Physics 10

Fatma BİLDACI

PALME PUBLICATIONS

Author
Fatma BİLDACI

ISBN: 978-605-282-139-8

Koza Matbaacılık
Sertifika No: 45553

GENEL DAĞITIM
PALME YAYINEVİ

Merkez:
A. Adnan Saygun Caddesi,
No.: 10/1 Sıhhiye/ANKARA
Tel.: 312 433 37 57
Faks: 312 433 52 72

Şube 1:
Olgunlar Sokak, No.: 4/5,
Bakanlıklar/ANKARA
Tel. : 312 417 95 28
Faks : 312 419 69 64

Şube 2:
Kazım Dirik Mahallesi, Ankara Caddesi,
No.: 259/C, Bornova/İZMİR
Tel. : 232 343 10 77
Faks : 232 343 10 78

bilgi@palmeyayinevi.com www.palmeyayinevi.com

"I am not leaving a spiritual legacy of dogmas, unchangeable petrified directives. My spiritual legacy is science and reason."

Mustafa Kemal ATATÜRK

Atatürk's Address to the Turkish Youth

You, the Turkish youth! Your primary duty is to forever protect and defend the Turkish independence and Turkish Republic. This is the mainstay of your existence and of your future.

This foundation is your most precious treasure in the future, as well, there will be malevolents, within and abroad, who will seek to deny your birthright. If one day you are compelled to defend your independence and the republic, you shall not reflect on the conditions and possibilities of the situation in which you find yourself, in order to accomplish your mission. These conditions and possibilities may appear unfavorable. The adversaries who scheme against your independence and your Republic may be the representatives of a victory without precedent in the world. By force or by ruse, all citadels and all arsenals of our dear fatherland may have been taken; all of its armies may have been dispersed and all corners of the country may have been physically occupied. More distressing and more grievous than all these, those who hold and exercise the power within the country may have fallen into gross error, blunder, and even treason. These holders of power may have even united their personal interests with political ambitions of the invaders. The nation itself may have fallen into privation, and may have become exhausted and desolate.

You, the future sons and daughters of Turkey! Even under such circumstances and conditions, your duty is to redeem Turkish independence and the Republic! The strength you shall need exists in the noble blood flowing through your veins.

October 20, 1927
Mustafa Kemal ATATÜRK

CONTENTS

3 WAVES

4 OPTICS

ABOUT THE BOOK

This part of the book will guide you through the book and give an understanding about the color coding of the tables and the symbols used in the book.

UNIT # — UNIT NAME

The concepts that will be covered in this unit are:

	CONTENTS	PAGE
Unit & subtitle #	Subtitle	#

Each unit starts with the *number* and the *name* of the unit. The *concepts* that will be covered throughout the unit is given in the box below. A *table* showing the *number of the subtitle number, subtitle itself*, and *page number* follows them.

Dependent quantity (unit)	Independent quantity (unit)
data	data
data	data

Tables in blue color appear by the graphs. They show the dependent and independent variables with their symbol and unit. Heading is followed by data.

Symbol	Quantity	Unit

Tables in green color are related to the equations given in the topic. They show the symbol, name of the quantity and its unit.

Material	Quantity (unit)

Tables in orange color give the literature values of the quantity mentioned in the topic.

Tables in pink color appear whenever needed to give a list of information of similar type, such as the volume of regular shaped objects or to make a comparison of quantities of similar type.

TITLE

In these blue boxes you can find the lives of some physicists, interesting facts about animals or how stuff works related to the topic.

Physics Extra

In the *Physics Extra* title some extra information is given for students who are curious to learn more about the topic.

Note for the teachers: The extensions of the curriculum for Science High Schools are given in this box.

Example

An *Example* question about the recent topic is given in this box.

Solution

The *Solution* to the example question is given here.

Check Your Understanding

Conceptual questions about the topic are asked in *Check Your Understanding.*

Conceptual Questions

Conceptual questions from the unit are asked in this box.

True - False

True - False questions from the unit are asked in this box.

Fill in the Blanks

Incomplete sentences which can be a definition or a conceptual sentence are given in *Fill in the Blanks* box.

Solved Questions

Question | **Solution**

Additional questions and their answers are given in *Solved Questions.*

Problems

1. The ....

(answer key)

Open ended questions are asked in *Problems* box. The answer to the question is given below the page and a blank space to write the answers left in the column.

The questions written in navy blue are about physics extra part.

Multiple Choice Questions

(answer key)

Multiple Choice Questions are asked in these boxes. The answer key is given below the page.

The questions written in navy blue are about physics extra part.

My Notes

This box is found at the back of the book for your extra notes.

You will receive the book with the *EXPERIMENTS AND ACTIVITIES WORKBOOK.*

Activity

NAME OF THE ACTIVITY

Brief explanation on the activity and list of materials

Directions:

1. ....

quantity	measurements

Questions about results of the activity

An *Activity* is an easy to perform hands on work that you can perform with your friends.

Experiment

NAME OF THE EXPERIMENT

A. DESIGN

Research Question:

How will .....

Selecting Variables

Independent variable: ....

Dependent variable: ....

Controlled variables: .....

Materials:

....

Methods for controlling the variables:

.....

Procedure:

1. .....

B. DATA COLLECTION AND PROCESSING

Data collection:

Trial	1	2
quantity (unit)		
quantity (unit)		

Observations:

Data processing:

Trial	1	2
quantity (unit)		

Data presentation:

Plot a graph showing ....

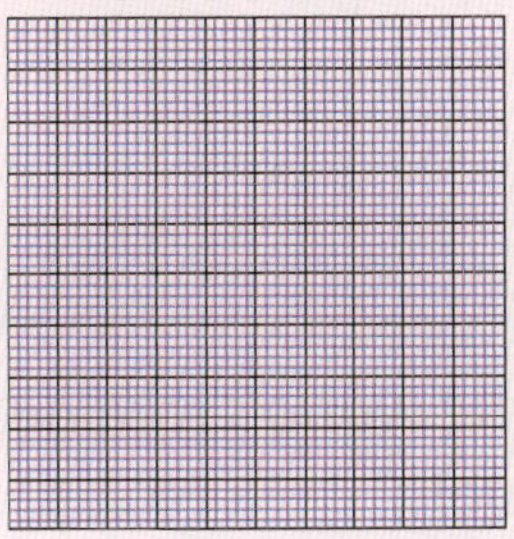

C: CONCLUSION AND EVALUATION

An experiment is a controlled procedure which is carried out to discover, test or to demonstrate something. We perform experiments in order to prove that our hypothesis is correct. In preparing an experiment there are three kinds of variables that are needed to be identified beforehand: independent, dependent, and controlled. The data collected in the experiment should be recorded and organized to have a conclusion statement. The project conclusion is a summary of the results of the project experimentation and a statement of how the results relate to the hypothesis.

An *experiment report* has three main parts: Design, Data Collection and Processing, Conclusion and Evaluation. With these lab reports you will take your first steps towards a scientific research.

UNIT 1

ELECTRICITY AND MAGNETISM

The concepts that will be covered in this unit are:
Electric current, potential difference, resistance, ohm's law and magnetic field.

CONTENTS PAGE

"Electric power is everywhere present in unlimited quantities and can drive the world's machinery without the need of coal, oil, gas or any other of common fuels."
Nikola TESLA

ELECTRICITY AND MAGNETISM

1.1. Current, Potential Difference, Resistance

1.1.1 Electric Current, Resistance and Potential difference

At 9th grade you had studied about charges at rest. Now we will consider charges in motion, that is to say, the *electric current*. Our technology today depends on the electric current. If electric current passes through a light bulb, you can illuminate your room, you can get warm if it passes through an electric heater, and so on...

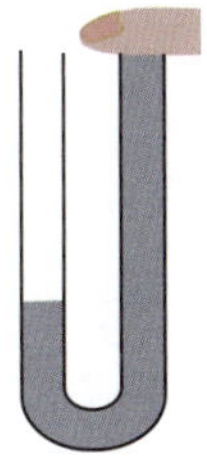

If you take your finger away from the right side of the U-tube given in the figure, the level of liquid on the right arm will move down, and the level of liquid on the left arm moves up, until the level of liquid on both arms is the same. The higher gravitational potential energy on the right arm is converted to kinetic energy of the liquid molecules until both arms have the same potential energy.

What if we connect two oppositely charged plates with a connecting wire? Can we use the analogy of hydrostatic equilibrium for this case?

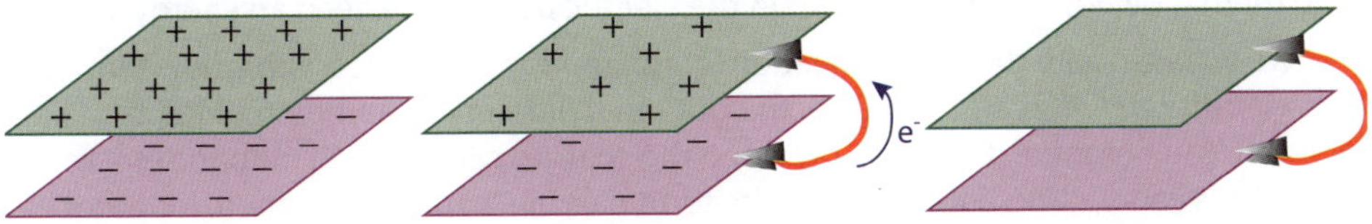

The positively charged plate is usually referred as having a *high potential energy* and the negatively charged plate is usually referred as having a *low potential energy*. There is a *potential difference* between these so called high and low potential plates so that the negative charges flow from one plate to another. This flow continues until both plates become electrically neutral or in other words until both of them have the same potential energy.

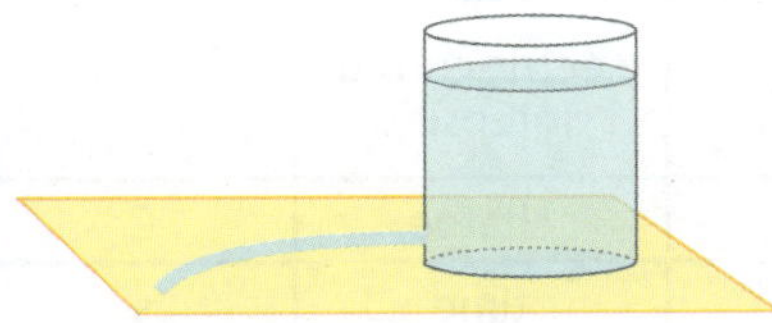

If you punch a hole at the bottom of a container filled with water, the water will continue to flow until it is finished. Here, the potential energy of water molecules stored within the container is transformed to the kinetic energy of the water molecules spurting out of the container. For having a continuous flow of water, you must somehow continue to fill the container, like pouring water in it.

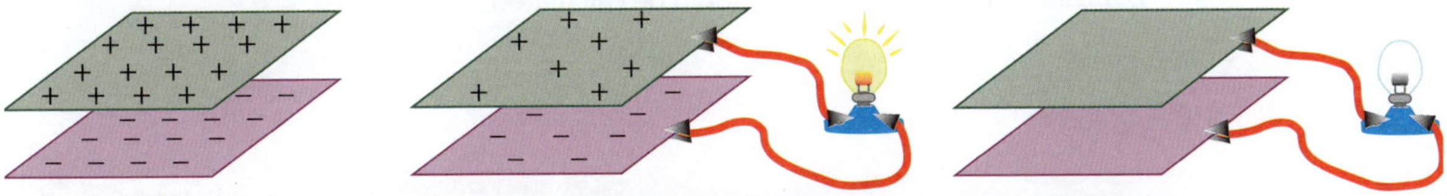

If we connect a light bulb between these oppositely charged plates, we can observe that it gives light for a short time. Just like the spurting water analogy, we should somehow maintain the flow of charges.

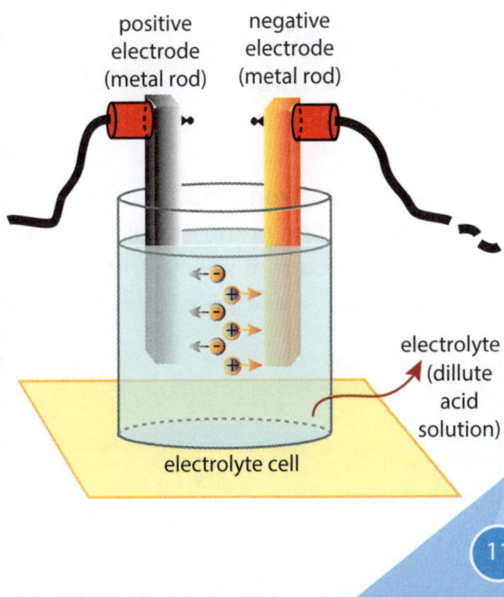

We have mentioned that only the electrons of a solid are allowed to move. *Electrons* are the *charge carriers* in solids. For *fluids*, since their molecules are not fixed at their position like solids', *positive charges (ions) can move as well as the negative charges (ions).*

Pure water, which is a good insulator, becomes a conductive medium when some amount of salt or dilute acid is poured in it. This solution is called an electrolyte. There are positive and negative ions in an electrolyte which are free to move. If we immerse two oppositely charged metal rods, electrodes, in this solution, we have an electrolyte cell. The electrodes have opposites signs. Thus an electric field rises from the positively charged rod to the negatively charged one. The presence of the electric field forces the positive ions to move along the electric field, whereas the negative ions move in the opposite direction of the electric field. The motion of the charged particles produces a *flow of charges.* Thus an electric current flows in the electrolyte cell. The electric current does not begin or end in the electrolyte cell. It comes to the electrolyte cell and flows through it.

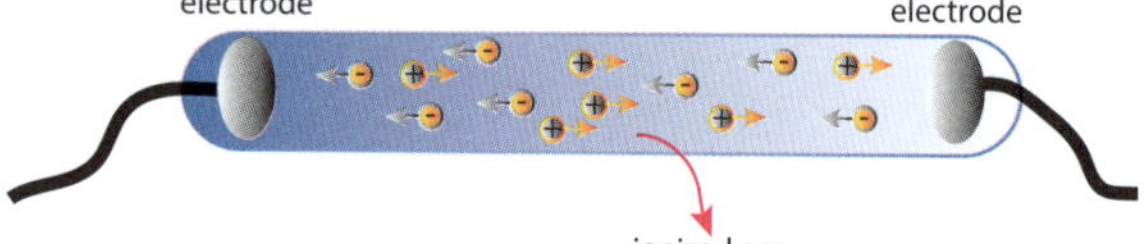

An ordinary gas is an insulator because it does not have free electrons. But once it becomes an ionized gas (now it is a plasma), free electrons are produced and it behaves like a conductor. The negative and positive ions within an ionized gas moves towards the oppositely charged electrodes,

Electric current needs a conduction medium like a wire, an electrolyte or an ionized gas. It exists when there is a motion of both positive and/or negative charges. Electric current is the name given to the motion of the charges. It is the charges that flow, not the electric current. Just like the water molecules in a river flow, not the stream.

As we have recently explained, if the charges on a conductor are at rest; then there is no electric field within the conductor. If an electric field is present in the conductor, the charges are forced to move. The electric field sets the charges in motion, and keeps them moving. If we consider the motion of electrons within a conducting wire, we must also consider the collisions of the electrons with the atoms of the metal. These collisions slow them down as their kinetic energy is converted to heat energy because of the collisions. If we don't continue to push them, the motion of the electrons ceases as they move. To keep them moving, we should maintain the electric field, so we need an energy source.

We will associate the energy needed for the motion of charges with a new term; the *electric potential difference*. The electric potential difference can be provided by a battery. With the potential difference provided the charges can be set in motion. If there is a continuous conducting path between the terminals of a battery, then there is an electric circuit. You can connect a light bulb, a radio... across the ends, the *terminals*, of the battery. Once the continuous path is set, the charges begin to flow through the wires from one terminal of the battery to the other one.

The electric current intensity in a wire is defined as the net amount of charge through the full cross-section of the wire in unit time:

Symbol	Quantity	Unit
I	electric current intensity	$\frac{C}{s} = A$
q	charge	C
t	time	s

$$I = \frac{\Delta q}{\Delta t}$$

Electric current intensity is measured in terms of coulombs per second (C/s), The unit has its own name, ampere (A), which is named after French physicist André Ampére.

For the electric current intensity of an ionized gas or an electrolyte the total charge of positive and negative ions should be calculated.

Example

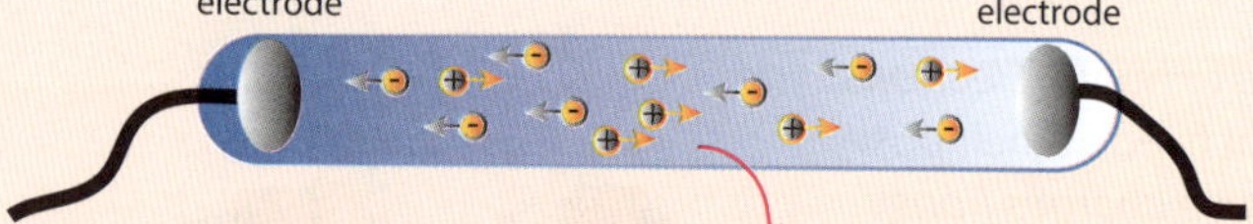

5×10^8 electrons and 1×10^8 H^+ ions are flowing through the cross-section of the ionized gas tube in 5 s.

Find the current in amperes.

$(q_{e^-} = q_{H^+} = e = 1.6 \times 10^{-19}$ C$)$

Solution

H atom consists of a proton and an electron. H^+ ions are actually protons. This is why the charge of one electron and one H^+ are the same.

The net charge flowing through the cross-section in unit time is the sum of number of electrons and H^+ ions.

For the current intensity.

$$I = \frac{q}{t} = \frac{n.e}{t}$$

$$I = \frac{(n_{e^-} + n_{H^+}).e}{t}$$

$$I = \frac{(5 \times 10^8 + 1 \times 10^8) \times (1.6 \times 10^{-19})}{5}$$

$$I = 1.92 \times 10^{-11}\ A$$

The circuit should have a continuous conducting path for a current to flow through it. A *closed circuit* is a circuit with no breaks or gaps in the conducting path. If there is a gap or a cut wire in the circuit then it is an open circuit. In an *open circuit*, the conducting path is not continuous, thus the current can't flow. Notice that in the open circuit aside, even though a battery is present in the circuit the lamp does not give light. The connecting wire is not connected to the negative terminal of the battery.

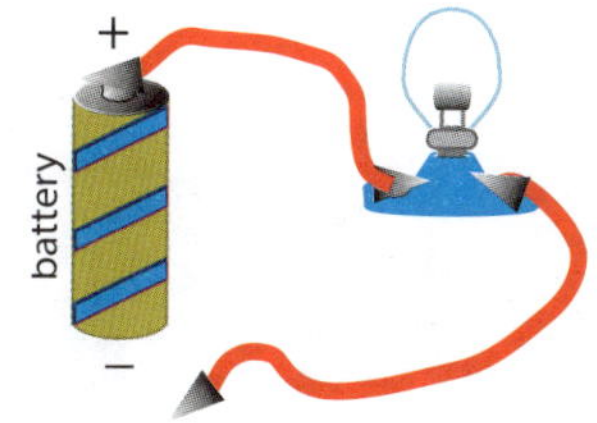

open circuit

Charge is conserved whether it is moving or not. At any instant, the magnitude of current at point A is equal to the magnitude of current at point B. A battery does not create or destroy a net charge. The electric field formed within the battery between the positive and negative terminals forces electrons of the electrodes to enter the circuit from point B. The potential difference provided by the battery produces an electric field within the conducting wires also. Thus the electron moves through the wires and the bulb. The free electrons of the conducting wires and the bulb are attracted towards the positive terminal of the battery. The bulb uses the energy of the electrons to give light, not the electrons. The flow is continuous and begins as soon as the wires are connected to both terminals of the battery. The number of electrons leaving the negative terminal is equal to the number of electrons entering the positive terminal of the battery at any instant. The charge is conserved.

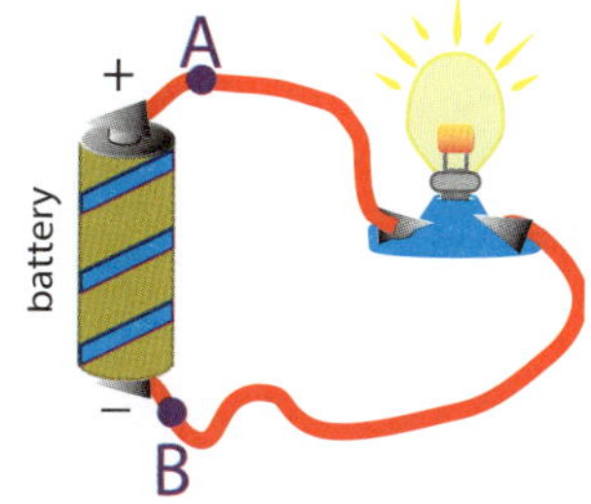

closed circuit

Today we know that the electric current is provided by the flow of electrons through a wire. But the studies about the electric current were done before the discovery of electron. Two centuries ago, physicists were thinking that the positive charges were flowing in a circuit. The direction of the conventional current is from the positive (high potential) terminal to the negative (low potential) terminal of the battery. When somebody speaks about *the flow direction of the current, a flow beginning from the positive terminal of the battery and ending at the negative terminal of the battery* is meant. What we are talking about here is the flow direction of the electric current, not the electrons. This choice for the flow direction of current makes difference only in microscopic level. Although we are talking about a direction for current, it is not a vector quantity. In a wire current is always parallel to the wire no matter how the wire curves.

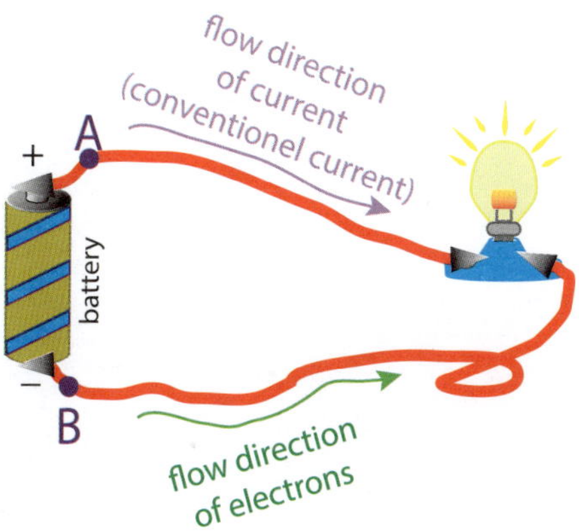

It is an interesting phenomenon that the moment you turn the light on, the bulb gives out light. Are the electrons moving with the speed of light, so that the bulb can give light at that instant?

Only light energy travels with the speed of light. Even though the mass of the electrons is very tiny, they have it. Electrons are matter; not energy. In a typical electric circuit, the drift velocity of electrons is a very small value such as 0.1 cm/s. But the electric field within the wire travels almost with the speed of light. All of the electrons in the conducting path react simultaneously to the electric field and start moving through the wire. This flow is just like the flow of water through a pipe. If one drop of water is forced to enter to a completely full pipe from one end, another drop of water must be pushed out of the pipe from the other end. The drops of water flowing through or taps are not the same individual drops coming from the water supply.

When you plug in a device such as a portable desk lamp, energy flows from the power outlet to the lamp. Power utilities do not sell electrons, they sell the electrical energy. The wires of your device is the supply of electrons. When you get an electric shock, the current is made up by the electrons in your body. They don't flow from the wire to your body. The electric energy causes the free electrons of your body to move.

The main task of a battery in a circuit is to provide a continuous potential difference for the charges to flow. Batteries do not produce the current. They provide the energy. Materials show an opposition to the flow of charges through them. This opposition of the wire to the flow of current is called the resistance of a wire. Resistance is sometimes a wanted property like for light bulbs or electric heaters or hair driers. We can control the amount of current in an electric circuit by using the resistors. The resistance of the wire depends on its physical properties.

1.1.2 Factors affecting the resistance of a solid conductor

A long queue at a passport check point

Long queues are sometimes formed at passport check points if the number of check points is not enough. As the queue gets longer, the time you spend gets longer for your turn to come. Using the same analogy, as a conductor gets longer, number of interactions of electrons with the atoms of the conductor gets more. One individual electron will face more resistance to its motion as it is travelling through the conductor. The steel wire with length 2L has more resistance when compared to the one with a length of L.

The resistance of a wire is directly proportional to the length of the wire.

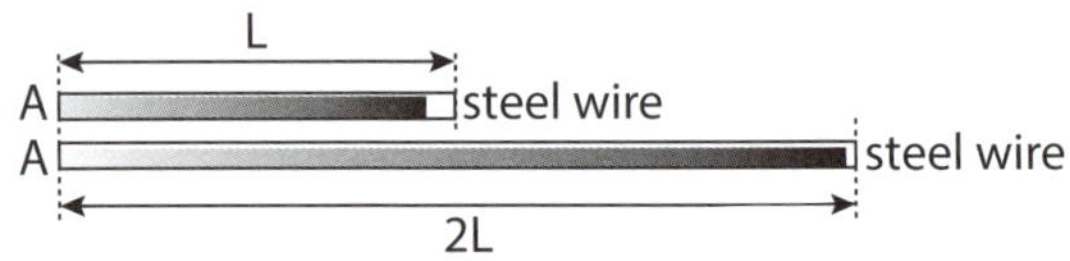

$$R \; \alpha \; L \; \textit{(directly proportional)}$$

If there are more desk at the check points, the number of people passing through them in total is going to be larger. If a conductor gets wider, number of interactions of electrons with the atoms of the conductor gets less. As a conductor gets wider, its resistance gets smaller. The steel wire with cross-section area A has more resistance when compared to the one with cross-section area 2A.

The resistance of a wire is inversely proportional to the cross-section area of the wire.

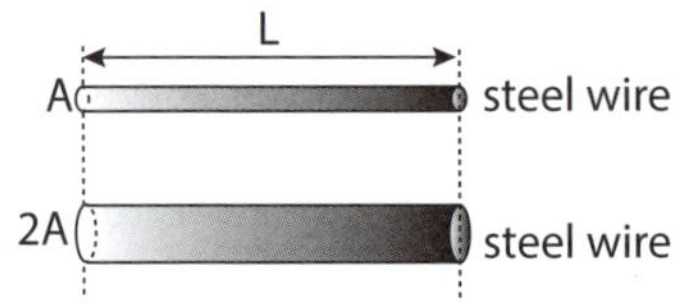

$$R \; \alpha \; \frac{1}{A} \; \textit{(inversely proportional)}$$

Not all the officers at the check point will complete the procedures at the same time. Each of them have their own working style. Just like the officers, each material has its own resistance to the flow of charges. A copper wire and a steel wire will not show the same resistance to the flow of electrons. The resistance of a wire in unit length having a uniform unit cross-sectional area is called resistivity.

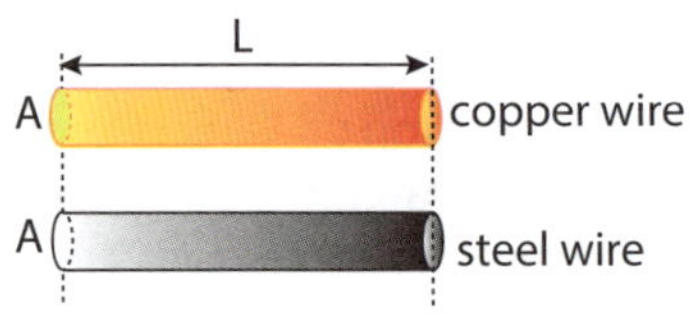

$$R \; \alpha \; \textit{resistivity (directly proportional)}$$

Resistivity of some materials

Material	Resistivity (ρ) (Ω.m)	Temperature coefficient per °C
Silver	1.59×10^{-8}	0.0061
Copper	1.68×10^{-8}	0.0068
Gold	2.44×10^{-8}	0.003715
Aluminum	2.65×10^{-8}	0.00429
Tungsten	5.60×10^{-8}	0.0045
Iron	9.71×10^{-8}	0.005671
Platinum	10.6×10^{-8}	0.003927
Mercury	98×10^{-8}	0.0009
Carbon	$3\text{-}60 \times 10^{-5}$	-0.0005

The list aside is for the resistivity of some materials. Silver, copper and gold are the first three. *Silver* is expensive and tarnishes in air so quickly. That's why it is not broadly used. *Gold* is quite expensive also, but because it is marvelously resistant to corrosion (gradual destruction of materials by chemical reaction with their environment), it is used to plate connectors and such. Most of the conducting wires are made of *copper* because it isn't too expensive. Although the resistivity of *aluminum* is greater than copper's, it is preferable to copper in some situations because of its smaller density. Its resistance for same mass of copper is smaller. For transmission lines aluminum is widely preferred.

To summarize, the resistance of a solid conductor is directly proportional to its length and inversely proportional to its cross-section area. The kind of the material affects the resistance of the conductor also.

The unit of resistance, *ohm*, is named after the German physicist George Simon Ohm. Its abbreviation is the Greek letter omega, Ω.

Physics Extra

The mathematical expression for the resistance of a wire is:

$$R \; \alpha \; \frac{L}{A} \Rightarrow R = \rho\frac{L}{A}$$

The unit of resistivity is ohm.meter (Ω.m). The symbol used for resistivity is another Greek letter, rho, ρ.

Symbol	Quantity	Unit
R	resistance	Ω (ohm)
L	length of wire	m
A	cross-section area of wire	m^2
ρ	resistivity	Ω.m

Example

Two cylindrical copper wires of lengths L and 2L respectively have the same mass.

If the resistance of the wire with length L is 5 Ω, find the resistance of the wire with length 2L.

Solution

If both wires have the same mass although they have different lengths, then their cross-section areas should be different. If:

$$m = d.V = d.h.A$$

then

$$d.L.A_1 = d.2L.A_2 \Rightarrow A_1 = 2A_2$$

$$R = \rho\frac{L}{A_1} = 5\ \Omega \Rightarrow$$

$$R' = \rho\frac{L}{A_2} = \rho\frac{L}{\frac{A_1}{2}} = 2.\rho\frac{L}{A_1} = 2 \times 5 = 10\ \Omega$$

The resistivity of all substances changes to some degree with changes in temperature. The change in the resistance of the conductor is roughly proportional to the change in the temperature and to the original resistivity, ρ_o which is measured at a reference temperature t_o.

$$\rho_t = \rho_0\,[1 + (t-t_0)]$$

If the temperature change is too large this equation is not adequate.

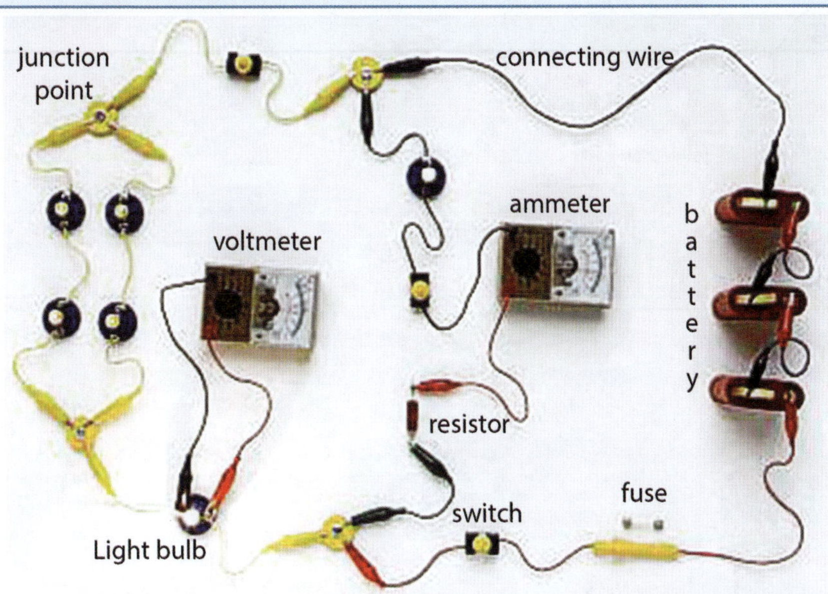

Above is a photo of a circuit prepared by using light bulbs, wires, dry cells, potential and current measuring devices. If you have a close look, you can identify the mentioned devices in the circuit. Even if the devices are labelled in the photo, trying to understand what is connected to what can be annoying sometimes. Instead of real appearance drawings or photos, we prefer to use circuit diagrams. A circuit diagram is a logical blueprint showing which device is connected to which device by using symbols. Before we draw a circuit diagram for the aside photo, let's learn the symbols for parts of an electric circuit.

Name	Symbol	Description
Dry cell	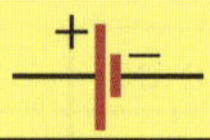	Supplies energy to the circuit.
Battery	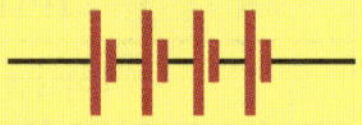	Dry cells connected in series.
Switch	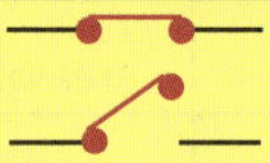	Controls the flow of current by making or breaking the current.
Ammeter		Measures the electric current in a circuit. It has a very low resistance in order not to change the magnitude of the current. It is always connected in series so that all of the charges passes through it.
Voltmeter		Measures the potential difference in a circuit. It has a very high resistance therefore the current flowing through the voltmeter is zero in practice. It is always connected in parallel so that all of the charges pass through the circuit element.
Resistor	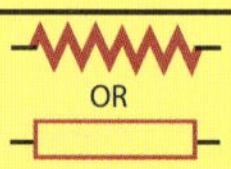	Slows down the flow of electrons.
Rheostat	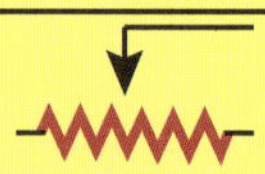	Resistor whose resistance value can be varied.
Light bulb	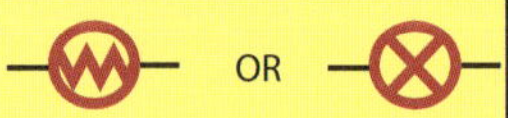	Transforms the electrical energy to light (and heat) energy.
Connecting wires		Lengths of copper wires for connecting the parts of the circuit. They have very low resistance, thus do not affect the current in a circuit.
Fuse	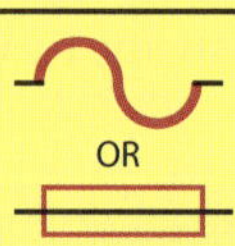	Disconnects the circuit when the current is above a certain value.

Check Your Understanding

The picture is a simple network prepared by lamps, dry cells, voltmeter and ammeter.

Draw the scheme of this circuit using the symbols of electric circuits.

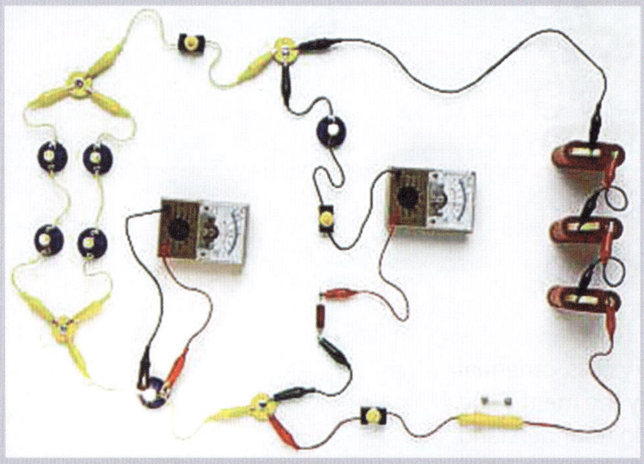

1.2. Electric Circuits

1.2.1. The Relationship Between Current, Resistance and Potential Difference

Georg Simon Ohm worked on the relationship between the electric current and the potential difference. He had performed a series of experiments. He measured the values of current and the potential difference across the ends of a conductor.

We will repeat his experiment with a resistor connected to a battery. An ammeter is connected in series to the circuit for reading the values of current through the resistor. A voltmeter is connected across the ends of the resistor to measure the potential difference across its ends. The potential difference is increased and the current is read. The measurements are:

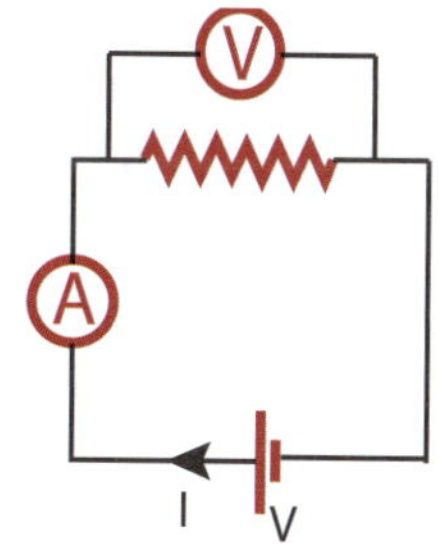

V (volt)	6	12	18	24	30
I (ampére)	2	4	6	8	10

According to the measurements, as the potential difference between the ends of the resistor increases, the ammeter readings also increase. The ratio of potential difference to ammeter reading is:

$$\frac{6}{2} = \frac{12}{4} = \frac{18}{6} = \frac{24}{8} = \frac{30}{10} = 3V/A$$

The ratio remains constant. When you change the resistor, the numerical values change but the empirical results remain the same, that is, the ammeter reading increases but the ratio is always equal to a constant number. Also reversing the terminals of the battery does not affect the magnitude of the current. Only the flow direction of the current is reversed when the terminals are shifted.

The results tell us that the ratio of potential difference to the current is a constant value. The constant value in this ratio is equal to the resistance of the wire. This leads us to the Ohm's Law of Resistance. According to Ohm's law:

$$\frac{V}{I} = R \text{ or } V = I.R$$

Symbol	Quantity	Unit
V	potential difference	V (volt)
I	electric current	A(ampére)
R	resistance	Ω (ohm)

According to Ohm's Law the current is not only dependent on the potential difference. The resistance of the wire is also important for the magnitude of the current. The main intend of a battery is to maintain a constant potential difference for the circuit. Different values of current can be produced by the same battery depending on the resistance of the device.

Ohm's law is not a fundamental law. It is a description of metal conductors. Materials or devices which do not obey Ohm's law are called non-ohmic. Diodes, vacuum tubes, transistors are among the non-ohmic devices.

Example

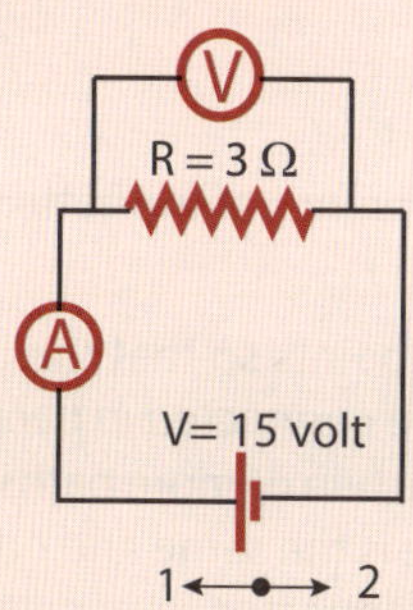

An electric circuit is prepared by using a 15 V battery, 3 Ω resistor, ammeter and voltmeter.

a. What is the flow direction of current?

b. What is the value read on ammeter?

c. What is the value read on the voltmeter?

Solution

a. From positive terminal of the battery of the dry cell to its negative terminal, that is direction 1 in the figure.

b. Using Ohm's law of resistance:

$V = I \,.\, R$

$15 = I \,.\, 3$

$I = 5\,A$

c. $V = I \,.\, R = 3 \times 5 = 15\,V$

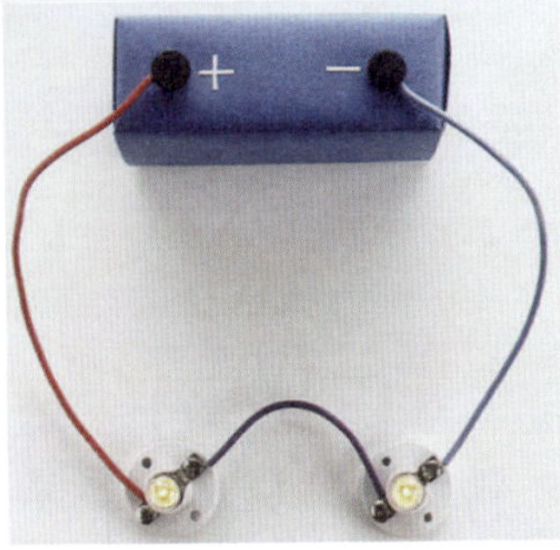

Light bulbs connected in series

Electric circuits usually consist of more than one device. The main two connection for these devices is series connection and parallel connection. There can be circuits using the combinations of series and parallel connections.

If two or more electric devices are connected one after another, then they are said to be *connected in series* to each other. For a series connection of electric devices there is only one path for the current to flow. So the current passing through each device is the same.

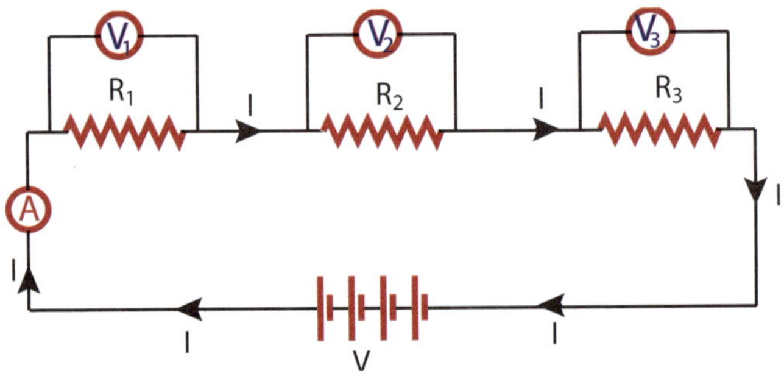

The sum of potential drops through each resistor is equal to the potential difference provided by the battery to the circuit.

$V = V_1 + V_2 + V_3$

If we substitute Ohm's Law of resistance:

$I \cdot R_{eq} = I \cdot R_1 + I \cdot R_2 + I \cdot R_3$

$I \cdot R_{eq} = I \cdot (R_1 + R_2 + R_3)$

$R_{eq} = R_1 + R_2 + R_3$

R_{eq} is the equivalent resistance. This is the *one* resistor value which causes the same magnitude of current to flow through the circuit. The resistance value of the circuit increases if the resistors are connected in series. There are some limitations in a series circuit. If one device in the circuit fails to operate, the current ceases in the whole circuit. The failing device acts as a gap for the electric current. The path is not complete anymore.

Example

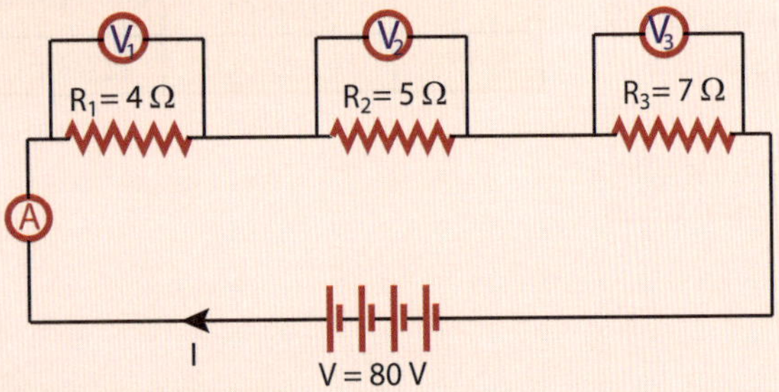

For the electric circuit given in the figure:

a. Find the equivalent resistance of the circuit.

b. Find the value read on the ammeter.

c. Find the values read on the voltmeters V_1, V_2 and V_3.

d. What is the relationship between V_1, V_2, V_3 and V which is the potential difference of the battery?

Solution

a. The equivalent resistance of the circuit is:

$R_{eq} = R_1 + R_2 + R_3 = 4 + 5 + 7 = 16\ \Omega$

b. Using Ohm's law of resistance:

$$I = \frac{V}{R} = \frac{80}{16} = 5\ A$$

c. For the potential differences between the ends of each resistor, we will use Ohm's law again.

$V_1 = I \cdot R_1 = 5 \cdot 4 = 20\ V$

$V_2 = I \cdot R_2 = 5 \cdot 5 = 25\ V$

$V_3 = I \cdot R_3 = 5 \cdot 7 = 35\ V$

d. $V_1 + V_2 + V_3 = 20 + 25 + 35 = 80\ V$

Their sum is equal to the potential difference of the battery.

If two or more electric devices are connected between two points so that the current splits between them and then rejoins, then they are said to be *connected in parallel* to each other. For a parallel connection of electric devices there are two or more paths for the current to flow. The potential difference between each device is the same and equal to the potential difference of the battery if all of the devices are connected in parallel to each other.

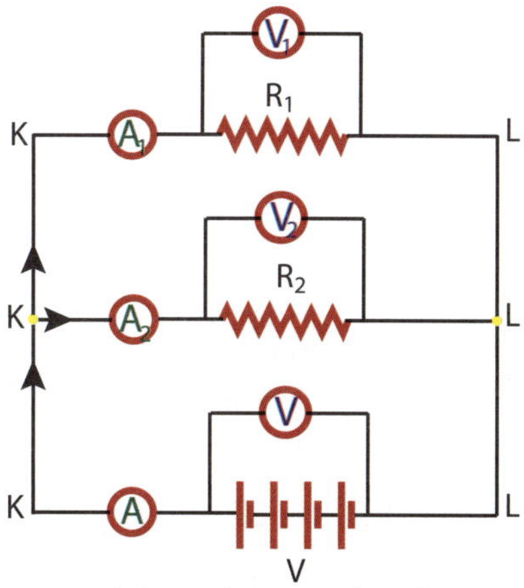

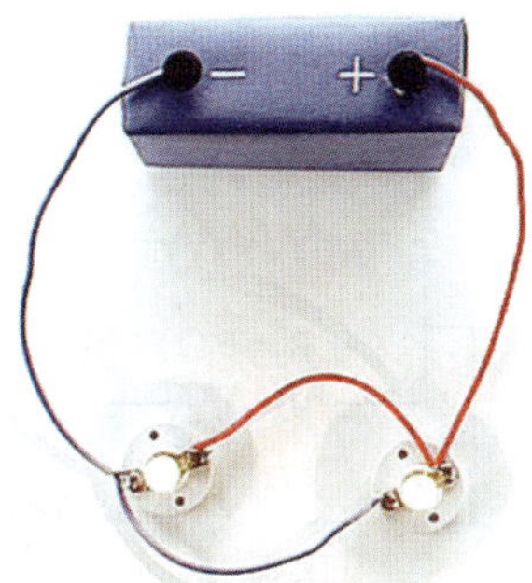

Light bulbs connected in parallel

The sum of currents on each branch is equal to the main current of the circuit.

$I = I_1 + I_2$

If Ohm's Law of resistance is substituted:

$$\frac{V}{R_1} + \frac{V}{R_2} = \frac{V}{R_{eq}}$$

$$V.\left[\frac{1}{R_1} + \frac{1}{R_2}\right] = \frac{V}{R_{eq}}$$

$$\frac{1}{R_1} + \frac{1}{R_2} = \frac{1}{R_{eq}}$$

R_{eq} is the equivalent resistance. According to the equation the resistance value of the circuit decreases if the resistors are connected in parallel. This is an expected result since you are giving additional paths to current to flow.

Example

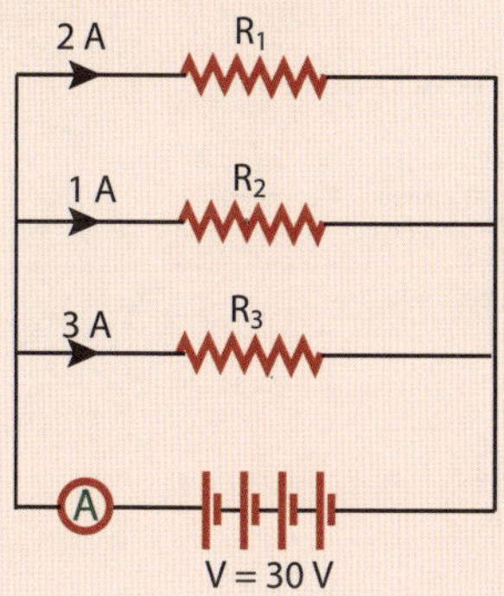

An electric circuit is prepared as given in the figure. The current passing through each resistor is given in the figure.

a. Find the value read on the ammeter.

b. Find the value of the resistances, R_1, R_2 and R_3.

c. What is the equivalent resistance of the circuit?

Solution

a. $I = I_1 + I_2 + I_3 = 2 + 1 + 3 = 6$ A

b. Using Ohm's law: $R_1 = \frac{V}{I_1} = \frac{30}{2} = 15\ \Omega$

$R_2 = \frac{V}{I_2} = \frac{30}{1} = 30\ \Omega$

$R_3 = \frac{V}{I_3} = \frac{30}{3} = 10\ \Omega$

c. 1st method: Using Ohm's law $R_{eq} = \frac{V}{I} = \frac{30}{6} = 5\ \Omega$

2nd method: Using the equation for R_{eq}

$$\frac{1}{R_{eq}} = \frac{1}{R_1} + \frac{1}{R_2} + \frac{1}{R_3} = \frac{1}{15} + \frac{1}{30} + \frac{1}{10} = \frac{2+1+3}{30} = \frac{6}{30} = \frac{1}{5}$$

$R_{eq} = 5\ \Omega$

Since the current is splitting into separate branches in a parallel circuit, each device on each branch is working independent of the one on the other branch. So if one lamp on one branch is burnt out, the other lamps can continue to work on the other branches without being interrupted. The lamps in our houses are connected parallel. If a lamp in a room of your house is burnt out, the lamps in the other rooms will continue to give out light. If a circuit has series and parallel connections at the same time, then the circuit is a simple network. The equivalent resistance and the current in a simple network can be calculated by using the equations we derived for series and parallel connections and also Ohm's law of resistance.

Example

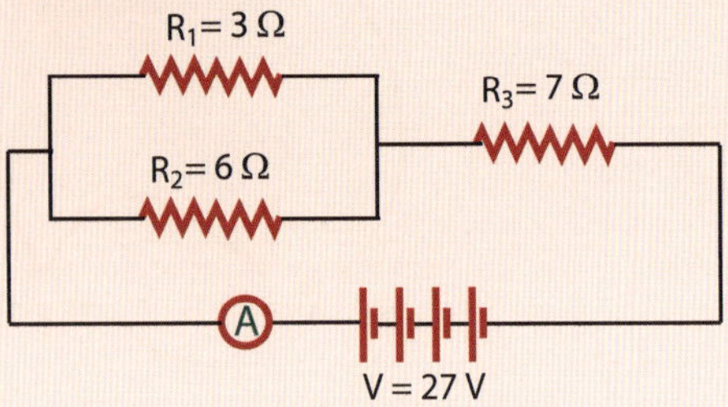

For the electric circuit given in the figure:
a. Find the total resistance of the circuit.
b. Find the current drawn from the battery.
c. What is the current through each resistor?
d. Find the potential difference across each resistor.

Solution

a. For R_1 and R_2:

$$\frac{1}{R_{12}} = \frac{1}{R_1} + \frac{1}{R_2} = \frac{1}{3} + \frac{1}{6} = \frac{2+1}{6} = \frac{1}{2}$$

$$R_{12} = 2\ \Omega$$

For the whole circuit:

$R_{eq} = R_{12} + R_3 = 2 + 7 = 9\ \Omega$

b. Using Ohm's law of resistance:

$$I = \frac{V}{R_{eq}} = \frac{27}{9} = 3\ A$$

c. Since R_1 and R_2 are parallel to each other, their potential differences are the same. So:

$$I_1.R_1 = I_2.R_2$$
$$I_1.3 = I_2.6$$
$$I_1 = 2I_2$$
$$I = I_1 + I_2 = 2I_2 + I_2 = 3I_2 = 3\ A$$
$$\Rightarrow I_2 = 1\ A\ ;\quad I_1 = 2\ A$$

d. Voltage drops for R_1 and R_2 are the same since they are parallel to each other.

$$V_1 = V_2 = I_1 . R_1 = 2 . 3 = 6\ V$$
$$V_3 = I . R_3 = 7 . 3 = 21\ V$$

Example

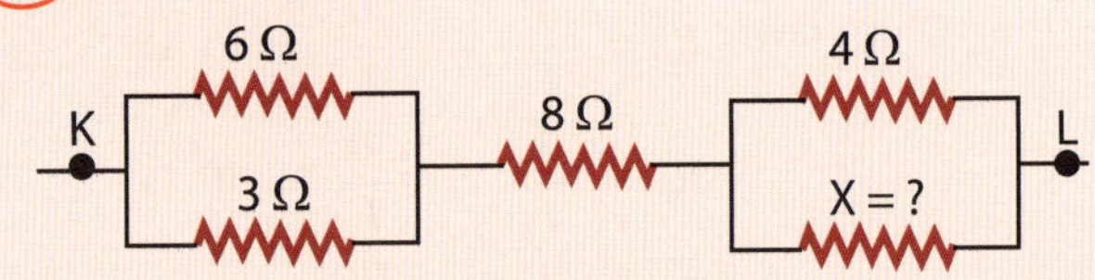

The equivalent resistance between points KL is 13 Ω. Find the value of unknown resistance, X.

Solution

6 Ω and 3 Ω are in parallel with each other.

$$\frac{1}{R_1} = \frac{1}{6} + \frac{1}{3} = \frac{1+2}{6} = \frac{3}{6} = \frac{1}{2}$$

$$R_1 = 2\ \Omega$$

Now they are in series with 8 Ω.

$$R_2 = 2 + 8 = 10\ \Omega$$

4 Ω and X are parallel to each other. They are in series with R_2. The equivalent resistance of 4 Ω and X should be:

13 - 10 = 3 Ω

Thus for X:

$$\frac{1}{3} = \frac{1}{4} + \frac{1}{X}$$
$$\frac{1}{X} = \frac{1}{3} - \frac{1}{4} = \frac{4-3}{12} = \frac{1}{12}$$
$$X = 12\ \Omega$$

Example

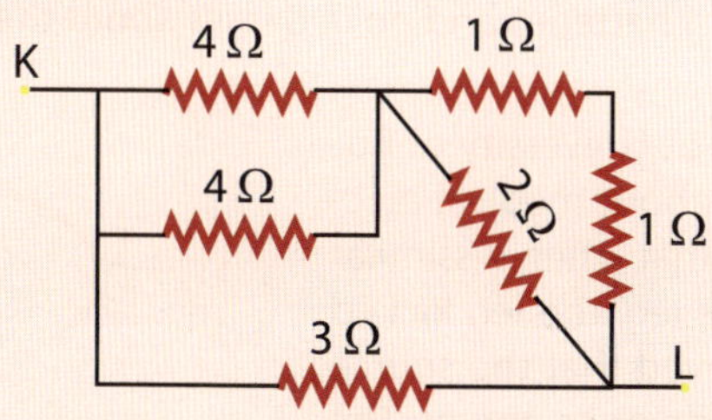

Some part of an electric circuit is given in the figure.

Find the value of the equivalent resistance between points KL.

Solution

It is not very clear in the diagram whether the resistors are in parallel or in series with each other. For such circuits labelling is a practical method.

The rules are:

1. Wires without a device have the same potential value, so the label of all points on the empty wire is the same.

2. If there is a device on the wire, then the point just after the device has a new label

3. A new label is given at a junction where at least 3 or more wires are intersecting.

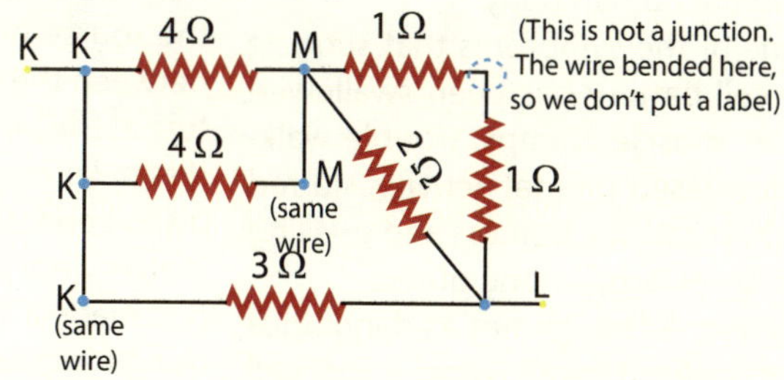

$$\frac{1}{R_{KM}} = \frac{1}{4} + \frac{1}{4} = \frac{2}{4} = \frac{1}{2}$$

$R_{KM} = 2\ \Omega$

$R_{ML} = 1+1 = 2\ \Omega$

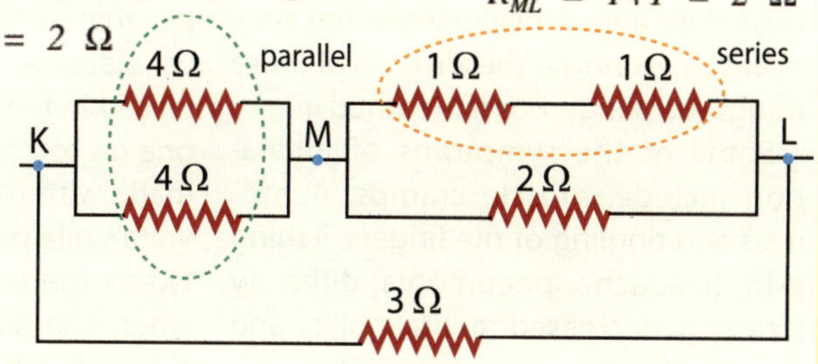

$$\frac{1}{R_{ML}} = \frac{1}{2} + \frac{1}{2} = \frac{2}{2} = 1$$

$R_{ML} = 1\ \Omega$

$R_{KL} = 2+1$

$R_{KL} = 3\ \Omega$

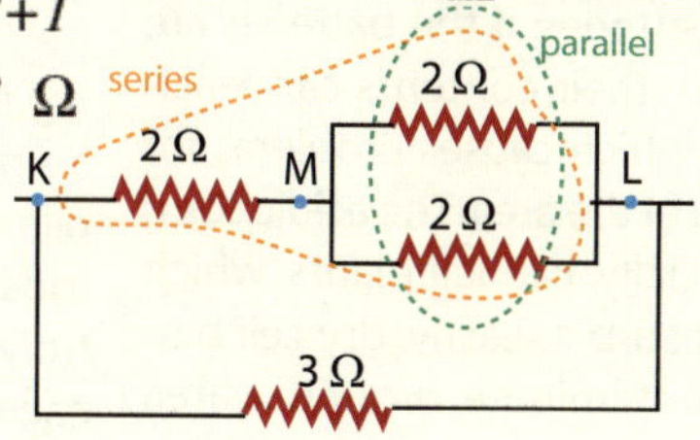

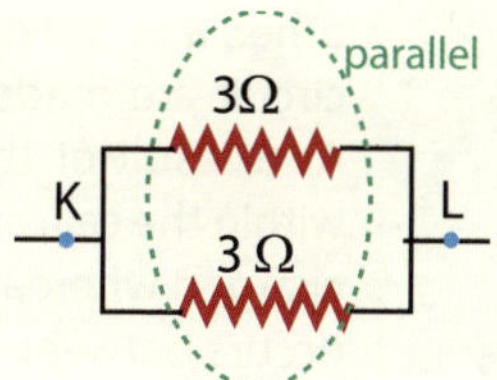

$$\frac{1}{R_{KL}} = \frac{1}{3} + \frac{1}{3} = \frac{2}{3}$$

$R_{KL} = 1.5\ \Omega$

NEGATIVE EFFECTS OF A DRY CELL BATTERY

Swallowing: A dry cell battery is extremely dangerous when swallowed, whether or not it possesses an electric charge at the time. The electrolyte compound is powerfully acidic, and the metals that make up the alternating layers of the battery are also dangerous to have in the human body.

Some of the symptoms that suggest a dry cell battery has been swallowed include: muscle cramps, trouble walking, decreased mental activity, slurred speech, spasticity, tremors and swelling in the lower legs, among others.

Inhaling: Although not as dangerous as out-and-out swallowing a dry cell battery, inhaling the vapors that come from them is still dangerous. If dry cell batteries are opened up, they usually release dust, and, if they're set on fire, they release vapors into the air. Both of these things are dangerous if breathed in.

Some of the symptoms of inhalation include: muscle cramps, numbness and tingling of the fingers, itching skin, headache, pneumonia, difficulty sleeping, decreased mental ability and slurred speech.

Environmental Poisoning: The contents of dry cell batteries are equally dangerous to the environment. Animals might mistake the batteries for food and swallow them, getting poisoned by the dry cell batteries. If the batteries are cracked open, their contents can leach into the soil, which causes problems for the local flora of the area. The acidic compound, especially, can kill plants, which may try to absorb it. Lastly, dry cell batteries can contaminate ground water, which makes any water pumped from the region of their dumping unsafe for anyone to drink.

Read more: http://www.ehow.com/list_5846105_negative-effects-dry-cell-battery.html

1.2.2. Dry Cells in Series and Parallel

We had mentioned that the batteries provide the energy for the charges to flow in a circuit. Now we will learn the history of batteries and go in details about their functions in a circuit.

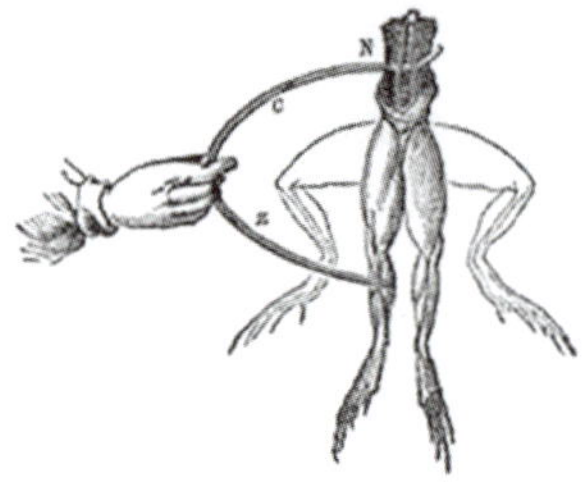

In 1780, Luigi Galvani, professor at University of Bologna, was performing experiments about medical electricity. He and his assistant observed that when dissimilar metals were inserted in a dead frog's leg, the leg kicked as if the frog was alive. Galvani believed that the source of electric charge was the frog's muscle or its nerves. The metal was merely transmitting the charge to convenient points. He called this phenomenon the *animal electricity*.

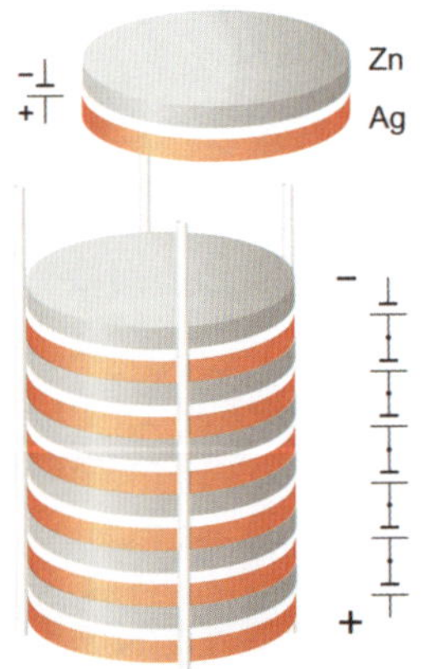

Alessandro Volta, professor at University Pavia, was not confident about Volta's results. He was thinking that the source of electricity was not the animal, the contact between the dissimilar metals was responsible of it. Moisture at the point of contact was an important factor for the dissimilar metals to conduct the electricity. He performed his experiments and found that certain combinations of metals were producing a greater effect than the others. He had made a list about the order of effectiveness of metals using his measurements. This list is still used by the chemists. Another outcome of his measurements was carbon could be used instead of one of the metals.

Volta published results of his experiments in 1800. The pile that he had constructed was consisting of zinc and silver plates. A piece of cloth soaked in brine or dilute acid is placed in between the metals. He placed these couplings one on top of another. Thus the first electric battery was made with this pile of couplings. The layered structure of Volta's pile provided a continuous potential difference between the two ends. This potential difference was small when compared with the static electric producing machines, but it was able to produce a large amount of electric charge. The silver end had excess positive charge and the zinc end had excess negative charge. The potential difference was much increased and a steady flow of electric charge was supplied. When strips of metal were brought close together a spark was produced.

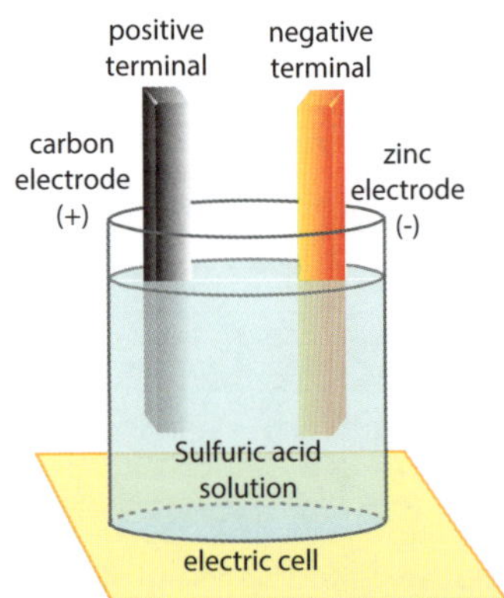

A battery produces electricity by *transforming the chemical energy into electrical energy*. For a simple battery we need two plates like Volta's pile or rods of dissimilar metals or carbon. These rods, the electrodes, are immersed in an electrolyte; brine or dilute acid solution. The part of each electrode out of the electrolyte is called a terminal. Such a set up is called an electric cell. If two or more cells are connected in series to each other, a battery is formed. The connections to a circuit are made from the terminals of the cell. As a result of the chemical reactions occurring within the cell, one electrode becomes negatively charged whereas the other one becomes positively charged. Thus a potential difference occurs between the two terminals. When the terminals are connected to the wires, the charges flow in the circuit. After a time the electrodes get used up, less energy is provided to the circuit. The cell becomes *dead*. The potential difference of a cell depends on the electrodes and their ability to give electrons to the circuit.

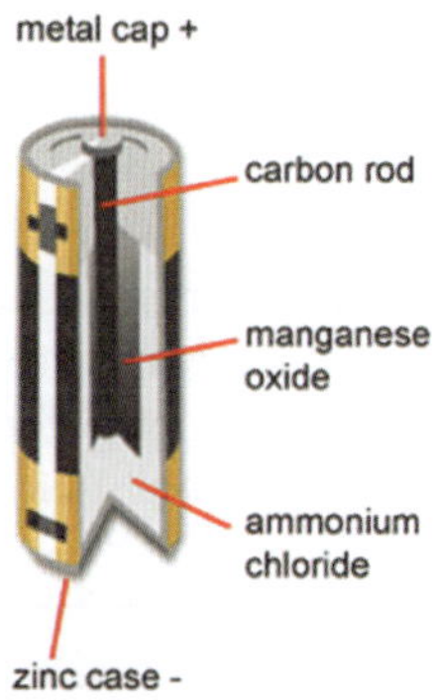

The flashlight batteries or the dry cells use carbon and zinc as their electrodes. The zinc case is the negative terminal and the carbon rod at the center is the positive terminal. The space between the two terminals contain a paste like mixture of manganese dioxide, ammonium chloride and carbon.

Zinc-Carbon batteries contain varying amounts of mercury and potassium hydroxide and lead. Most of these batteries are disposable lead-acid cells which are not recycled. Lead from disposed dry-cells leaches into the ground, contaminates the soil and water.

Old fashioned dry cells

When two or more cells are connected so that the positive terminal of one is connected to the negative terminal of the next, then they are said to be connected in series. The net potential difference of the circuit is the sum of their individual potential differences.

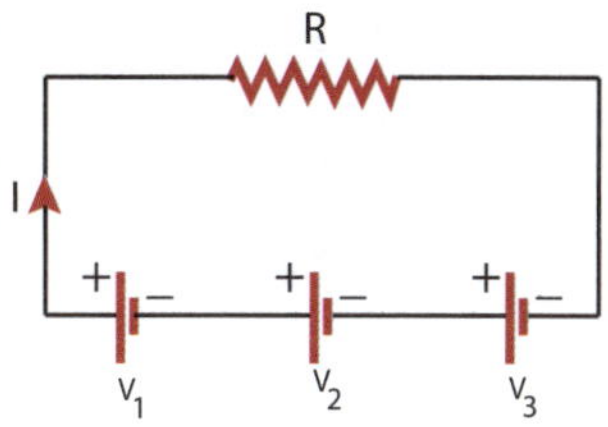

dry cells connected in series

$$V_{net} = V_1+V_2+V_3+....$$

$$I = \frac{V_{net}}{R_{eq}} = \frac{V_1+V_2+V_3+....}{R}$$

We use series connection of cells for having a greater potential difference.

For some circuits the negative terminal of one cell is connected to the negative terminal of the next. These cells are said to be oppositely connected. In this case the net potential difference of the circuit reduces. This might seem unreasonable but, such an arrangement is actually how a battery charger works. The cell with a higher potential difference (V_{high}) charges the other cell with a lower potential difference (V_{low}). Because of its greater potential difference V_{high} forces the electrons back into the negative terminal of V_{low}.

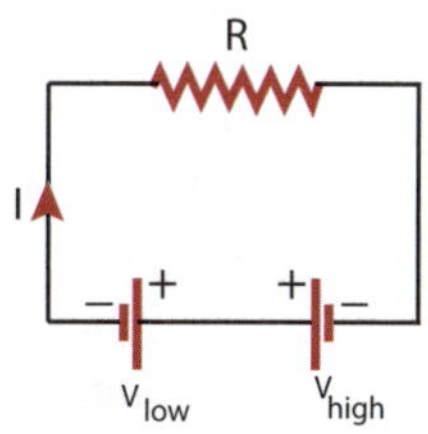

If the potential differences of the dry cells are equal to each other, than a parallel connection of cells is available. A parallel connection is not used to increase the potential difference of the circuit. The potential difference of the circuit is equal to one of the dry cell's potential difference.

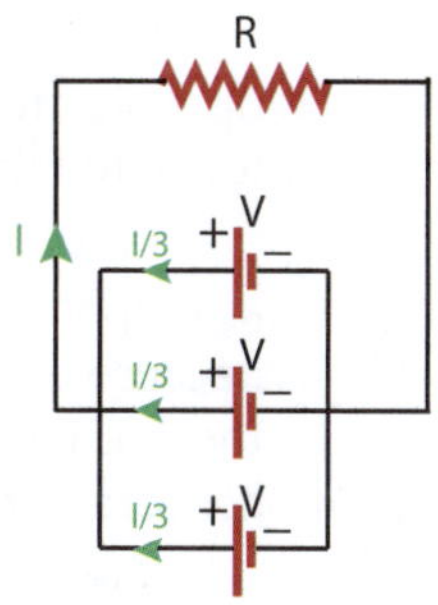

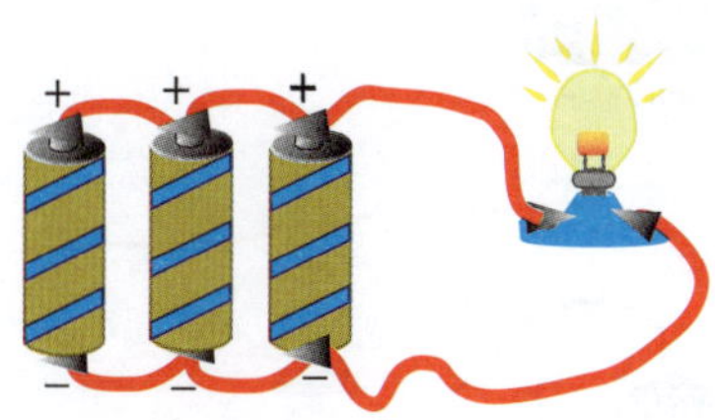

dry cells connected in parallel

$$V_{net} = V$$

$$I = \frac{V_{net}}{R_{eq}} = \frac{V}{R}$$

When large currents are needed parallel connection is used to increase the energy provided to the circuit. Each cell in the parallel connection produces a fraction of the main current. Thus the cells go dead in a longer time.

A battery holder

Being long lasting is one of the most important properties of the dry cells. Working life of a cell depends on the current drawn from it. If q is the charge on the cell, and I is the current drawn from the battery, then:

For the lasting time for a cell:

$$q = I.t \Rightarrow t = \frac{q}{I}$$

CIRCUIT 1

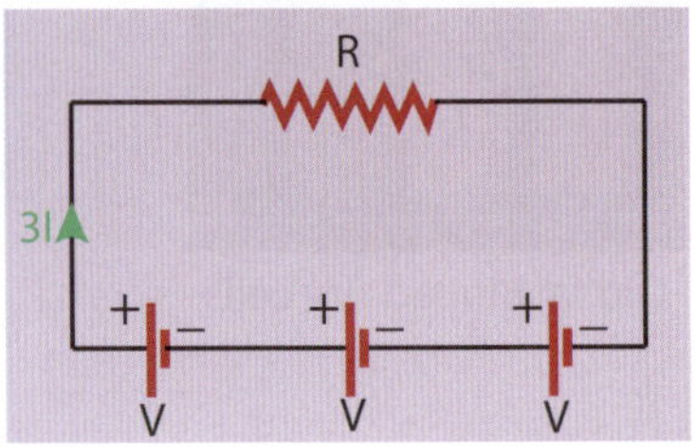

CIRCUIT 2

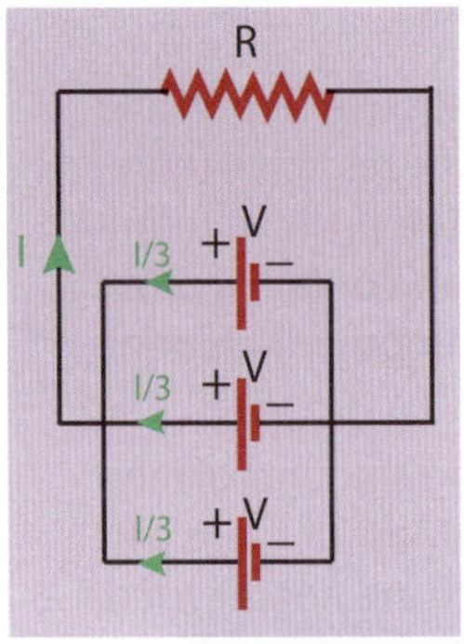

The circuits aside are prepared by using identical dry cells and resistors. Let's compare the lasting times of the dry cells in the circuits:

For having a comparison of lasting time, we will compare the times spent for drawing equal amount of charge, Q, from the batteries.

In CIRCUIT 1 three cells are connected in series with each other and in CIRCUIT 2 three cells are connected in parallel to each other. Below table gives the net potential difference (V_{net}), current (I) and the lasting time (t). In the table the ratio Q/I is assumed equal to t.

	V_{net}	I	t
CIRCUIT 1	V + V + V = 3V	$I_1 = \frac{3V}{R} = 3I$	$t_1 = \frac{Q}{3I} = \frac{t}{3}$
CIRCUIT 2	V	$I_2 = \frac{V}{R} = I$	$t_2 = \frac{Q}{I/3} = \frac{3Q}{I} = 3t$

Even though the main branch current in CIRCUIT 2 is I, the current drawn from each of the parallel cells is I/3. That is why when comparing the lasting time of the dry cells in CIRCUIT 2, the current is taken as I/3. According to the example circuits, CIRCUIT 2, which has dry cells in parallel, lasts 9 times longer than CIRCUIT 1, which has dry cells in series.

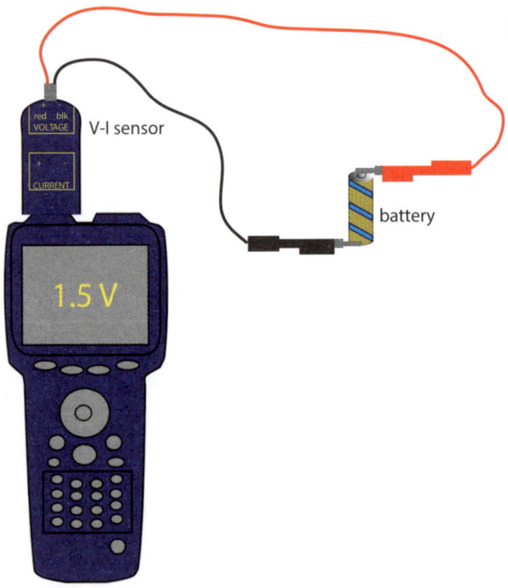

data logger

Until now we have assumed that the batteries that we use are *ideal*. This means that the potential difference between the ends of the battery will remain the same even when it is powering a device in the circuit. In a real circuit, the ability of a battery to act as an ideal battery is limited due to battery size, chemical properties, age and temperature.

When a battery is dead, it is usually because of the *internal resistance* it had built up. The battery cannot supply a useful amount of power to operate an external load. In such a case a voltmeter is not a useful tool to have an idea about the degree of life left in a battery.

If you measure the potential difference across the ends of a battery which is disconnected from the circuit (open circuit), you will usually read a value equal to the terminal voltage of the battery.

Now connect the same battery to an external resistance and wait for a while as it operates. If you touch it, you will feel that the battery is now hot. This is because of the current moving through the inside of the battery. As a result of the internal resistance of the battery some of the electrical energy produced is converted to heat energy as the electrons move through it.

data logger

If you measure the potential difference of the battery when it is connected to an external resistance, the value that you will read on the voltmeter will be less than the value you had measured for the open circuit.

According to the measurements aside, for the open circuit:

$V_{OC} = 1.5\ V$

For the closed circuit:

$V_{CC} = 1.2\ V$

Potential drop across the ends of the battery as a result of its internal resistance:

$\Delta V = 1.5 - 1.2 = 0.3\ V$

The difference between the two measurements is the potential drop across the ends of the battery as a result of electrical energy used by its internal resistance.

1.2.3. Electrical Energy and Electrical Power

Blades of a blender use the electrical energy to rotate. A mixer uses electrical energy when it is whisking the egg whites. For both cases, the electrical energy is converted to mechanical energy. When you turn on the light, the electric energy is transformed to heat and then light energy by a filament lamp. Electric toasters, hair dryers, electric heaters, kettles use the electrical energy to obtain thermal energy.

As the electrons are moving through a conductor, they collide with the atoms of the wire. During the collisions, some part of the kinetic energy of electrons is transferred to atoms that they collide. The kinetic energy of the atoms increase. An increase of kinetic energy of the atoms of the wire results an increase in the temperature. The increased thermal energy is transferred as heat by conduction and convection to the air (or to the water molecules in a kettle), radiated as light like in a filament bulb.

The electrical devices use the energy given and transform it as heat, kinetic energy. Power is the work done or energy used / transferred in unit time.

$$power = \frac{work}{time} = \frac{energy\ transferred}{time}$$

The electrical energy transferred is actually the work done by the charges. If the energy provided by the battery for 1 coulomb of charge is V, then the energy for the whole charge is:

$$W = V \cdot q$$

$$I = \frac{q}{t} \Rightarrow q = I.t$$

$$W = V.I.t$$

The work done in unit time by the electric charges is:

$$P = \frac{V.I.t}{t} = V.I$$

If Ohm's law is substituted for potential difference, V:

$$P = I.R.I = I^2.R$$

If Ohm's law is substituted for electric current, I:

Symbol	Quantity	Unit
P	power	W (watt)
V	potential difference	V (volt)
I	electric current	A (ampére)
R	resistance	Ω (ohm)

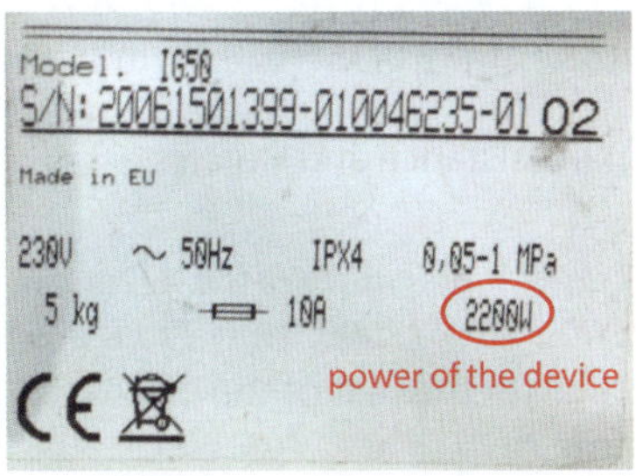

Electric devices have a sticker on them which gives the information about its electrical properties such as the potential difference, the current, etc. The power of the device is also written on that sticker. For the aside picture, 2200 W represents the power of the device. The device uses 2200 joules in one hour.

WARNING: As you had learnt at the Energy unit of 9th grade, *power* is the work done in unit time. *Watt* is the unit of power which is also equal to *joules* per *seconds*. For the electrical appliances, as its power is being defined, the time unit used is <u>hour</u>. As you are reading the power of the appliance keep in mind that the value given is the energy used in one hour.

Brightness of a bulb is dependent on its power. On the package usually numbers like 40 W, 60 W are written. It means that 40 W are used for one hour. If the lamps in a circuit are identical, then their resistors are the same. So in an electric circuit, depending on the connections of these identical light bulbs only the current through the bulb ($P \alpha I^2$ then) or only the potential difference across the ends of a bulb ($P \alpha V^2$ then) can be checked to compare the brightness as a practical method.

If you check your electric bill, you will see a *kWh usage* section. kWh is the abbreviation for kilowatt-hours. A kWh is the energy used in one hour. In that section you can see how many kilowatts of electrical energy you have used for that period of the bill. You pay for the energy that you have used, not the power.

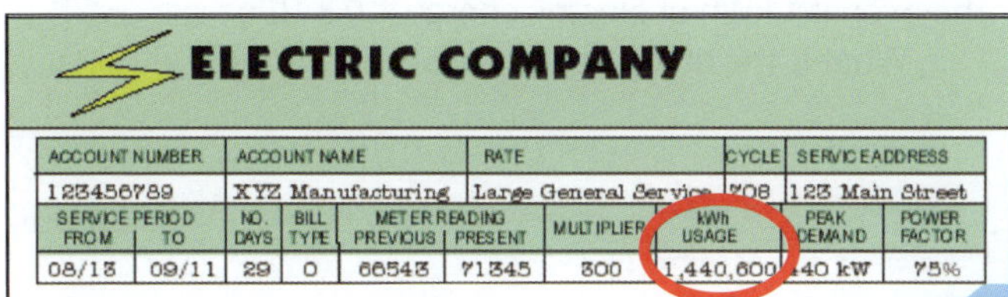

ACCOUNT NUMBER	ACCOUNT NAME	RATE	CYCLE	SERVICE ADDRESS
123456789	XYZ Manufacturing	Large General Service	708	123 Main Street

SERVICE PERIOD FROM	SERVICE PERIOD TO	NO. DAYS	BILL TYPE	METER READING PREVIOUS	METER READING PRESENT	MULTIPLIER	kWh USAGE	PEAK DEMAND	POWER FACTOR
08/13	09/11	29	0	66543	71345	300	1,440,600	440 kW	75%

Example

The power of a blow dryer is 2000 watts per hour.

a. How much electrical energy will be consumed when it operates for 6 minutes?

b. How much will be paid for the energy consumption of the dryer if 1 kWh of electric energy is 0.27 TL.

Solution

a. Blow dryer uses 2000 W in 1 hour. We have used it for 6 min.

$$6\ min = 6\ min\ x\ \frac{1\ h}{60\ min} = 0.1\ h$$

$W = P\ .\ t = 2000\ x\ 0.1 = 200\ Wh = 0.2\ kWh$

b. We will pay:

$0.27\ x\ 0.2 = 0.54\ TL = 54\ krs$

Example

A 5 Ω resistor is connected to a 30 V power source. Then it is placed in an isolated beaker which has 0.5 kg of water at 20 °C in it.

a. What is the power of the resistor?

b. How much heat is needed to have water at 100 °C?

c. In how many minutes will the water reach to 100 °C? (c_{water} = 4186 J/kgC°)

Solution

a. $P = \frac{V^2}{R} = \frac{30^2}{5} = 180\ W$

b. $Q = m.\ c.\ \Delta t = 0.5\ x\ 4186\ x\ (100 - 20) = 167440\ J$

c. $P = \frac{energy}{time} = \frac{Q}{t}$

$$t = \frac{Q}{P} = \frac{167440}{180} = 930.2\ s \simeq 15.5\ min$$

If you turn off your TV set by using the remote control, there will be a tiny light on the TV. This light shows you that your device is ready to be switched on from the remote control any time you wish. But as it is waiting, it is still consuming power. Stand-by power, sometimes referred as the *vampire power*, is the electrical energy consumed by electrical devices while they are switched off and not performing their function. You may think a device has gone to sleep when you switch off. But in reality its power consumption is just a little less than the power it consumes when it's active. If a device has an external power supply, remote control, continuous display (including an LED), or charges batteries power will be drawn continuously. Sometimes there is no obvious sign of continuous power consumption, but you can be sure about the vampire energy there.

Up to 80% of the energy consumption is used just to maintain a network connection by the games consoles. If regulators, organizations, service providers and device makers pay attention to energy efficiency there is an opportunity of 60 % of savings.

Example

An average microwave will use around 1200 watts per hour. In addition to using energy while cooking or heating, a microwave will also use 4 watts of power while in standby mode. This energy is used to display the clock and to wait for a user command. Suppose that the microwave oven is used for cooking 20 minutes a day, and for the rest of the day it is in stand by mode.

a. Calculate the energy used by the oven in a month (30 days).

b. How much will be paid for the energy consumption of the oven if 1 kWh of electric energy is 0.3 TL.

c. What is the efficiency of the oven?

Solution

a. $t_{output} = 20\ min/day\ x\ 30\ days = 600\ min = 10\ h$

$W_{output} = P_{output} \cdot t_{output} = 1200\ x\ 10 = 12000\ Wh = 12\ kWh$

But the oven is left in stand by for the rest of the day.

For one day the stand by time is:

$(24\ h\ x\ 60\ min/h) - 20\ min = 1420\ min$

The stand by time for the whole month is:

$t_{stand\ by} = 1420\ min/day\ x\ 30\ days = 42600\ min = 710\ h$

$W_{stand\ by} = P_{stand\ by} \cdot t_{stand\ by} = 4\ x\ 710 = 2840\ Wh = 2.8\ kWh$

Energy used by the oven for the whole month is:

$W_{net} = W_{input} = 12 + 2.8 = 14.8\ kWh$

b. We will pay:

$0.3\ x\ 14.8 = 4.44\ TL$

c. $e\ \% = \frac{W_{output}}{W_{input}} x100 = \frac{12}{14.8} x100 = 81\ \%$

19 % of the energy consumed is the stand by energy.

While waiting for the manufacturers to pay attention and do something about vampire energy, you can reduce the standby power at your home by:

- Using a power strip with on/off switches as a central "turn off" point. If you plug all of your products into a power strip and flip off the power strip when these items are not in use, they are truly off.
- Unplugging your chargers: cell phone chargers, camera chargers, battery chargers or power adapters, etc. These are drawing some amount of energy even when not in use (and even when not connected to an end-use product).
- Looking for ENERGY STAR when shopping. All ENERGY STAR qualified products are among the lowest power consuming in their category in standby mode. Enable the ENERGY STAR power management settings on your computer and monitor, so they go into power save mode when not in use.

1.2.4. Safety Precautions Against Electricity

The utility companies provides the electrical energy for you. The electrical supply is connected to outlets in your house. In a simple two-wire electrical power system, the conductor connected to ground is called the *neutral*, and the other conductor is called the *hot*, also known as the *live* or the *active*. It is named *hot* because of its potential for shock hazard. It will be dangerous to touch unless voltage is secured by proper disconnection from the source.

A fuse

The electricity runs through the wires in the walls of your house. Even though it is very small, these wires have resistance. A large enough current may cause these wires to heat up. They may become so hot that they can start a fire. If a wire carries more current than it is safe, then it is said to be overloading.

A fuse box: Each button controls a different circuit in the house

To prevent the overloading on a wire, fuses or circuit breakers are used. They operate like a switch in the circuit and open the circuit (create a gap in the conducting path) when the current exceeds a certain value. If a circuit repeatedly opens a circuit breaker or burns out a fuse then there may be two reasons.

1. Too many devices are connected to the circuit: Electrical appliances in a house are in parallel to each other. Each device connected the circuit, draws its own current and adds up to the current of the main branch. Thus the current of the main branch increases with each device plugged in. If the current on the main branch is greater than the current of the fuse or the circuit breaker, some of the devices should be unplugged to lower the current on the main branch. There are several circuits in a house or apartment, like a circuit for kitchen, a circuit for the living room, etc. One of the devices may be plugged to another circuit within the house. An electrician chooses the proper fuse size according to the wires used in the construction of the building. NEVER replace a properly rated fuse with a higher one. You may start a fire!

Plugs used in various countries

2. There may be a short circuit: Two wires that shouldn't be touching to each other may come in contact. That may be because the insulation of the wires had worn through. This shortens the path for the current. Thus the resistance of the circuit becomes very small and a very large current can be drawn. Short circuits should be repaired immediately.

Lets consider a mini-oven with a conductive metal case. If it is operating properly, then there should be no shock hazard. The wires conducting power to the oven's heating element are insulated from touching the metal case (and each other) by rubber or plastic. However, if one of the wires inside the oven accidently come in contact with the metal case, the case will be made electrically common to the wire, and touching the case will be just as hazardous as touching the bare wire. Whether or not this presents a shock hazard depends on which wire accidentally touches:

Left: A socket with no grounding
Right: A grounded socket

- If the "hot" wire contacts the case, the user of the oven is in danger.
- If the neutral wire contacts the case, there is no danger of shock.

50 % of danger is a very high rate. Some engineers solve this problem simply by making the outside case of the device nonconductive, thus it is double-insulated. Some engineers solve this problem of safety by maintaining a conductive case, but using a third

conductor to firmly connect that case to ground. The third connection on the power cord provides a direct electrical connection from the device case to earth ground, making the two points electrically common with each other. If they're electrically common, then there cannot be any voltage dropped between them. If the hot conductor accidently touches the metal appliance case, it will create a direct short-circuit back to the power source through the ground wire, tripping any over current protection devices. The user will remain safe.

0.2 – 2mA	Slight tingling
5mA	Faint shock
6 – 16mA	Muscles freeze, painful shock
30 – 99mA	Respiration and muscle paralysis
100 – 2000mA	Ventricular fibrillation (erractic heart functioning)
2000mA and above	Cardiac arrest, burning of organs and tissues leading to death

Electric current passing through the human body can heat the tissues and cause burns. It may also stimulate nerves and muscles and we may feel a shock. The results of an electric shock depends on the magnitude of the electric current, how long it acts on the tissue and which part of the body it passes through. The physical conditions of the human body is also important.

The resistance of human body depends on its physical conditions. It may vary from 1000 Ω to 100000 Ω for dry skin. If the skin is wet then the resistance lowers instantly. Different parts of our body have different resistance values. Currents having a path outside the skin will cause considerably less damages. If the path is from hand to hand or to leg, the electric current would directly pass through the path of the heart. This is an internal path, and the results could be lethal. The threshold value of current for human body is about 10 mA. But sometimes low levels of current may also cause involuntary contraction of the muscles. The person may be pushed away from the circuit and fall from an elevation. The fall may result with extreme injuries or even death. So, in this case, he will not die of electrocution, but due to the fall.

Birds on an electric wire, resting safely

To receive a shock, there must be a potential difference between one part of your body and another part. Current always chooses the easier path, so it will pass through the less resistant path connecting these two parts of your body. This also explains how birds can stand on electric wires harmlessly. As long as the bird stays on one cable, all of its body parts will have the same potential value. Thus there won't be any potential difference and no harm will be caused on the bird. But if the bird accidentally touches to the other wire, it gets highly electrocuted.

The first thing to do to rescue a person having an electric shock is cutting the current off. If switching off is not possible, then getting the person away from the electric supply by using a dry wooden stick or some other insulator may help. Then if necessary apply artificial respiration and call paramedic team.

1.3. Magnets and Magnetic Fields

1.3.1. Magnetic Fields Produced by Magnets

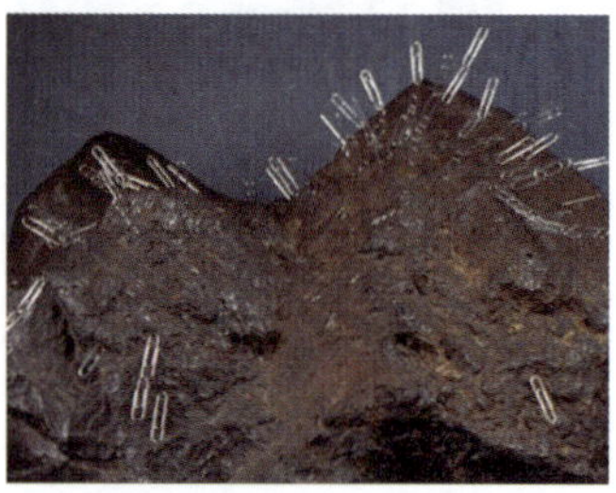

Paper clips attracted by an iron ore lodestone

Magnetite

Although there are various stories about the history of magnets, the most common one tells that it was first found at western part of Asia Minor by the Greek. Some sources say that it was a shepherd named Magnes who noticed that some black stones were naturally attracting his iron staff, and named after him. Some other sources say that it was first discovered at Magnesia (Manisa of today) and the name comes from the city. Magnet was an iron ore lodestone or magnetite, a natural compound of iron (Fe_3O_4). It was attracting some certain substances like iron fillings.

We can clearly say that people knew the attractive properties of magnets before the electric charges. Chinese people made use of the magnets to find their direction at 11th century. People made use of magnets but first systematic studies about magnets were done by William Gilbert in 1600. He came up with the theory that Earth itself is a giant magnet. His studies are followed by a number of physicist like H. C. Oersted, C. F. Gauss, M. Faraday, A. M. Ampére.... Our technology today not only depends on electricity. The studies and researches about magnetic properties of matter lead to discoveries using the electrical and magnetic properties of matter, in other words to electromagnetism. In the nineteenth century, James Clerk Maxwell founded the studies about electromagnetism. Here at this point, it will be good to emphasize that electricity and magnetism are two different characteristics of matter. They work together or they are present together in many cases, but they are not the same concepts.

If you suspend a bar magnet from the middle so that it can rotate freely in the horizontal plane, you will observe that its one end will always *seek* for the north. The ends of a magnet is usually called the poles of the magnet. The north seeking pole of a magnet is its *north pole* and likewise, the south seeking pole of a magnet is its *south pole*. Chinese people observed this property of magnets and used them as an aid for finding their direction. A small, lightweight magnet, called the needle, is balanced on a nearly frictionless pivot point. One end of the needle is colored or marked as N for North Pole. This simple device is called a compass. No matter where you are, the compass needle always points the North Pole.

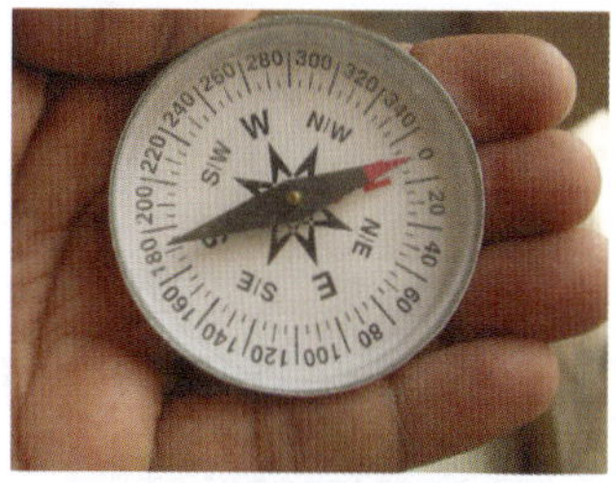

A compass

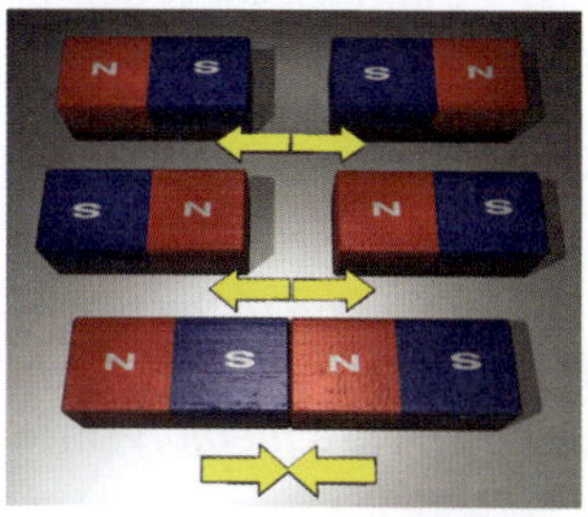

Like the electric charges, like poles of two magnets will repel each other and unlike poles of two magnets will attract each other. As in electrostatic forces, the force applied by the magnets is also a force acting at a distance. If any pole of a magnet is brought close enough to paper clips they will begin to move towards it even though they are not touching. This behavior of substances leads us back to the field concept. A field is present around a magnet. This field is called the magnetic field.

Different shapes of magnets

The magnetic field exists all around a magnet but it is strongest at its ends.

It is easy to find an isolated positively or negatively charged particle. But in the case of the magnets, if you break a magnet in two pieces, you will have two smaller magnets with two poles. If you continue to break the magnets in two, at all times the resulting magnets will have N and S poles even at the microscopic scale. No magnetic monopoles (one-pole) exist.

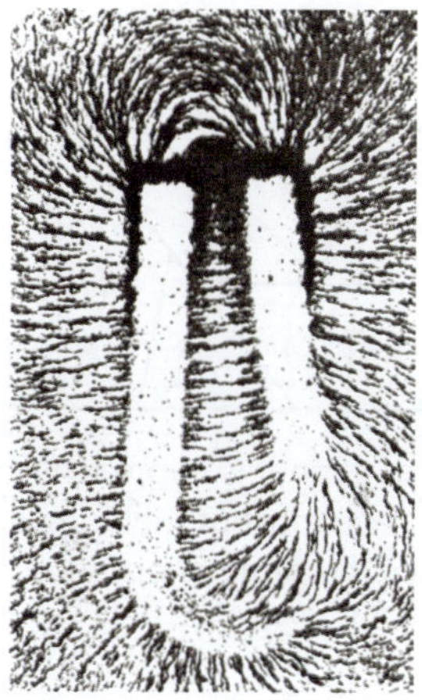

The iron fillings are more condensed at the poles of a horse shoe magnet.

Electrically neutral objects have equal numbers of positive and negative charges. All neutral objects are attracted by charged objects as a result of induction. A difference rises at this point between electricity and magnetism. Not all objects are attracted by magnets. Iron, cobalt, nickel, gadolinium and some of their oxides and alloys show strong magnetic effects. Other materials' magnetic effect is very weak that can only be detected by very delicate instruments. You cannot attract a piece of paper with a magnet even if you have a magnet as big as a car. To understand the reason why not all substances are attracted by a magnet, we must investigate the substances in atomic scale.

We know that atoms consist of a nucleus which has protons and neutrons and there are the electrons in the energy levels. The electrons rotate around the nucleus and themselves just like the Earth rotating around the sun and itself. The rotational motion of an electron around itself is called the spin of the electron. The spin of the electrons causes a magnetic field. We can assume an electron as a tiny magnet then. The quantum physics offers that in every energy level there are two electrons with opposite spins referred as up and down. In a filled shell, the spins cancel out each other. If an atom has odd number of electrons, at least one electron's spin is not coupled, a magnetic field rises.

Substances such as iron are called ferromagnetic (ferrum = iron in Latin) materials. In ferromagnetic materials, the atoms have unpaired electrons thus a net spin. These tiny electron magnets line up and reinforce each other. If there is no net spin and/or they don't line up and reinforce each other, then the substance is not ferromagnetic.

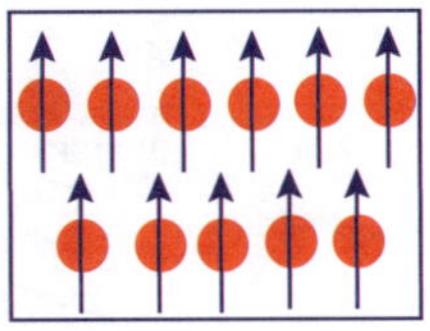

Electrons in a magnetic domain are lined up in the same direction and produce a net magnetic field.

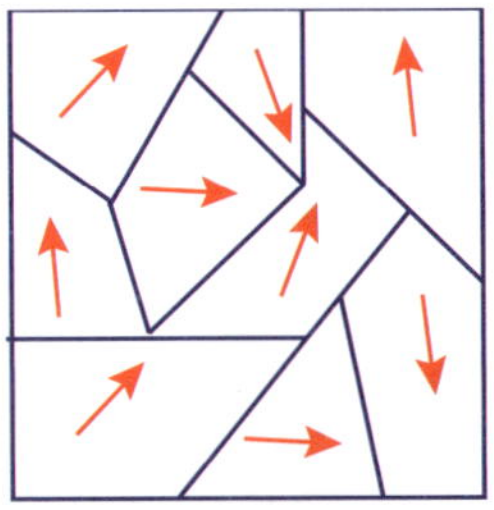

Not all the atomic magnets in the domains have the same alignment, they are randomly oriented.

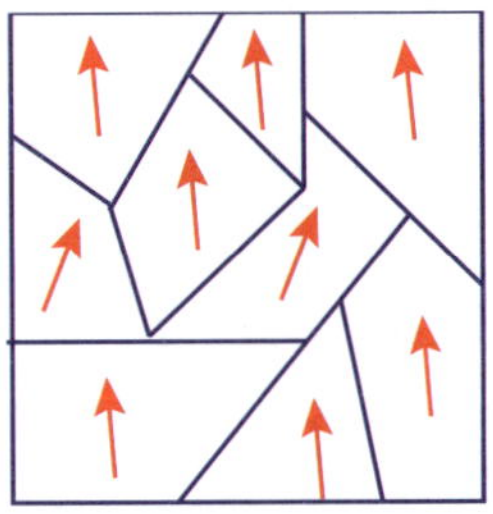

When a magnet is approached to iron piece, most of the domains have the same orientation

If we consider a very small piece of iron, spins of electrons for all atoms of that tiny region are lined up in the same direction. We will name that small region a magnetic domain. You can consider a magnetic domain as a region which has a size of 1 mm length or width. We can say that there is a net magnetic field in that domain. If you consider a larger piece of iron, again all the domains will have a net magnetic field individually, but as a whole these magnetic fields do not have the same direction. They have their own orientation which doesn't result a net magnetic effect throughout the iron piece. If a magnet is brought near that piece of iron, the magnetic field of the magnet penetrates into the iron and rotates the atomic magnets of the domains within the iron. Most of the domains align with the external magnetic field and predominate. The iron will be attracted towards the magnet. Because the N pole of the tiny magnets within the iron piece faces the S pole of the magnet or vice versa.

In the presence of an applied magnetic field the iron becomes magnetized, but this is not for a very long time. The magnetization is temporary. But if a strong enough magnetic field is applied for a sufficient time or if a magnet is rubbed on a piece of iron always in the same direction (like always rubbing the magnet from the beginning to the end of an iron rod), then the iron may become permanently magnetized for a long time. The iron is a permanent magnet now. Ferromagnetic substances are said to have a *memory* since they remain magnetized after the external field is removed.

If you drop the magnet on the floor or hit it with a hammer, you will cause the magnetic domains to loose their alignments. The magnet may loose some or all of its magnetization. If a magnet is heated, the increase in the temperature increases the random motion of atoms. This randomizes the alignment of the domains and the magnet begins to loose its magnetization. If a critical temperature is exceeded (Curie temperature) a magnet cannot be made at all.

A second type of magnetic materials, paramagnetic substances, like aluminum, copper, antimony, do also have magnetic dipoles. But these magnetic dipoles attract so weakly with each other. In the absence of an external magnetic field these dipoles are randomly oriented. Once a paramagnetic substance is placed in an external magnetic field, the magnetic dipoles tend to line up with the field. But at the same time, the thermal motion of the atoms or molecules increase. The magnetic orientation process competes with the thermal motion. The magnetization of a paramagnetic substance is increases with an increasing external magnetic field and decreasing temperature.

Diamagnetic substances like water, plants, your skin, almost have no magnetic dipoles. When they are placed in an external magnetic field, the magnetic dipoles that they have are oriented opposite to the field direction.

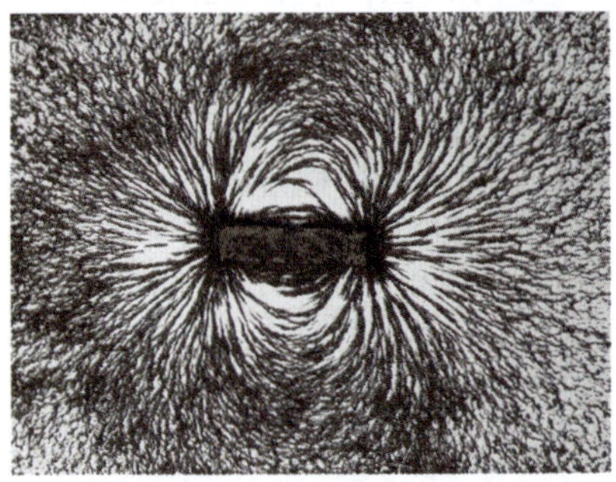

A magnet is surrounded by a magnetic field. If iron fillings are sprinkled over a bar magnet, magnetic field lines can be observed. Since there are no magnetic monopoles, magnetic field lines leave the magnet from one pole and enter to it from the opposite pole. They continue inside the magnet and always form closed loops unlike the electric field. The electric field extends outwards from a positively charged particle and they enter to a negatively charged particle. The magnetic field lines leave the magnet from the N pole and enter to it from the S pole.

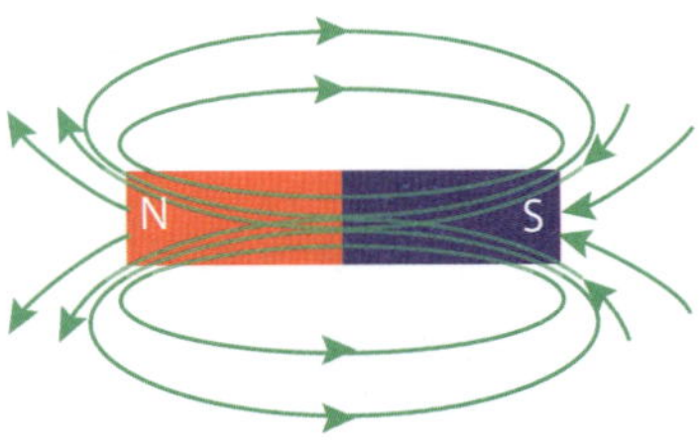

The number of field lines in unit area is proportional to the strength of the field at that region. The direction of the magnetic field at a point on a field line is tangent to that field line. It is defined as the direction that N pole of a compass needle would point if it was placed at that point.

The repulsion and attraction forces between the poles of magnets can be represented just like we did for electric charges. In the case of like poles, the field lines will expel each other. For the unlike poles, the field lines will leave the pole of one magnet and enter to the pole of the other magnet.

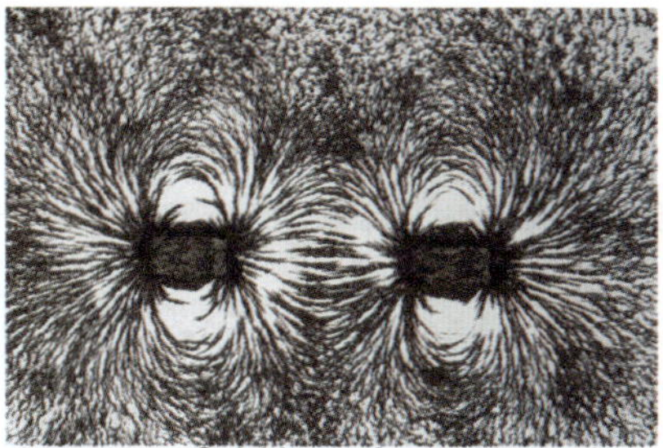

Magnetic field lines for unlike poles

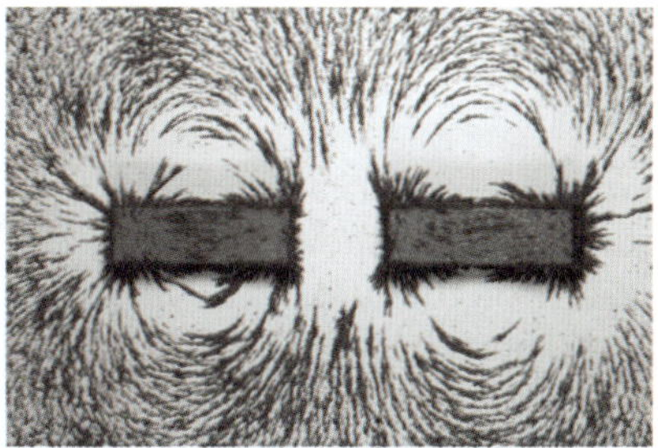

Magnetic field lines for unlike poles

The force present between magnets have the characteristics of the electrostatic force which was first studied by Coulomb. Likewise, the force applied by the magnets on each other is directly proportional to the magnetic intensity of their poles and inversely proportional to the distance between the magnets.

1.4. The Relationship Between Current and Magnetic Field

1.4.1. Magnetic Field Around a Current Carrying Wire

In 1820 Hans Christian Oersted, a Danish physicist and chemist, was preparing a lecture on Voltaic cell. As he was setting up his materials he noticed that a compass needle near the experimental set up was being deflected when the electric current was flowing in the circuit. A compass needs a magnetic field to be deflected from the Earth's North Pole. This deflection convinced him that magnetic fields radiate from all sides of a wire carrying an electric current. This was a very important observation since it was the first to see a direct relationship between electricity and magnetism.

Oersted with his students

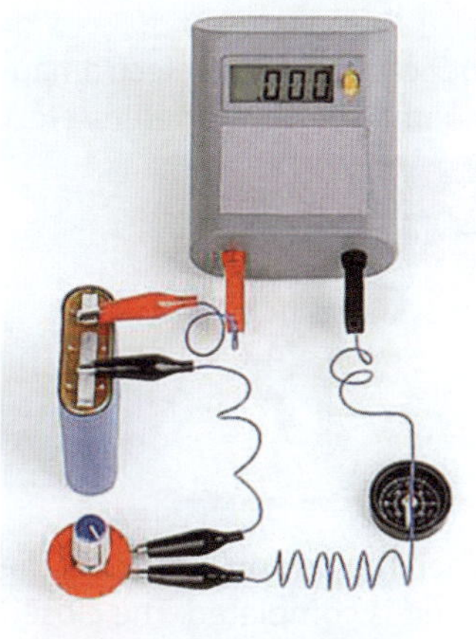

No deflection in the compass needle when the switch is off.

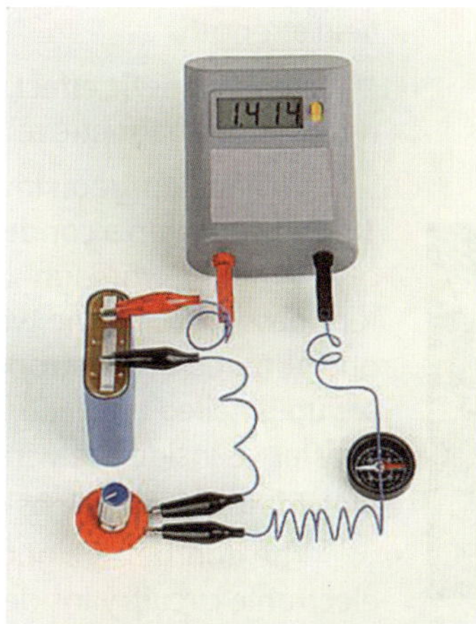

The compass needle deflects when the switch is on and the current is flowing through the wire.

Compass needles placed near a straight section of a current carrying wire experiences a force on it. This force causes the needle to align tangent to a circle near the wire. The magnetic field lines produced by a current in a straight wire are in the form of circles with the wire at the center. The direction of these lines is indicated by the north pole of the compasses.

Physics Extra

To find the direction of the magnetic field around a current carrying wire, the right hand rule is used. The thumb of your right hand shows the flow direction of current. If you wrap your right hand around the current carrying wire, the tip of your four fingers points the flow direction of the magnetic field.

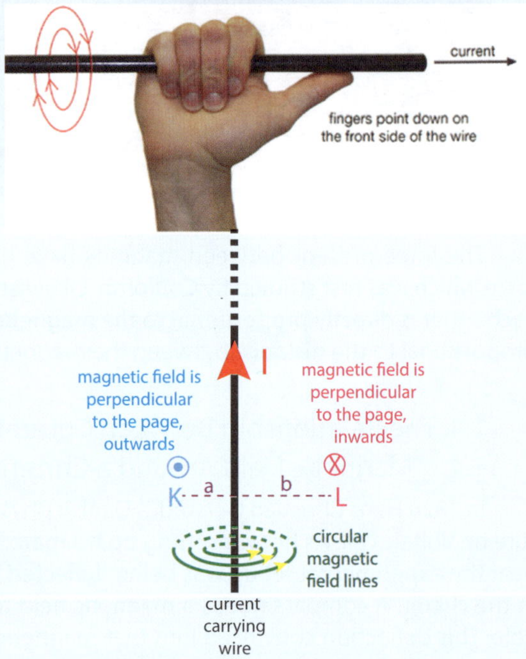

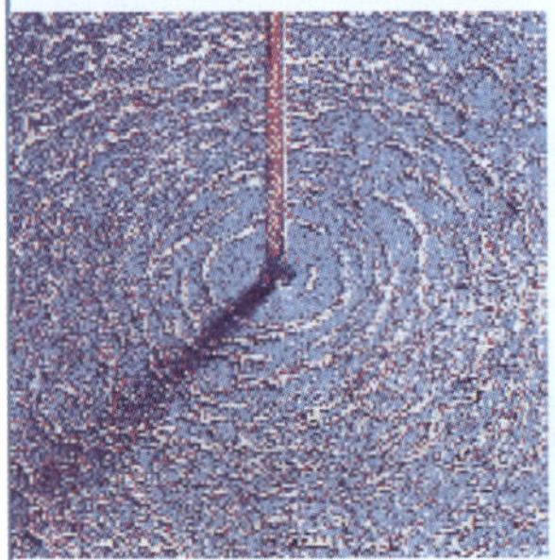

The magnetic field lines formed around a current carrying wire

It is usually so hard to illustrate the magnetic field of a point in two dimensional analysis. The magnetic field vector is perpendicular to the page and point outwards (towards you) at point K. For such a magnetic field direction, a small circle with a dot inside is used. You can imagine that the tip of an arrow is approaching to you. Likewise at point L, the magnetic field vector is perpendicular to the page and point inwards (into the page). This time for the direction of the magnetic field, a small circle with a cross sign inside is used. You can imagine that the tail of an arrow is getting away from you.

As you get away from the current carrying wire the strength of the magnetic field will decrease.

Another observation depends on the magnitude of the current flowing through the wire. More current means more charge and more charge means a stronger magnetic field strength.

The magnetic effect of an electric current has many applications, e.g. in electromagnets, electromagnetic relays, loudspeakers, electric motors, ammeters, voltmeters, etc.

William Sturgeon invented the electromagnet in 1825. If you wrap a conducting wire around a iron rod or any other ferromagnetic substance and let the current pass through the wire, you can produce almost one hundred times stronger magnetic field. This easy set up is called an electromagnet. The ferromagnetic substance inside the wire coils (a solenoid) is a magnet as long as the current flows through the solenoid.

An electromagnet

The doorbell is a good example of electromagnets. When the button is pushed, the electronic circuitry inside the door bell closes. Thus the circuit is completed. The closed circuit allows electricity to flow. The flowing current creates a magnetic field and causes the clapper to become magnetized. The hardware of most doorbells consist of a metal bell and metal clapper. When the magnetic charges causes them to clang together, you hear the chime inside. The bell rings, when the button is released, the circuit opens and the doorbell stops ringing. Thus the bell doesn't ring always. It rings only when a current passes through the circuit.

Old fashioned door bell

Strong electromagnets are used to control charged particle beams in high energy accelerators such as the super colliders. The maglev trains also levitate and reach to high speeds with the use of electromagnets.

Maglev train

Toroidal magnets of the huge detector ATLAS at CERN

1.4.2. Magnetic Field of the Earth

The magnetic fields exists everywhere around us. We live in a magnetic field that is created by the Earth's core. Many of the technological devices that we use have their own magnetic fields.

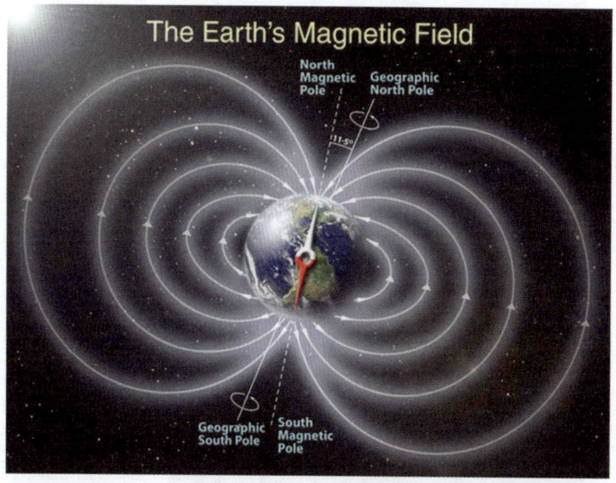

The pattern of field lines around the Earth are like as if there were a giant bar magnet inside the Earth. N pole of a compass needle is attracted towards the geographic north, so the S pole of that giant magnet should be at the northern hemisphere, whereas its N pole should be in the southern hemisphere since the south pole of the compass needle points towards the geographic south. Of course there is not a giant magnet inside the Earth. But the Earth has large deposits of iron ore deep beneath its surface. The temperature is very high below the surface. This keeps the iron ore to become permanently magnetized. But the iron is molten and there are convection currents in the Earth's core. The charged ions or the electrons circle in the liquid, and produce magnetic field just like a current loop produces a magnetic field.

If you suspend a compass needle from the middle so that it can rotate either in vertical and horizontal planes, you will observe that its alignment will change continuously as you move from the equator to northward or from the equator to southward. The needle is horizontal to the ground only at the equator. As you get away from the equator, the needle rotates such that it points more and more towards the surface of the Earth. In 1832, James Ross located a point at Hudson Bay in Canada. At this point the compass needle was directly pointing downward. That was the location of Earth's south seeking pole, in other words, north pole. This point was approximately 2100 km away from the geographical north of the Earth. North seeking pole of the Earth, the south pole was approximately 1900 km away from the geographical south pole of the Earth. This point was to the south of Australia. The North magnetic pole has been moving in the Canadian artric. At 2001 it was 600 km away from the nearest town Resolute Bay, Canada. The pole kept going during the 20th century, north at an average speed of 10 km per year, lately accelerating to 40 km per year. At this rate it will exit North America and reach Siberia in a few decades. The magnetic south pole changes its position just like the north pole but at a slower rate.

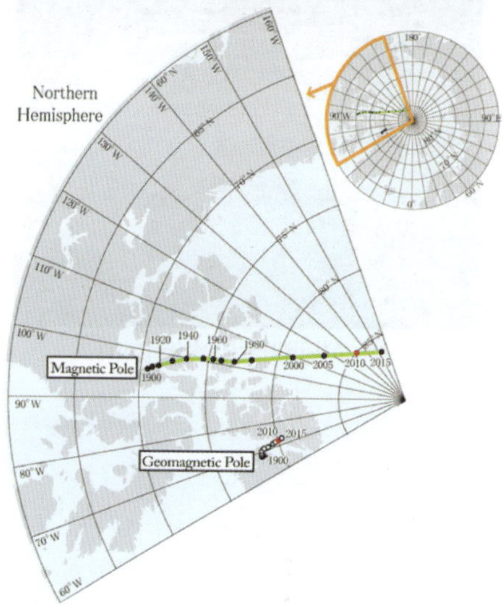

We can conclude that, the magnetic and geographical poles of the Earth do not coincide. The geographic poles are defined with respect to the rotational axis of the Earth. The angular difference between the geographic and magnetic poles of the Earth is called magnetic declination. The angle that the compass needle makes with the horizontal at any point is referred to as the magnetic inclination or the angle of dip.

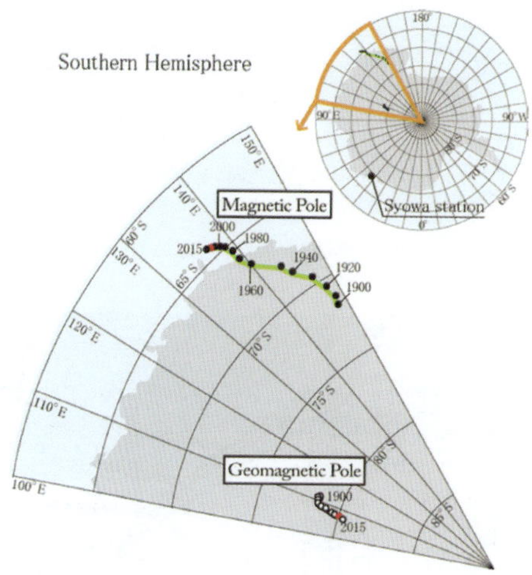

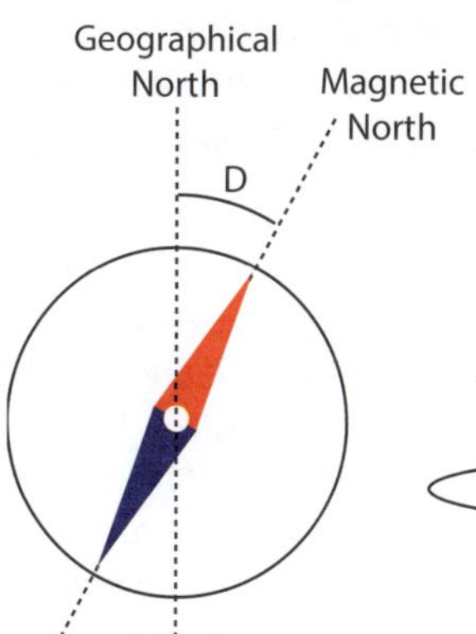

Decliniation

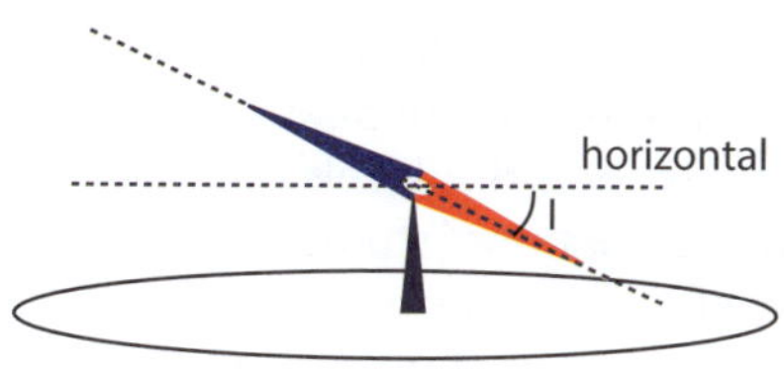

Incliniation

The position of the magnetic poles varies from year to year, point to point. The direction of the magnetic poles have been reversed throughout the geological time. A careful analysis of rock strata provides the evidence for this change. Molten iron atoms do not have a magnetic orientation because of the thermal motion. But a very small amount of the iron atoms will align with the Earth's magnetic field. When these iron atoms cool

and solidifies in igneous rocks, this predominance records the direction of the Earth's magnetic field. Magnetic domains of grains of iron in the sedimentary rocks also settle themselves with the Earth's magnetic field. By using sensitive instruments, the slight magnetism can be measured. If samples of rock from different strata are tested throughout the geological time, the magnetic field of Earth for different periods can be charted. There had been times when the magnetic field of the Earth had diminished to zero and the poles reversed themselves. More than 20 reversals had taken place in the past 5 million years. The most recent reversal had taken place approximately 730000 years ago.

The strength of Earth's magnetic field also changes according to location and season. Varying ion winds in the Earth's atmosphere cause more rapid but much smaller changes in its magnetic field. The energetic interactions of the atoms in the atmosphere with the solar UV rays and X rays produce the ions in this region. The motion of these ions produces a small but important part of the Earth's magnetic field.

We are always under the influence of the cosmic rays from the outer space. Cosmic rays are charged particles departed from a star such as the Sun. They are deflected away by the Earth's magnetic field and make up the Van Allen Radiation Belts. Van Allen Belts are two doughnut shaped rings. The inner belt is about 3000 km away from the Earth and the outer belt is 15000 km away. Astronauts stay in orbit below these belts at safe distances from the radiation. Most of the particles trapped by the outer ring comes from the sun. The cosmic bombardments are strongest at the poles. This is because the charged particles that hit the Earth there travel along the magnetic field lines and they are not deflected. As you get away from the poles, cosmic ray bombardments get less and less. It is the smallest at the equator.

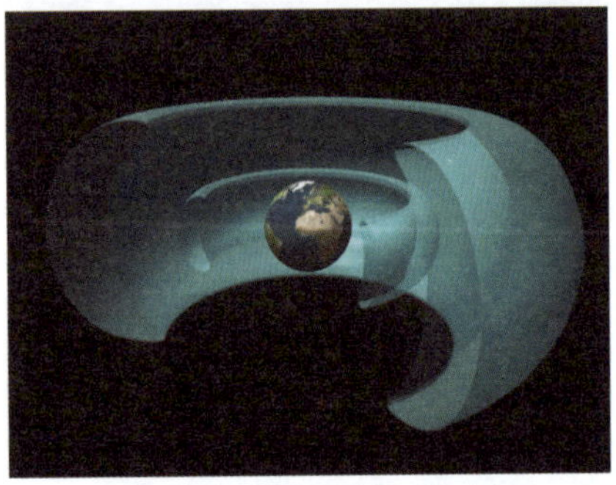

Van Allen Radiation Belts

Not only Earth has a magnetic field in the solar system. The strength of a planets magnetic field is related to the rate of rotation. Jupiter rotates faster then the Earth so it has a larger magnetic field. Venus rotates slower thus it has a smaller magnetic field.

Whether it is a living thing or non-living thing, everything has either a weak or a strong magnetic field. We, humans do have a magnetic field and this magnetic field is coherent with the Earth's magnetic field. When the first astronauts returned to earth they experienced what was called "space sickness"..nausea; headaches; disorientation; sleep disorder. At first it was thought to be diet; and weightlessness. But then it was discovered that a lack of magnetic field was the problem. Each space craft was fitted with a magnetic field generator. The problems stopped.

We cannot live without some form of magnetic field. But with the recent changes in our technology, there had been an enormous increase in the electric and magnetic fields, in other words, electromagnetic fields around us. Now we are under the influence of the magnetic fields produced by the electrical devices like mobile phones, televisions, all electrical appliances and also the high voltage transmission lines. The high artificial electromagnetic fields may cause some health problems on humans, animals and plants.

Short term health problems that may occur are fatigue, anxiety, insomnia, prickling and/or burning skin, rashes and muscle pain.

Strong, artificial EMFs like those from power lines can scramble and interfere with your body's natural EMF, harming everything from your sleep cycles and stress levels to your immune response and DNA. After hundreds of international studies, the evidence linking EMFs to cancers and other health problems is loud and clear. Researchers found that children living within 200 m of power lines had a 70% greater risk for leukemia than children living 600 m away or more. Several studies have identified occupational exposure to extremely low-frequency electromagnetic fields (EMF) as a potential risk factor for neuro-degenerative disease. There is strong prospective evidence that prenatal maximum magnetic field exposure above a certain level (possibly around 16 mG) may be associated with miscarriage risk.

Even though navigational applications of the smart phones are used most commonly in our daily life and technology, a compass would be the most reliable device when you are out in the wilderness. A compass never looses the GPS signal and is never out of battery.

All kinds of compasses include a magnetized needle that orients itself to the magnetic fields in the Earth. As you are finding your direction with a compass don't forget to take the magnetic declination into consideration and be careful about ferromagnetic substances that are nearby. Compass needle will interfere with ferromagnetic substances and may loose track of the magnetic north pole.

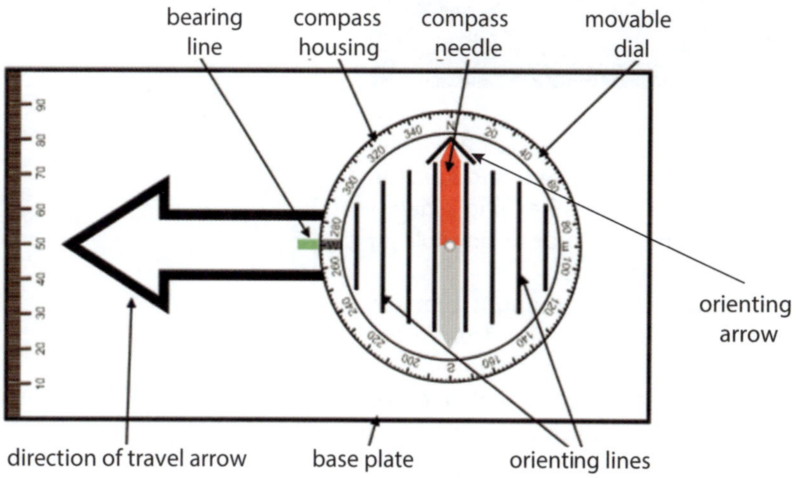

A most commonly used compass especially by hikers and in orienteering is the baseplate compass. The compass is embedded on the clear plastic baseplate. The arrow pointing away from the compass is the direction of travel arrow. The magnetized compass needle is placed a clear, plastic circle called the compass housing. The movable dial surrounds the compass housing that displays all 360 degrees of the circle. The magnetic needle is the compass needle. The red end of the compass needle is for north. The non-magnetic arrow within the compass housing is the orienting arrow. The orienting lines run parallel to the orienting arrow.

As you are finding your direction, begin with holding the compass flat in your palm and your palm in front of your chest. The red end of the compass needle will show north but usually it will not coincide with N on the movable dial unlike the figure above. Turn the movable dial until the magnetic compass needle and the orienting arrow line up. Now you can tell where you are travelling to by looking at direction of travel arrow. For the figure above the direction of travel is west. If the direction of travel is somewhere between North, East, South and West; look closely at the degree markers on the compass. If it intersects at 32, you're facing 32 degrees Northeast. Ignoring the magnetic declination may result reaching to a very distant place from your target. Don't forget to take off local magnetic variation by twisting the degree dial the correct number of degrees to the left or right, depending on the declination and check where the direction of travel arrow lines up with the degree dial. Remember to check your compass from time to time and be sure not to accidentally twist the degree dial from its current position. As you are walking to follow the direction of travel arrow accurately, look down at the arrow, then focus on a distant object like a tree, telephone pole, or other landmark, but not a huge object as a mountain, and use this as a guide. Once you reach each guide point, use your compass to find another.

People used to accept that migrating animals were using the sun and the stars as they were migrating. In 1957, Hans Fromme noticed that the robins he'd placed in a cage were keeping on moving to southwest in fall. The robins were in a closed room without the sun and the stars but they were still trying to migrate to southwest, to Spain. He was the first to suggest that animals were using Earth's magnetic field to navigate.

We, people are using compass to find our direction but there are several animals such as sea turtles, trout, pigeons, fruit flies, monarch butterflies and even magnetotactic bacteria have their own, in built, navigational system to find their way.

ANIMAL MAGNETIC SENSE

A team led by Can Xie's at Peking University in China has found a protein in fruit flies, butterflies and pigeons that they believe to be responsible for this magnetic sense.

Some scientists thought that the magnetic sense came from iron-binding molecules, and some others thought it came from a protein called cryptochrome, which senses light and has been linked to magnetic sense in birds. Xie's group was the first to guess these two were part of the same system. Xie's team first screened the fruit fly genome for a protein that would fit a very specific bill. The molecule had to bind iron, it had to be expressed inside a cell instead of on the cell membrane and do so in the animal's head – where animals tend to sense magnetic fields – and it also had to interact with cryptochrome.

"We found one gene fit all of our predictions," says Xie. They called it MagR and then used techniques including electron microscopy and computer modelling to figure out the protein's structure. They found that MagR and cryptochrome proteins formed a cylinder, with an inside filling of 20 MagR molecules surrounded by 10 cryptochromes. The researchers then identified and isolated this protein complex from pigeons and monarch butterflies.

In the lab, the proteins snapped into alignment in response to a magnetic field. They were so strongly magnetic that they flew up and stuck to the researchers' tools, which contained iron. So the team had to use custom tools made of plastic.

https://www.newscientist.com/article/dn28494-animal-magnetic-sense-comes-from-protein-that-acts-as-a-compass/

Solved Questions

Question

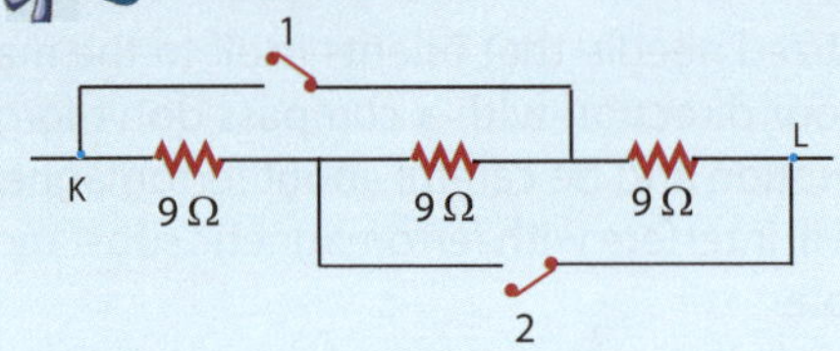

Find the equivalent resistance when

a. both switches are open.

b. switch 1 is open and switch 2 is closed.

c. both switches are closed.

Solution

a. All resistors are operating and they are in series to each other when both switches are open.

$R_{eq} = 9 + 9 + 9 = 27\ \Omega$

b. Only the first resistor operates when switch 2 is closed. The current prefers to flow from the low resistance branch. The last two resistors have a short circuit around them.

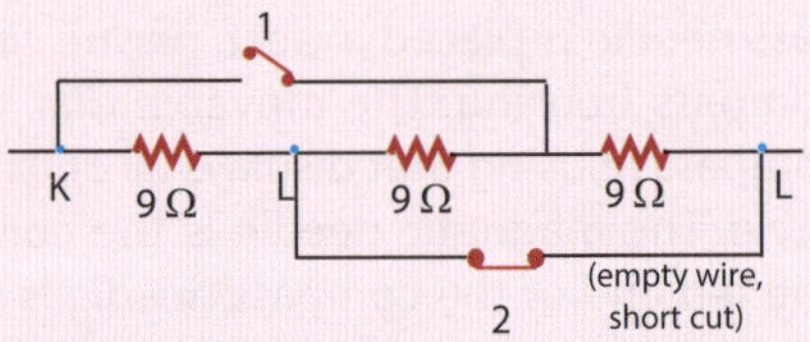

$R_{eq} = 9\ \Omega$

c. All resistors are connected between points K and L now when both switches are closed.

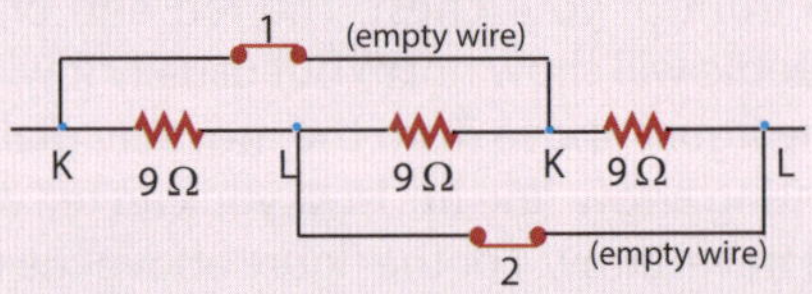

They are parallel to each other now.

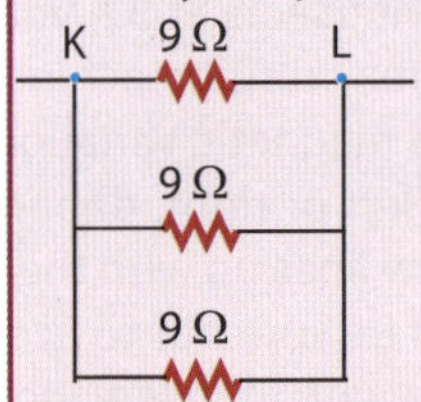

$$\frac{1}{R_{eq}} = \frac{1}{9} + \frac{1}{9} + \frac{1}{9}$$

$$\frac{1}{R_{eq}} = \frac{3}{9} = \frac{1}{3}$$

$$R_{eq} = 3\ \Omega$$

Question

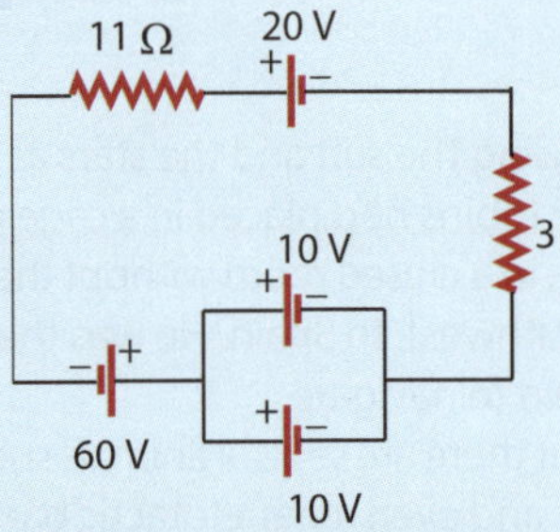

Find the current of the circuit.

Solution

10 V batteries are connected in parallel to each other so the net voltage they provide to the circuit is 10 V.

20 V and 60 V batteries are backing up each other whereas the 10 V batteries are opposing to the current they are providing to the circuit.

Then the net voltage of the circuit is:

$V_{net} = 20 + 60 - 10 = 70\ V$

The equivalent resistance is:

$R_{eq} = 11 + 3 = 14\ \Omega$

The current is:

$$I = \frac{V_{net}}{R_{eq}} = \frac{70}{14} = 5\ A$$

Question

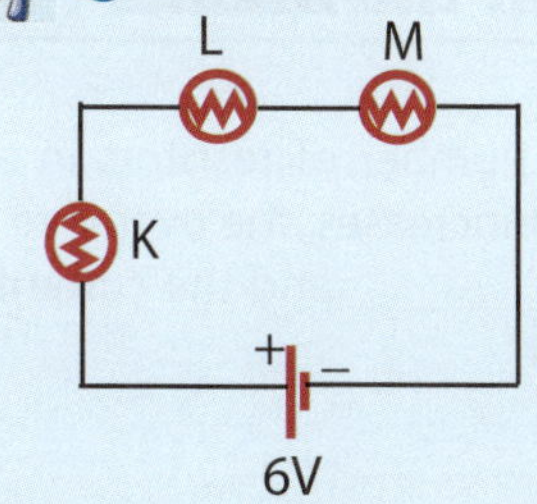

Figure 1

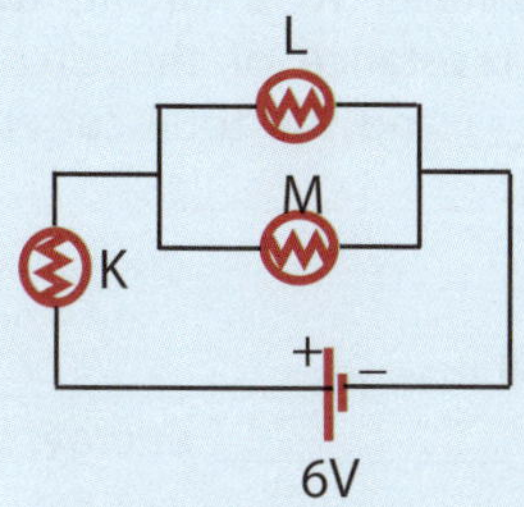

Figure 2

Identical bulbs are connected as given in figures 1 and 2.

a. Compare the brightness of the bulbs in figure 1.

b. Compare the brightness of the bulbs in figure 2.

c. How do the brightness of the bulbs compare with respect to figure 1 and figure 2.

Solution

Bulbs are identical so a comparison of the current or the potential difference will be enough.

a. Bulbs are connected series to each other and the current passing through them is the same. They have the same brightness.

b. Bulbs L and M are connected parallel to each other, the current passing through them is the same so they have the same brightness.

The current passing through K is twice that of the current passing through L or M. So K is brighter than L or M.

K > L = M

c. The equivalent resistance changes thus the current changes in circuits. But the battery of the circuits does not change so we can use the potential difference across the ends of each bulb.

	K	L	M
Circuit 1	2V	2V	2V
Circuit 2	4V	2V	2V
Change	Increases	Same	Same

Question

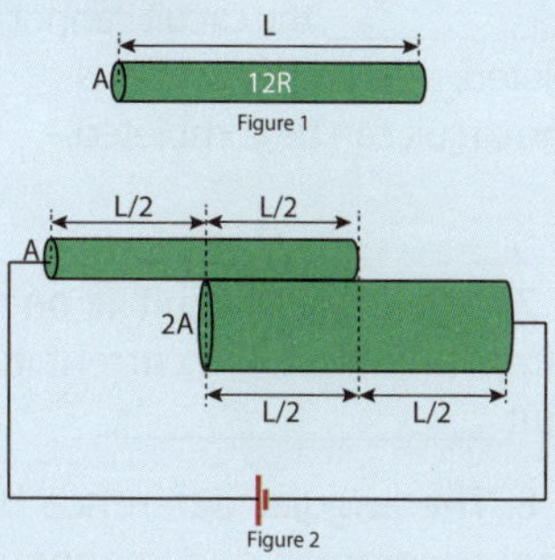

The resistance of a copper wire of length L and cross section area A is 12R.

Find the equivalent resistance of the network given in the figure.

Solution

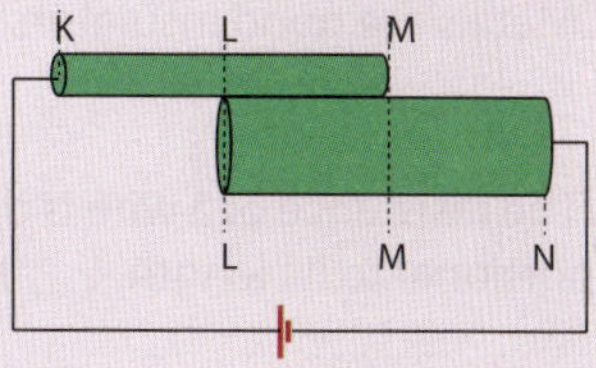

Resistors are connected parallel at points that they are touching to each other; L and M. For the resistances:

$$R_{wire} = \rho\frac{L}{A} = 12R \Rightarrow \rho\frac{L/2}{A} = 6R$$

$$\text{and} \quad \rho\frac{L/2}{2A} = 3R$$

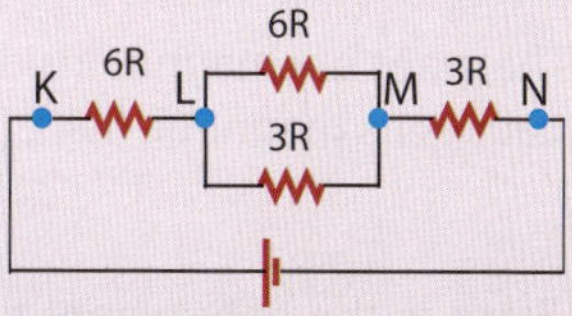

$$\frac{1}{R_{LM}} = \frac{1}{6R} + \frac{1}{3R} = \frac{3}{6R} = \frac{1}{2R}$$

$$R_{LM} = 2R$$

$$R_{KM} = 6R + 2R + 3R = 11R$$

MAGLEV TRAINS

Maglev is short for *magnetic levitation*, which means that these trains will float over a guideway using the basic principles of magnets to replace the old steel wheel and track trains.

A maglev train does not have the kind of engine used to pull conventional train cars along steel tracks. Instead of using fossil fuels, the magnetic field created by the electrified coils in the guideway walls and the track combine to propel the train.

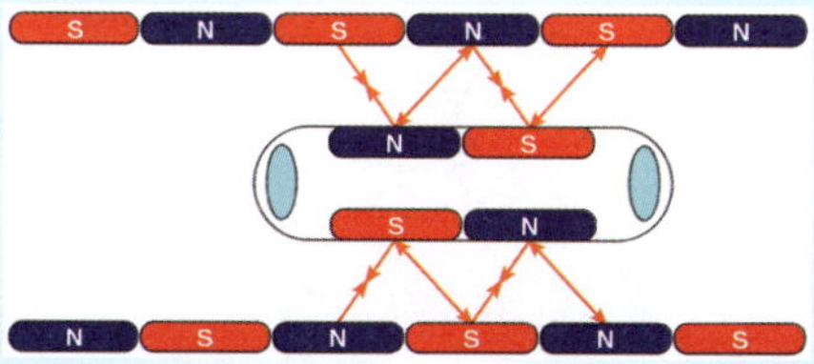

The magnetized coil running along the track, called a guideway, repels the large magnets on the train's undercarriage, allowing the train to levitate between 1 cm to 10 cm above the guideway. Once the train is levitated, power is supplied to the coils within the guideway walls to create a unique system of magnetic fields that pull and push the train along the guideway. The electric current supplied to the coils in the guideway walls is constantly alternating to change the polarity of the magnetized coils. This change in polarity causes the magnetic field in front of the train to pull the vehicle forward, while the magnetic field behind the train adds more forward thrust.

Maglev trains float on a cushion of air, eliminating friction. This lack of friction and the trains' aerodynamic designs allow these trains to reach unprecedented ground transportation speeds of more than 500 km/h. Developers say that maglev trains will eventually link cities that are up to 1700 km apart. At 500 km/h, you could travel from Paris to Rome in just over two hours.

Conceptual Questions

1. How will the light intensity of lamps change when more lamps are added in a series circuit?

2. What causes the overloading of electric circuits at your home? Suggest one way to prevent it.

3. If so many electrical appliances are working at the same moment, power is cut in a house. How and why does this happen?

4. Why is there a difference between the magnetic and geographical poles of the Earth? Why is there such a differentiation for the poles?

True - False

1. Charge becomes used up as it flows through a circuit. The amount of charge which exits a light bulb is less than the amount which enters the light bulb. (___)

2. Charge flows through circuits at very high speeds. This explains why the light bulb turns on immediately after the wall switch is flipped. (___)

3. The local electrical utility company supplies millions and millions of electrons to our homes everyday. (___)

4. An electrochemical cell can be a source of charge in a circuit. The charge which flows through the circuit originates in the cell. (___)

5. A battery is the dry cells connected in series. (___)

6. Metals are good insulators. (___)

7. The resistance of a wire is directly proportional to its length. (___)

8. For a continuous flow of electrons, a circuit without gaps is needed. (___)

9. A current causes a magnetic field around a current carrying wire. (___)

10. The giant bar magnet inside the Earth causes its magnetic field. (___)

11. Earth is the only planet with a magnetic field in our solar system. (___)

Fill in the Blanks

1. As the number of resistors in a series circuit increases, the overall resistance ___________ and the current in the circuit ___________.

2. As more and more resistors are added in parallel to a circuit, the equivalent resistance of the circuit ___________ and the total current of the circuit ___________.

3. A dry cell changes ___________ energy into ___________ energy.

4. The electric current of a circuit is measured by a ___________ and it must be connected in ___________ to the circuit.

5. Conductors try to stop the flow of electrons. This is called the ___________ of this conductor.

6. In an electric circuit; when the switch is ___________, the circuit cannot be completed, when the switch is ___________, the circuit can be completed.

7. In a series circuit if one of the lamps burn out, the current in the circuit ___________.

8. The angular difference between the geographic and magnetic poles of the Earth is called ___________ / ___________.

9. Magnetic field lines leave from the ___________ pole and enter from the ___________ pole of the magnet.

10. The magnetic filed lined formed around a current carrying wire are ___________ / ___________.

Conceptual Questions

1. How will the light intensity of lamps change when more lamps are added in a series circuit?

As the number of lamps increases the equivalent resistance of the circuit increases. This results as a decrease in the current of the circuit. As the current decreases, the brightness of the lamps also decreases.

2. What causes the overloading of electric circuits at your home? Suggest one way to prevent it.

For household circuits all elements are connected parallel to each other. If too many devices are connected, current will have more pathways to flow at the house and cause a overloading. As a result the insulation may melt because of overheating and cause fire. It can be prevented by operating less number of devices at the same time.

3. If so many electrical appliances are working at the same moment, power is cut in a house. How and why does this happen?

If too much current is drawn from the circuit, the electric wires of the house get heated up. The fuse shuts down the circuit in order to prevent a fire.

4. Why is there a difference between the magnetic and geographical poles of the Earth? Why is there such a differentiation for the poles?

The geographical poles are defined with respect to the rotational axis of the Earth, whereas the magnetic poles are defined to the location where a compass needle remains vertical to the surface of the Earth. The magnetic poles change their position in years and also swap in several hundred thousands of years. Defining the north and south with respect to the rotational axis is more dependable. But since the compasses use the magnetic poles for navigation, we also need the magnetic poles.

True - False

1. Charge becomes used up as it flows through a circuit. The amount of charge which exits a light bulb is less than the amount which enters the light bulb. (_F_)

2. Charge flows through circuits at very high speeds. This explains why the light bulb turns on immediately after the wall switch is flipped. (_F_)

3. The local electrical utility company supplies millions and millions of electrons to our homes everyday. (_F_)

4. An electrochemical cell can be a source of charge in a circuit. The charge which flows through the circuit originates in the cell. (_T_)

5. A battery is the dry cells connected in series. (_T_)

6. Metals are good insulators. (_F_)

7. The resistance of a wire is directly proportional to its length. (_T_)

8. For a continuous flow of electrons, a circuit without gaps is needed. (_T_)

9. A current causes a magnetic field around a current carrying wire. (_T_)

10. The giant bar magnet inside the Earth causes its magnetic field. (_F_)

11. Earth is the only planet with a magnetic field in our solar system. (_F_)

Fill in the Blanks

1. As the number of resistors in a series circuit increases, the overall resistance increases and the current in the circuit decreases.

2. As more and more resistors are added in parallel to a circuit, the equivalent resistance of the circuit decreases and the total current of the circuit increases.

3. A dry cell changes chemical energy into electrical energy.

4. The electric current of a circuit is measured by a ammeter and it must be connected in series to the circuit.

5. Conductors try to stop the flow of electrons. This is called the resistance of this conductor.

6. In an electric circuit; when the switch is open, the circuit cannot be completed, when the switch is closed, the circuit can be completed.

7. In a series circuit if one of the lamps burn out, the current in the circuit stops.

8. The angular difference between the geographic and magnetic poles of the Earth is called magnetic declination.

9. Magnetic field lines leave from the north pole and enter from the south pole of the magnet.

10. The magnetic filed lined formed around a current carrying wire are concentric / circles.

Problems

1.

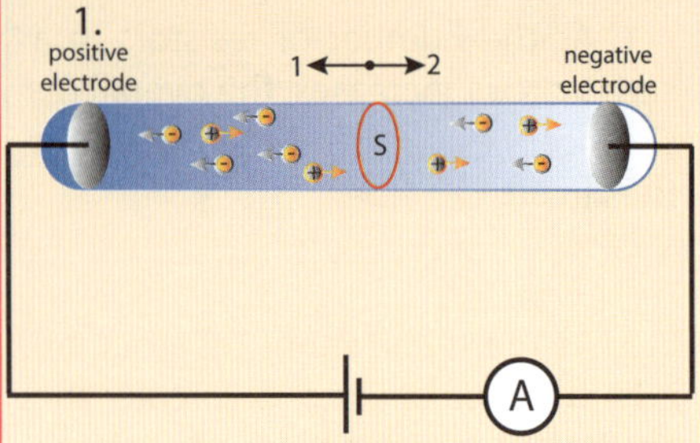

6×10^{19} hydrogen ions and 4×10^{19} electrons are passing through the cross-section S in 20 seconds.

a. What is the direction of the current?

b. What is the value read on the ammeter?

($1 e = 1.6 \times 10^{-19}$ C)

2.

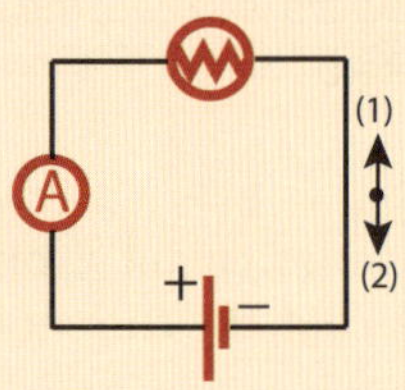

A charge of 120 C flows through the circuit in 1 min.

a. What is the value read on the ammeter?

b. What is the flow direction of current?

c. What is the flow direction of electrons?

3. Draw two lamps in parallel. An ammeter connected to one of the lamps and a switch to the other lamp. Then connect them to a battery formed by 3 cells.

4. A circuit consists of a 24 V battery formed by 4 cells and three resistors in series which are 8 Ω, 10 Ω and 12 Ω respectively. There is an ammeter between 8 Ω and 10 Ω resistors and a voltmeter is connected to the 12 Ω resistor. A switch is connected to the circuit after the 12 Ω resistor.

Draw the diagram of the circuit.

5. Two identical resistors are connected in parallel to each other. Three other identical resistors parallel to each other are connected in series to them. A switch is placed after the three parallel resistors. Then all of them are connected to two cells connected in parallel to each other.

Draw the diagram of the circuit.

6.

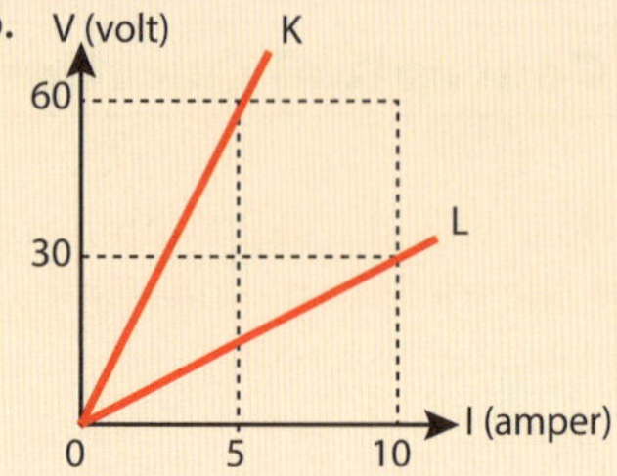

Potential difference versus current graph is given for the resistors K and L.

a. Find the resistance values of K and L.

b. Find the equivalent resistance when K and L are connected

i. in parallel to each other.

ii. in series to each other.

1. a. 2; b. 0.8 A / 2. a. 2 A; b. 2; c. 1 / 6. a. 12 Ω, 3 Ω; b. i. 2.4 Ω; ii. 15 Ω

7.

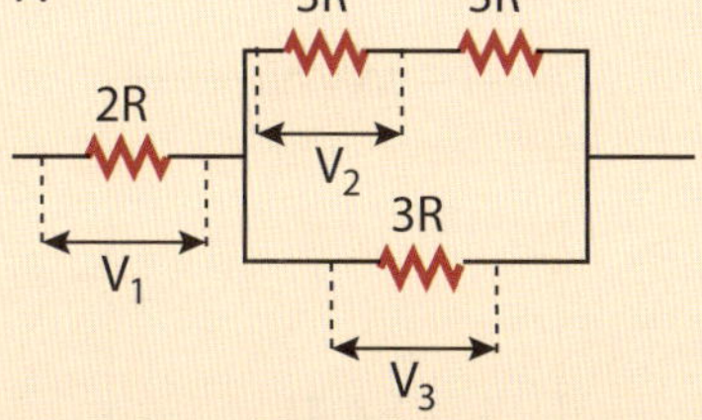

Find the relationship between the potential differences V_1, V_2, V_3.

8.

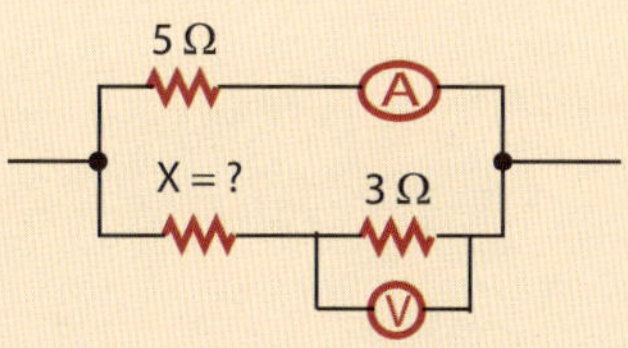

The value read on ammeter is 4 A and the value read on the voltmeter is 15 V.

Find the value of resistor X.

9.

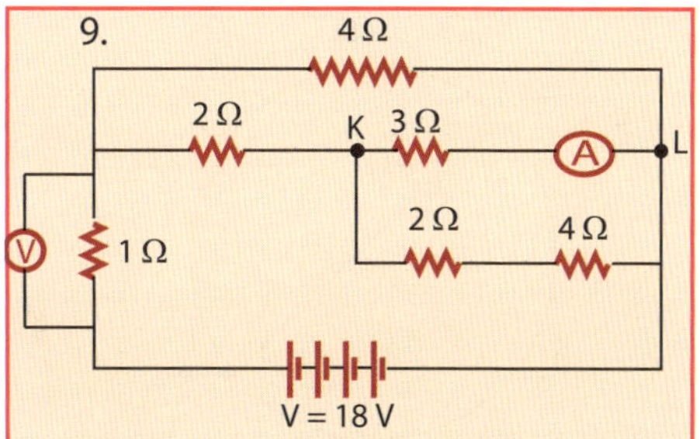

For the circuit given in the figure

a. Find the value read on the voltmeter, V.

b. The value read on the ammeter, A

c. The potential difference across points K and L.

10.

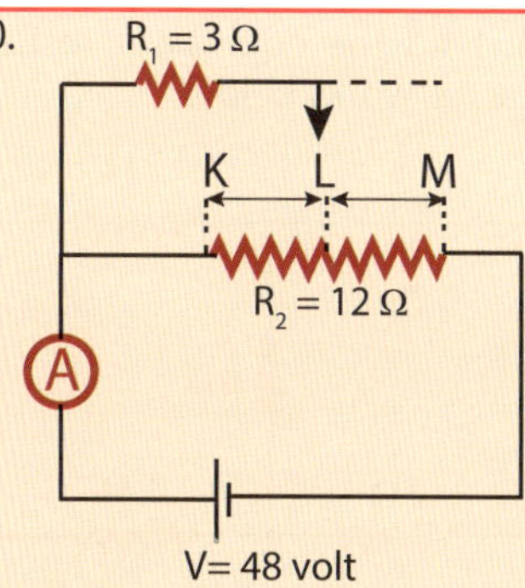

R_2 is a rheostat and L is its midpoint.

Find the value read on the ammeter when the rheostat arm is

a. at K.

b. at L.

c. at M.

7. $V_1 = V_3 > V_2$ / 8. 1 Ω / 9. a. 6 V; b. 2 A; c. 6 V / 10. a. 4 A; b. 6 A; c. 20 A

11. Find the equivalent resistance of the networks below.

a.

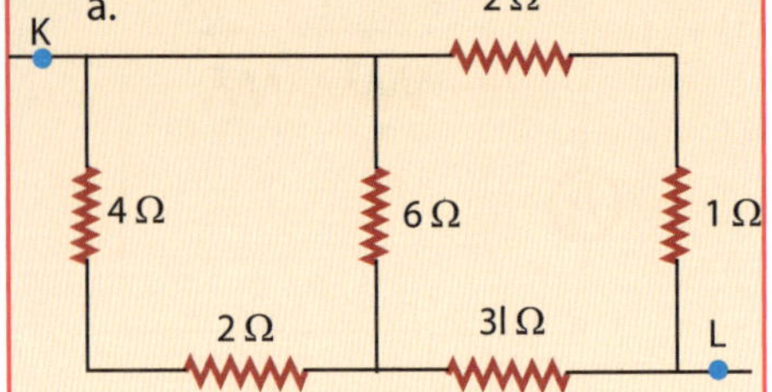

b.

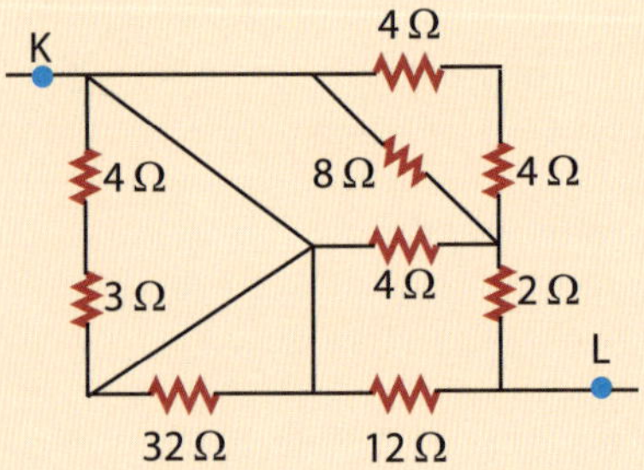

c.

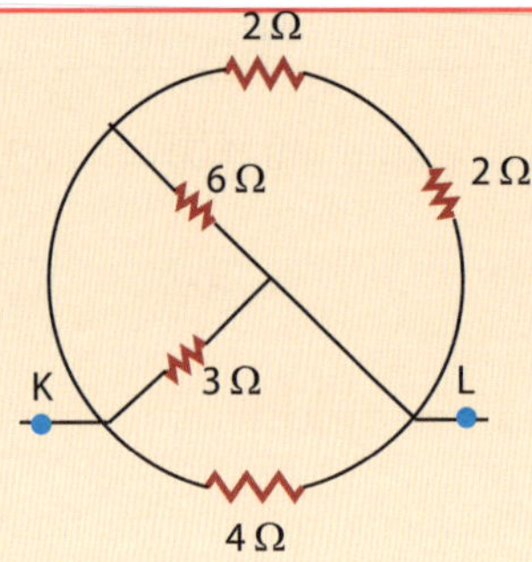

d.

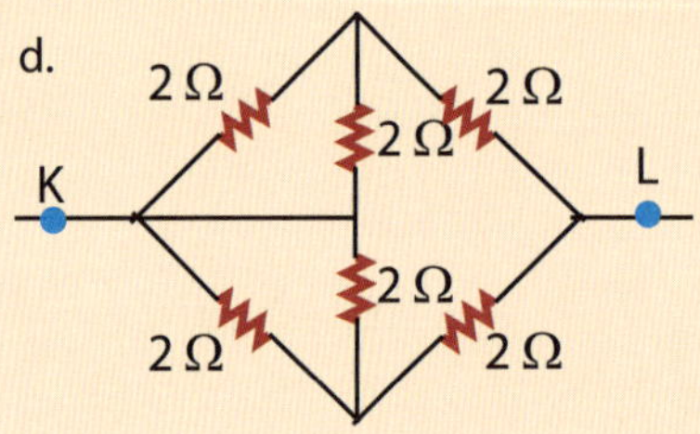

12.

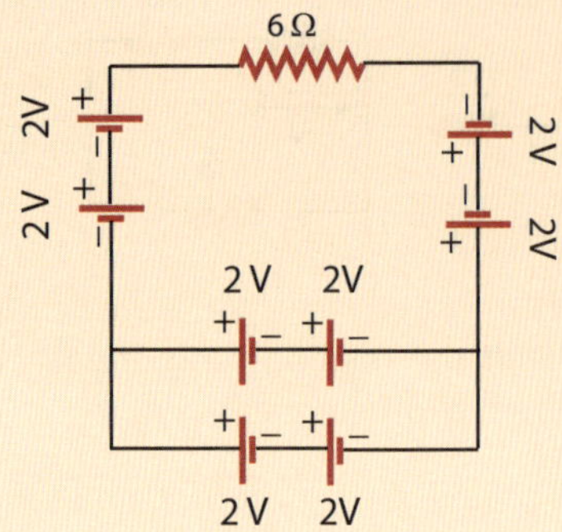

The potential difference of identical dry cells is 3 V. They are connected to a 6 Ω resistor and each other as given in the figure.

Find the current of the circuit.

11. a. 2 Ω; b. 3 Ω; c. 1 Ω; d. 1.5 Ω / 12. 2A

13.

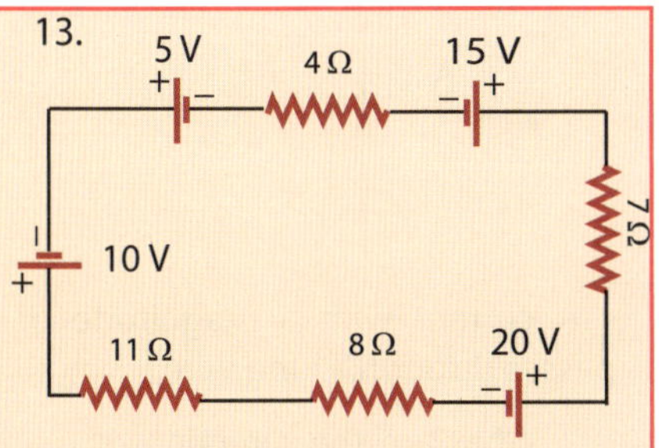

Find the current of the circuit.

14.

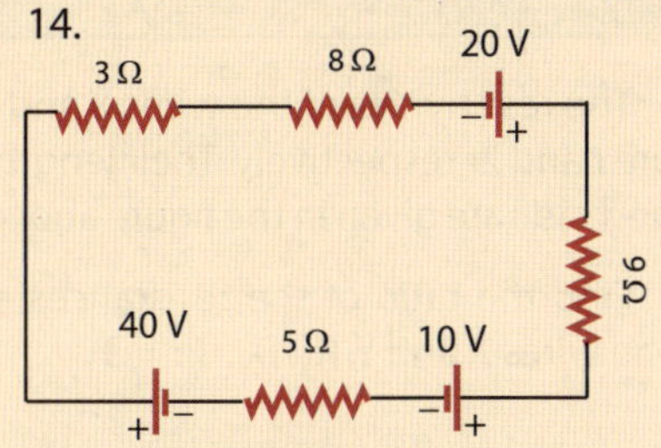

Find the current of the circuit.

15.

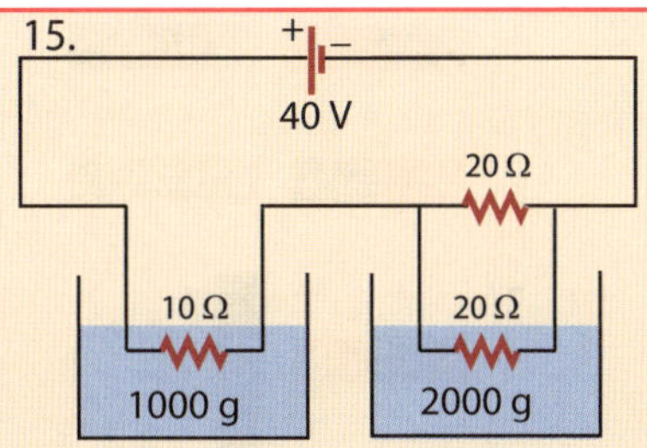

Two resistors are placed in containers which have 1000 g and 2000 g of water. The circuit works for 2 min.

Find the temperature change of water in the containers.

(c_{water} = 1 cal/g°C, 1 J = 0.25 cal)

16.

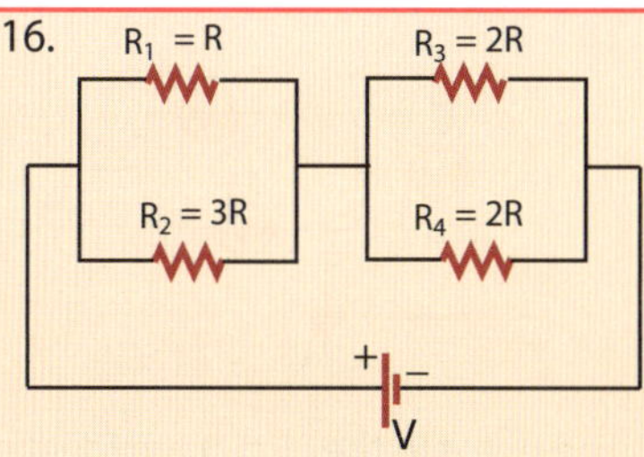

The power of the resistance R_1 is P.

Find the power of the circuit in terms of P.

17.

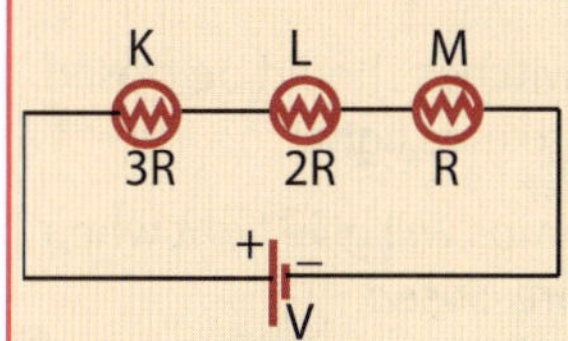

Figure 1

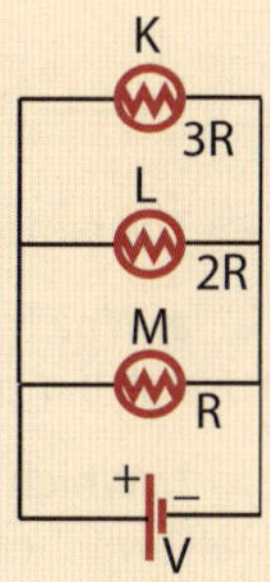

Figure 2

a. Compare the brightness of the bulbs in figure 1.

b. How will the brightness of the bulbs change when they are connected in parallel as given in figure 2.

13. 2/3A / 14. 2A / 15. 1.2 °C, 0.3 °C / 16. 28/9 / 17. a. K > L > M; b. M > L > K

18.

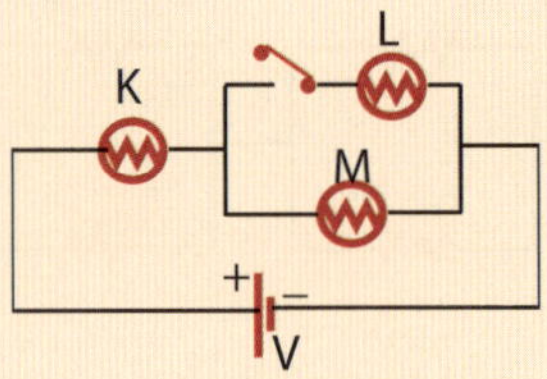

The bulbs in the circuit are identical.

Compare their brightness when the switch is

a. open.

b. close.

19.

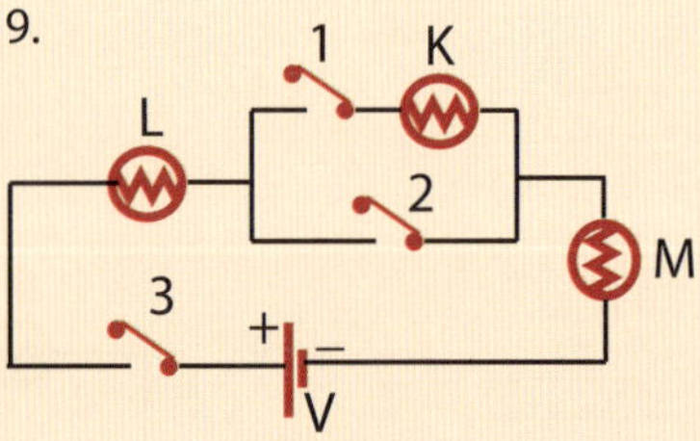

The lamps in the circuit are identical.

a. Which switches should be closed for all lamps to give light?

b. Which lamps will give light when all switches are closed.

20.

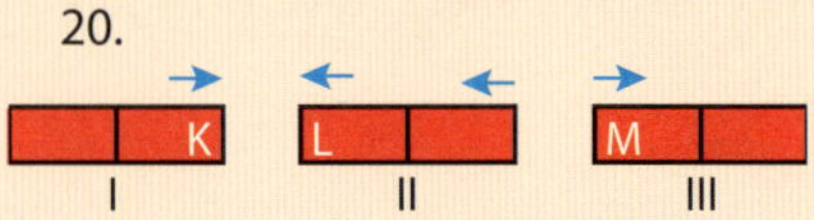

Magnets I and II are attracting each other whereas magnets II and III are repelling each other.

What can be said about the poles K, L, M?

21.

1. 

2. 

3. 

Strong magnets are approached to iron rods and magnetize them.

What can be the poles at the <u>ends</u> of the iron ores?

22. Draw the magnetic field lines of the following magnets.

a.

b.

c.

23.

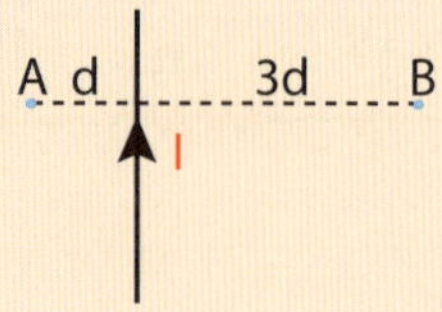

A current with a magnitude of I passes through a long wire.

a. What are the direction of magnetic fields at points A and B.

b. Compare the magnitude of magnetic fields at points A and B.

24.

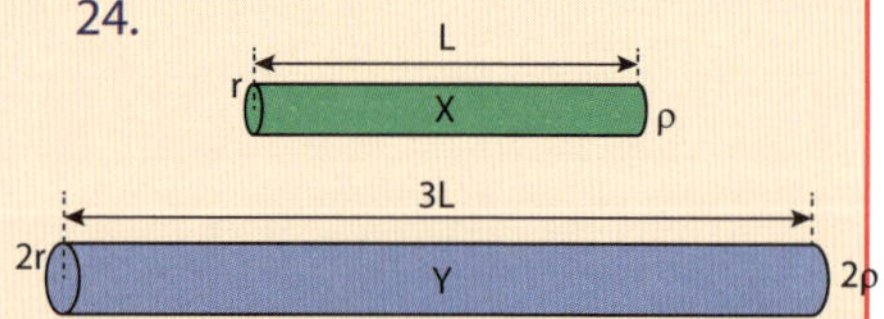

The resistivities of the wires X and Y are r and 2r respectively. Their lengths and radii are given in the figure above.

Find the ratio of the resistances of the wires X and Y, R_X/R_Y. ($\pi = 3$)

18. a. K = M; L = 0; b. K > L = M / 19. a. 1, 3; b. L, M / 20. N, S, N or S, N, S / 21. 1. SN, 2.NS, 3.N,SS,N / 23. a. A: ⊙, B: ⊗; b. A > B) / 24. 2/3

Multiple Choice Questions

1.

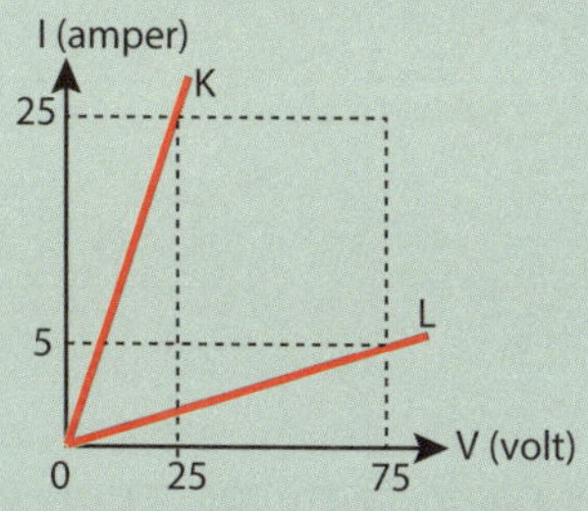

Current versus potential difference graph for resistors K and L is given in the figure. When K and L are connected in series the equivalent resistance is R_1 and when they are connected in parallel the equivalent resistance is R_2.

What is the ratio of R_1/R_2?

A. $\frac{16}{15}$ B. $\frac{1}{15}$ C. $\frac{256}{15}$

D. 1   E. $\frac{15}{256}$

2.

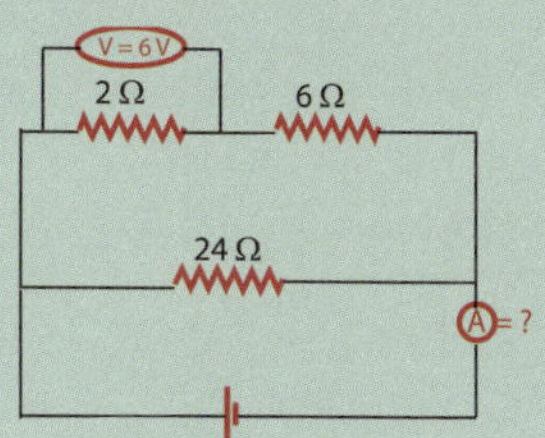

The value read on the voltmeter is 6 V.

Find the value that will be read from the ammeter in terms of amperes.

A. 4   B. 6   C. 8   D. 10   E. 12

3.

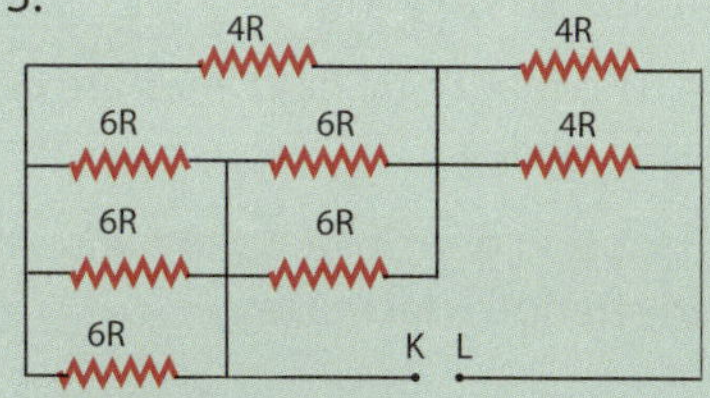

What is the value of equivalent resistance between points K and L in terms of R?

A. 1   B. 2   C. 3   D. 4   E. 5

4.

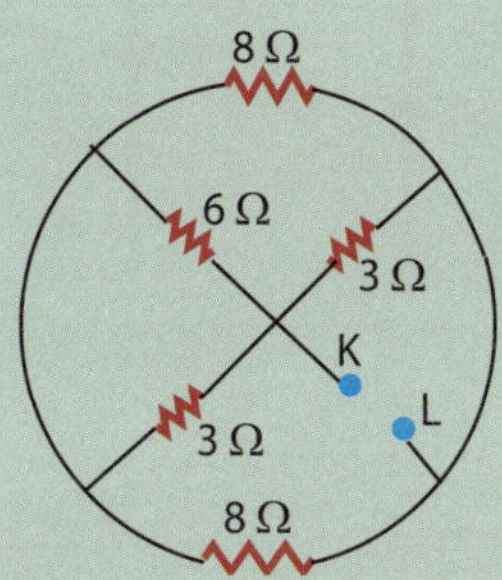

What is the value of equivalent resistance between points K and L in terms of W?

A. 6   B. 5   C. 4   D. 3   E. 2

Answer questions 5, 6, 7, 8 with respect to the given circuit. The lamps in the circuit are identical.

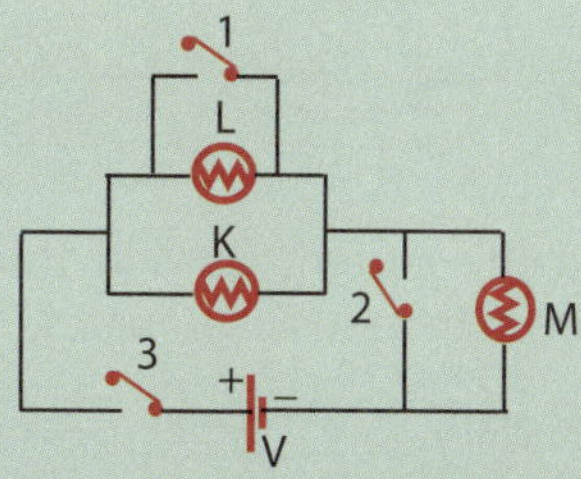

5. Which lamps will give light when all of the switches are closed?

A. Only K   B. Only L

C. K and L   D. L and M

E. None

6. Which switches should be closed so that all lamps give light?

A. None   B. Only 3

C. 1 and 2   D. 2 and 3

E. 1, 2 and 3

7. Which switches should be closed so that only lamp M gives light?

A. Only 1   B. Only 3

C. 1 and 2   D. 2 and 3

E. 1 and 3

8. What is the relationship between the brightness of the bulbs when only switch 3 is closed?

A. M < K = L   B. K < L < M

C. K = L < M   D. L < M < K

E. K = L = M

1.C 2.A 3.D 4.E 5.E 6.B 7.E 8.C

9.

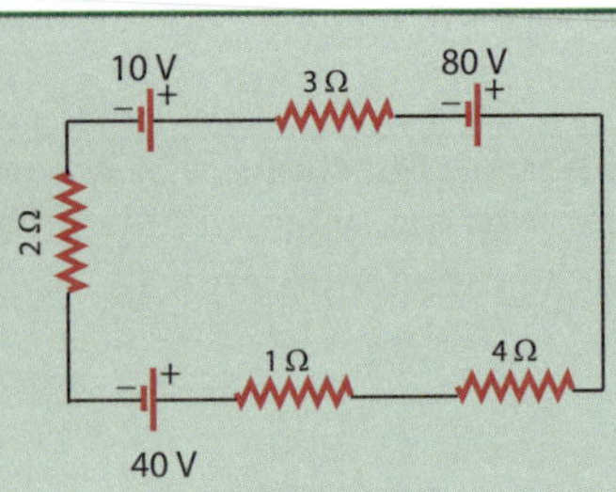

Find the current flowing through the circuit in terms of amperes.

A. 3 B. 5 C. 7 D. 11 E. 13

10. Which of the followings will be attracted by a magnet?

A. Iron nail

B. Glass marble

C. Nylon thread

D. Woolen cloth

E. Copper penny

11. I. If a magnet is divided in two, the resulting magnets will have their own N-S poles.

II. If a magnet is heated above a certain temperature, it looses its magnetism.

III. Magnets pull objects because of the electric field around them.

Which of the above statement(s) are true?

A. Only I B. Only III

C. I and II D. II and III

E. I, II and III

12. Which of the followings is *false* for the magnetic field produced by a current carrying wire?

A. The magnetic field lines are concentric circles.

B. As you get closer to the wire, the field strength decreases.

C. Changing the direction of current does not affect the magnitude of the produced magnetic field.

D. If more current passes from the wire, the magnetic field strength increases.

E. A compass needle deflects when a current passes through a wire.

13.

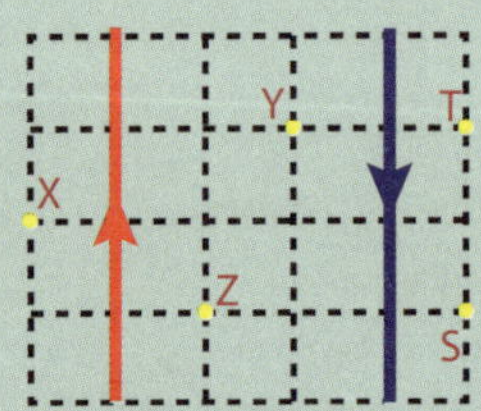

Two current carrying wires are placed parallel to each other as shown in the figure.

At which point(s) may the magnetic field strength be equal to zero?

A. Y and Z

B. X, Y and T

C. T and S

D. X, T and S

E. Y, T and S

Solve questions 14 and 15 with respect to the figure given below.

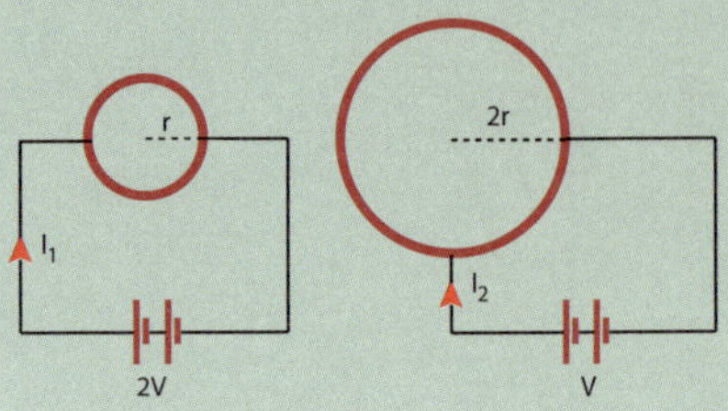

14. The circular resistances with radii r and 2r are formed by using a uniform wire. Then they are connected to batteries with potential differences 2V and V respectively.

Find the ratio of equivalent resistances, R_1/R_2.

A. $\frac{3}{2}$ B. $\frac{1}{2}$ C. $\frac{1}{3}$

D. 1 E. $\frac{2}{3}$

15. Find the ratio of currents through the circuits I_1/I_2.

A. $\frac{2}{3}$ B. 3 C. $\frac{1}{3}$

D. 2 E. $\frac{1}{2}$

9. B 10. A 11. C 12. B 13. D 14. E 15. B

UNIT 2 PRESSURE AND BUOYANCY

The concepts that will be covered in this unit are:
Solid pressure, pressure of fluids, Pascal's principle, Bernoulli's principle, Archimedes principle, buoyant force.

CONTENTS PAGE

The Antonov An-225 Mriya is an airlift cargo aircraft that was designed by the Soviet Union's Antonov Design Bureau in the 1980s. The An-225's name, Mriya means "Dream" (Inspiration) in Ukrainian. It is powered by six turbofan engines and is the longest and heaviest airplane ever built with a maximum take off weight of 640 tonnes. It also has the largest wingspan of any aircraft in operational service. A second airframe was partially built; its completion was halted because of lack of funding and interest.

PRESSURE AND BUOYANCY

2.1. Pressure

Walking on snow is usually a tricky thing in the winter time. You got stuck in it and with every other step it gets harder to get out of the snow. Or you can wear a pair of those special "snow shoes" which makes your walking easier. Walking with high heel shoes at the beach is not a good choice either. In both cases, a physical concept that we are going to talk about in this unit can be the explanation for these problems.

The ratio of force per unit area is called pressure in physics.

$$pressure = \frac{force}{area} \Rightarrow P = \frac{F}{A}$$

Symbol	Quantity	Unit
F	force	newton (N)
A	area	m^2
P	pressure	pascal (Pa)

According to this ratio, pressure is directly proportional to the force applied and inversely proportional to the area. Assume that force applied, which is the weight of the people is the same for both cases mentioned above. Snow shoes will provide a larger surface area thus a smaller magnitude of force per unit area (pressure) will be applied to the snow. This will result an "easy walking" on the snow.

Sometimes walking on snow is a hard thing to do

The same idea lies behind sharpening a knife. As the contact area of the knife gets smaller, force per unit area gets larger. A smaller area (thinner edge) means a smaller force to cut the same loaf. As you may imagine, preferring to use sharper needles of nails instead of dull ones is also a result of the force per unit area ratio.

Snow shoes help you to walk easier on snow

Last year you have learned that force is a vector quantity; a quantity with a certain direction. Although it is not mentioned so often, area is another quantity with direction. Its direction is represented by a normal vector, pointing outwards from the surface. Thus for the equation stated above for pressure:

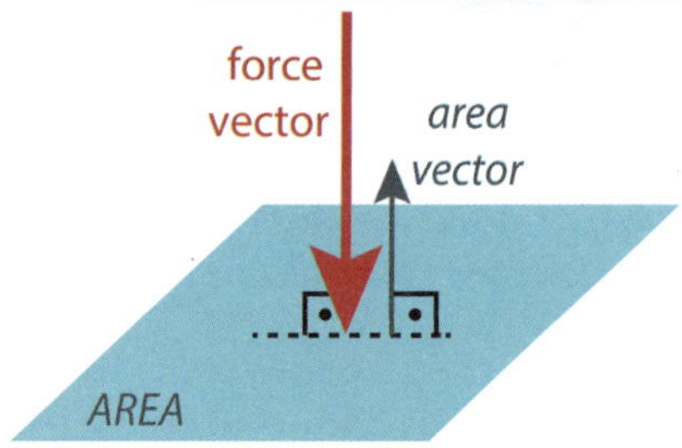

$$P = \frac{\vec{F}}{\vec{A}}$$

The division of two vector quantities gives a scalar quantity. So, pressure is a scalar quantity. It doesn't have a direction. It is given by a number and a unit. The unit of pressure N/m^2 is often called pascal (Pa) in the honor of Blaise Pascal; a French mathematician, physicist, inventor and writer.

High heels at the beach? Not a good choice....

Another important note about pressure is that the *force applied* and the *normal vector of the surface area* should be parallel to each other. For the drawing aside, although forces $\vec{F_1}$ and $\vec{F_2}$ have the same magnitude (length), their effects on unit area is not going to be the same. The second force is not acting perpendicular to the surface. Some of its effect is in the horizontal plane, not all of its effect is in the vertical plane.

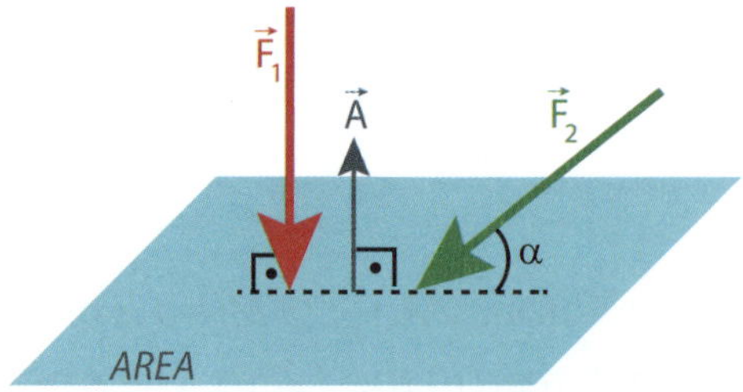

2.1.1. Pressure and Force of Pressure of Solids and Fluids

Earth pulls all the objects towards its center. This force of attraction between the earth and the objects is called the *force of gravity*, also known as the *weight* of the object. For most common situations weight of the object is the perpendicular force acting on the surface of contact. So we can conclude that solids exert a pressure due to their weight on the surface.

force applied on the contact area = weight of the object

Here we should state that the above equation is valid as long as no forces other than the weight of the object is acting on it. If the base area of the object (contact surface) is given as A, then the magnitude of the pressure acting on the base is:

$$P = \frac{F}{A} = \frac{W}{A}$$

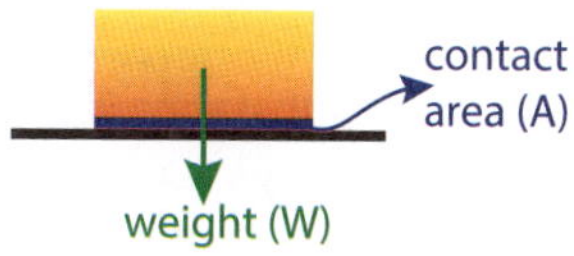

Example

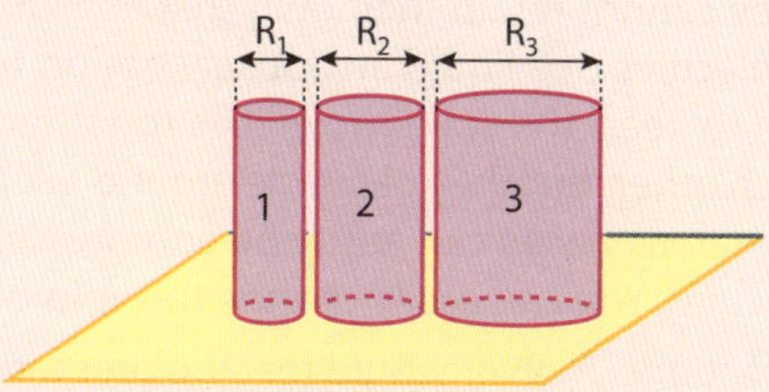

The cylinders given in the figure have the same weight. The relationship between their diameters is $R_1<R_2<R_3$.

What is the relationship between the pressures they exert on to the ground?

Solution

The area equation for a circle is $A = \pi.r^2$ where r is the radius, half of the diameter of the circle.

If the relationship between the diameters (also the radii) is $R_1<R_2<R_3$, then the relationship between the base areas of the cylinders is $A_1<A_2<A_3$.

The pressure of a solid is: $P = \frac{W}{A}$

According to the equation pressure is inversely proportional to the contact surface. The weight of the cylinders are the same. So, if $A_1<A_2<A_3$ then $P_1>P_2>P_3$.

Example

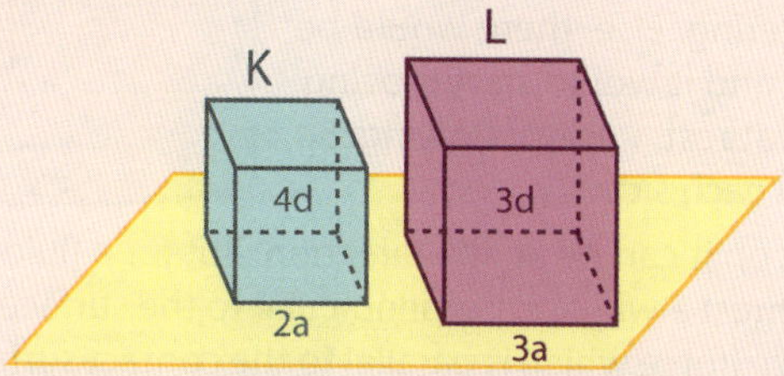

One edge of the cubes K and L are given as 2a and 3a as given in the figure. Density of K is 4d and that of L is 3d.

What is the ratio of the pressures they exert on to the ground; P_K/P_L ?

Solution

The pressure of a solid is: $P = \frac{W}{A}$

Base area of a cube is: $A = a^2$

Weight is: $W = m.g$ where m is the mass and g is the gravitational acceleration.

Mass is: $m = d.\ V$ where d is the density and V is the volume.

Volume of a cube is: $V = a^3$

So, $W = m.g = d.V.g = d.a^3.g$

If the substitutions are made:

$$P = \frac{W}{A} = \frac{d.a^3.g}{a^2} = d.a.g$$

$$P_K = 4d\ .\ 2a\ .\ g = 8dag$$

$$P_L = 3d\ .\ 3a\ .\ g = 9dag$$

$$\frac{P_K}{P_L} = \frac{8dag}{9dag} = \frac{8}{9}$$

Force of pressure is defined as the force causing that pressure on that surface area. For the magnitude of force of pressure:

$$P = \frac{F}{A} \Rightarrow F = P.A$$

So for a solid object that no other external forces are acting on it, the magnitude of force of pressure is:

$$F = P.A = \frac{P}{A}.A = W \Rightarrow F = W$$

Solids transmit the force without a change in magnitude and direction, thus produce a change in pressure because of the change in the area.

When a force of F is applied on top of the nail by the hammer, this force is directly transmitted to the tip of the nail. For the pressures at top and tip of the nail:

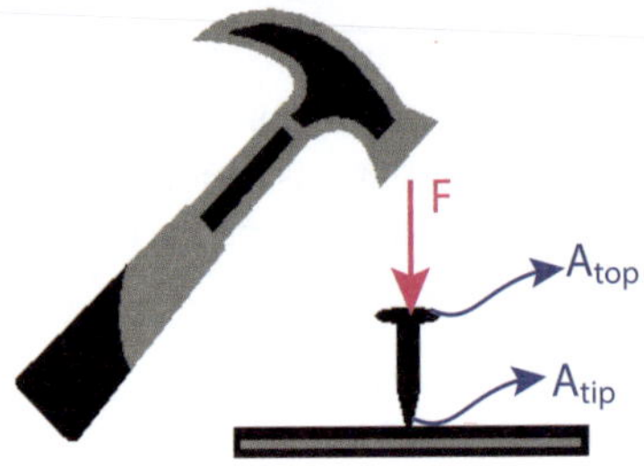

$$P_{top} = \frac{F}{A_{top}} \text{ and } P_{tip} = \frac{F}{A_{tip}}$$

$$A_{top} > A_{tip} \text{ so } P_{tip} < P_{top}$$

Any flowing substance is called a fluid. Liquids and gases are fluids. The physics of fluids is called fluid mechanics. At this point of our topic, we will study about fluids at rest, the fluid statics. Why a submarine can be crushed under the ocean, why you feel pain in your ears when you dive so deep, why hot air balloons float.... are studied under this title.

What keeps a submarine from crushing?

Since the fluids are flowing, they cannot retain their own shapes, but fill in or take the shape of their container. The molecules of a liquid are loosely bound to each other. But these bonds are strong enough to keep the solid molecules together. Liquid molecules cannot get far apart from each other, but they can slide around each other. Practically, liquids are incompressible whereas gases can easily be compressed due to the empty space between the gas molecules. You can easily change the volume of a gas.

Fluids exert a pressure in all directions. A diver would feel the pressure of water on every part of his body, not only his ears. Although we are not mostly aware of, the air surrounding us, exerts a pressure on our body also. If it is a fluid at rest, the pressure is the same for the same dept.

Yasemin Dalkılıç; Turkey's world record breaking most famous skin diver. How can she resist the pain in her ears when she dives to 100 m depth?

In the figure aside a cylinder is filled with a liquid. Let's consider an extremely tiny sphere of liquid within this liquid. The red arrows are representing the force of pressures acting on the surface of the sphere. The pressure on one side of the sphere should be equal to the pressure on the opposite side of the sphere that we are considering. Else, there would be a net force acting on the sphere and it would start moving. But the liquid in our container is at rest, it is not flowing. So the pressures should be equal on each side.

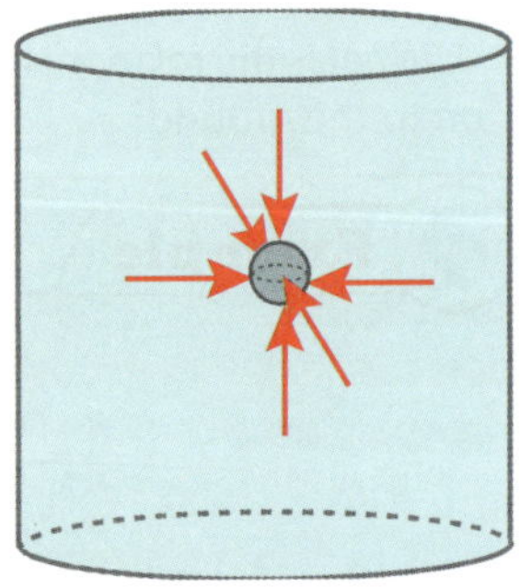

For a solid, the forces acting on it can be at any direction. But for a fluid at rest, the force acting on any solid surface must always be perpendicular to the surface of contact. For a fluid at rest, a tangential force (force which is parallel to the contact surface) cannot be sustained. If there were any tangential forces acting, the layers of the fluid would slide over each other. This results a *motion* for the fluid that we have been considering *at rest!* The inability of fluids to resist to these tangential forces gives them their characteristic ability to flow or to change their shape.

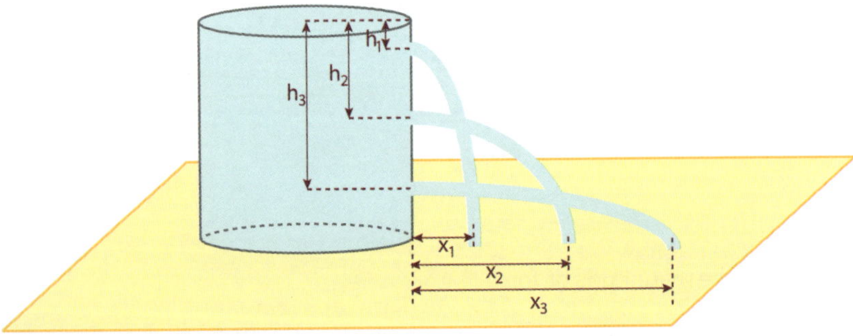

The fluid pressure produces a force. This force pushes on the walls of the container and also everywhere within the liquid. If you fill a container with water and then punch holes at different heights, you will observe that water spurts out from the holes. Another interesting observation would be that the horizontal displacements for the water spurting out of each hole are different for different depths of water. As the hole gets closer to the bottom of the container, the horizontal displacement of spurting water increases. The force of pressure of water behind each hole pushes the water forward trough the holes. According to our observation, as the hole gets deeper on the container, the pressure of the water increases.

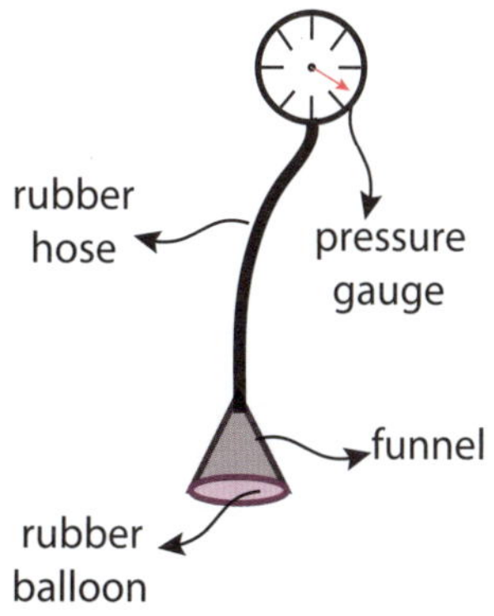

Stretch a rubber balloon to the mouth of a funnel, attach a rubber hose to the tube of the funnel. Attach the other end of the rubber hose to a pressure gauge. We can define a *pressure gauge* simply as a scale reading the pressure of air, especially the pressure of air within a tire.

If you lower the funnel into a container filled with a liquid, as you change the orientation of the funnel within the liquid, as long as you keep the face of the rubber balloon at the same depth, you will observe that the pressure gauge's scale will always show the same value. It doesn't matter whether you are pointing the rubber balloon up, down or aside. There is always a pressure exerted on it with the same value for the same depth.

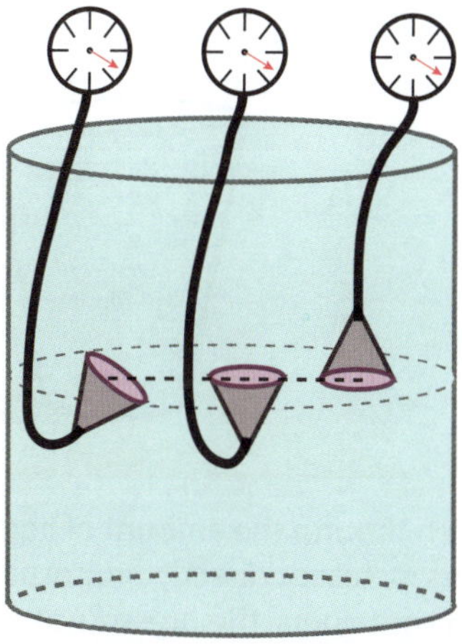

Physics Extra

Liquids, just like solids do, exert a pressure because of their weight. For a simpler approach, we will begin with the pressure of a point at the base of the container.

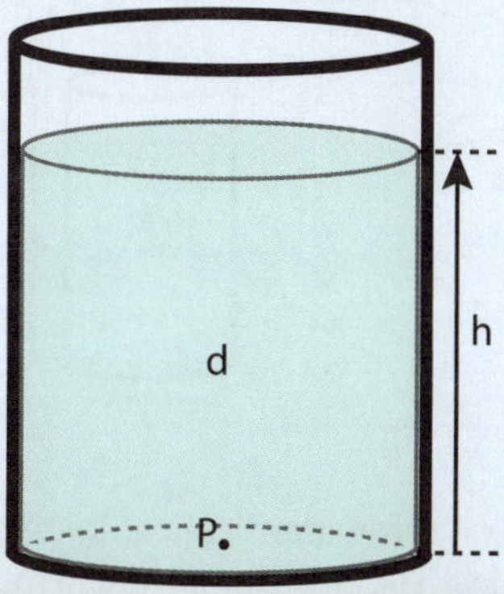

Weight of the liquid causes the force of pressure so the force acting on point P is:

$$W = m.g$$

where m is the mass of the liquid, g is the gravitational acceleration. Remember that mass of a liquid is dependent on its density and volume:

$$m = d.\ V$$

For a regular shaped container like the cylinder we have in our figure, the volume is dependent on the base area and the height of the container.

$$V = A.h$$

If we substitute for weight:

$$W = m.g = d.V.g = A.h.d.g$$

For the pressure exerted to the base of the container:

$$P = \frac{W}{A} = \frac{A.h.d.g}{A} = h.d.g$$

Symbol	Quantity	Unit
h	depth of liquid	m
d	density	kg/m^3
g	gravitational acceleration	m/s^2
P	pressure	Pa (pascal)

The equation above is valid for fluids whose density is constant and does not change with depth. For most common cases, just like calculating the pressure of water at the bottom of a swimming pool, the equation can be applied to liquids. But at great depths of the ocean, the density of water increases substantially because of the compression of the huge weight of water above it. So the pressure at these great depths is not going to be as we had calculated.

The fluid should also be *incompressible*. This equation cannot be used for calculating the pressure of a gas since the gasses can easily be compressed.

Example

What is the pressure at the bottom of a 1.5 km-deep oil well filled with oil of density 840 kg/m^3?
($g = 9.8\ m/s^2$)

Solution

The unit of depth should be in terms of meters.
$h = 1.5\ km = 1500\ m$
$P = h.\ d.\ g = 1500 \times 840 \times 9.8 = 12348000\ Pa = 12348\ kPa$

"h" in the given equation is the depth of the liquid. It is measured from the point considered to the surface of the liquid. This is understandable because as you swim up towards the surface of water, the pressure in your ears decreases. It is the *weight* of water *above* you which applies a pressure on your body, not the water *below* you.

If we have a careful consideration of the equation for the pressure of a liquid, there is nothing about the amount of the liquid within the container. Even if it is a thinner container or a larger one or a point with the same depth in a swimming pool, the pressure will have the same magnitude for the same depth and density of liquid.

Physics Extra

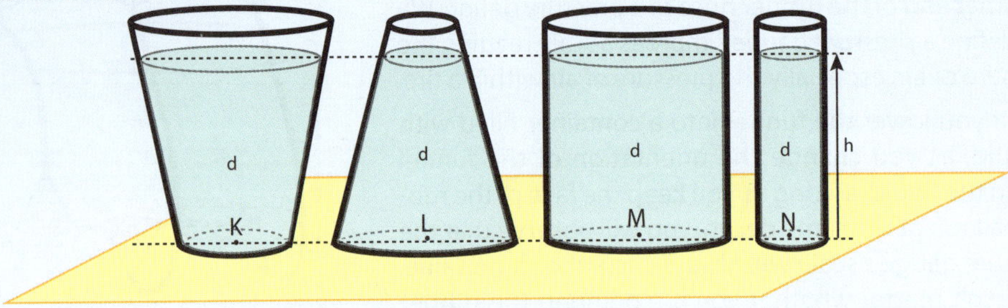

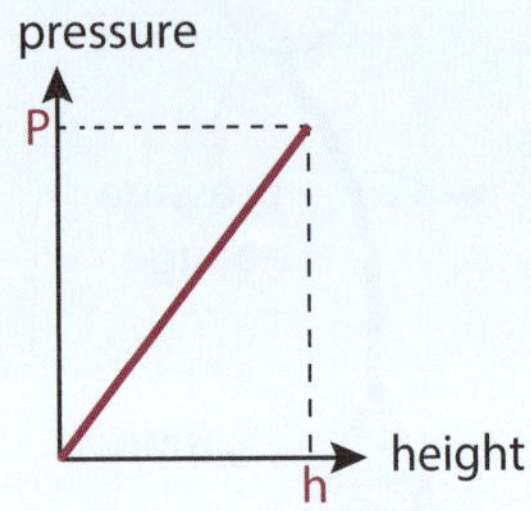

Even though the amount of liquid in the containers above are different, the pressure of the liquid above the base of the containers increases directly proportional to depth. Since the depth, density of liquid and the gravitational acceleration is the same for all containers, the pressure at points K, L, M and N are the same.

Example

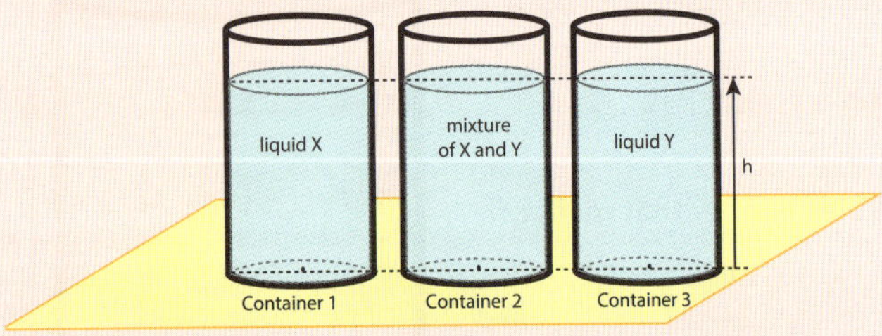

Three identical containers shown in the figure are filled with liquids X and Y and a mixture of X and Y up to the same height. The mixture in the container 2 was prepared by taking equal volumes from X and Y. The pressures at the base of the container 1 is P and that of container 2 is 5P.

What is the pressure at the base of the container 3 due to liquid Y?

Solution

$P_1 = d_X.h.g = P$

If the volumes of the liquids are equal then the density of the mixture is equal to the average of the densities of the liquids.

$$d_{mix} = \frac{d_X + d_Y}{2}$$

then

$$P_2 = \frac{d_X + d_Y}{2}.h.g = 5P$$

If we divide P_1 and P_2 side by side:

$$\frac{P_1}{P_2} = \frac{d_X.h.g}{\frac{d_X + d_Y}{2}.h.g} = \frac{P}{5P}$$

$$\frac{2d_X}{d_X + d_Y} = \frac{1}{5}$$

If we perform cross-product here:

$10d_X = d_X + d_Y$

$d_Y = 9d_X$

$P_3 = d_Y.h.g = 9d_X.\ h.g = 9P$

Now let's find the force of pressure for a liquid:

$$\left.\begin{aligned} F &= P.A \\ P &= h.d.g \end{aligned}\right\} F = h.d.g.A$$

This equation will be valid for a point at the base of the container. According to this equation we can say that the force of liquid pressure at the base of a container is equal to the *weight of the liquid column* above the base of that container.

What if we consider a point on the *lateral surfaces* of the container? Which depth should be considered when calculating the pressure acting on any lateral surface?

Each point on a lateral surface has its own depth. Instead of calculating the pressure at every point of the surface, we will prefer to calculate an average pressure acting on that surface. The average pressure acting on a lateral surface is:

$$P_{average} = \frac{P_{max} + P_{min}}{2} = \frac{h_{max}.d.g + h_{min}.d.g}{2} = \left(\frac{h_{max} + h_{min}}{2}\right).d.g$$

Physics Extra

Then for the force of pressure acting on that lateral surface:

$$F_{lateral} = P_{average} \cdot A_{lateral} = \left(\frac{h_{max} + h_{min}}{2}\right).d.g.A_{lateral}$$

Now, for the container given below; let's find the force pressures acting on the lateral surfaces and base of the container.

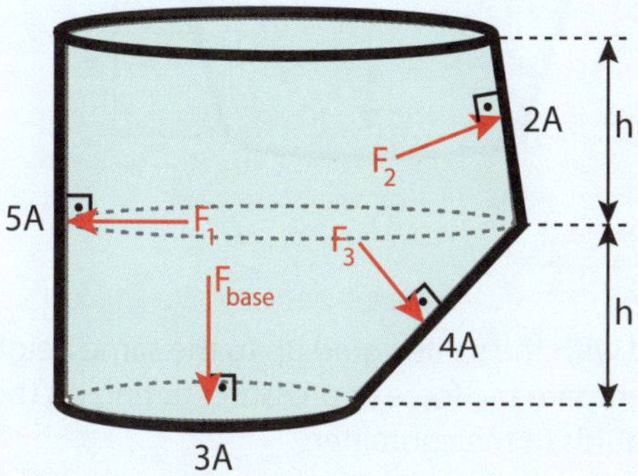

$$F_1 = \left(\frac{2h + 0}{2}\right).d.g.5A = 5hdgA = 5F$$

$$F_2 = \left(\frac{h + 0}{2}\right).d.g.2A = hdgA = F$$

$$F_3 = \left(\frac{2h + h}{2}\right).d.g.4A = 6hdgA = 6F$$

$$F_{base} = 2h.d.g.3A = 6hdgA = 6F$$

As you can see from the calculations, the force of pressure acting on the base of the container doesn't have to be the greatest force. The force of pressure is not only dependent on the pressure, it is also dependent on the surface area.

Example

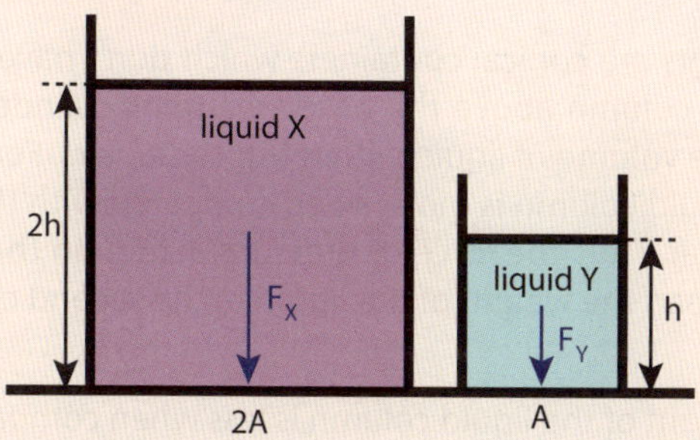

The containers given in the figure are filled with liquids X and Y up to the given heights. Forces of pressure acting on the base area of the containers are equal to each other.

Find the ratio of the density of the liquids dx/dy.

Solution

$$F_{base} = h.d.g.A$$

$$F_X = F_Y$$

$$2h.d_X.g.2A = h.d_Y.g.A$$

$$4d_X = d_Y$$

$$\frac{d_X}{d_Y} = \frac{1}{4}$$

Example

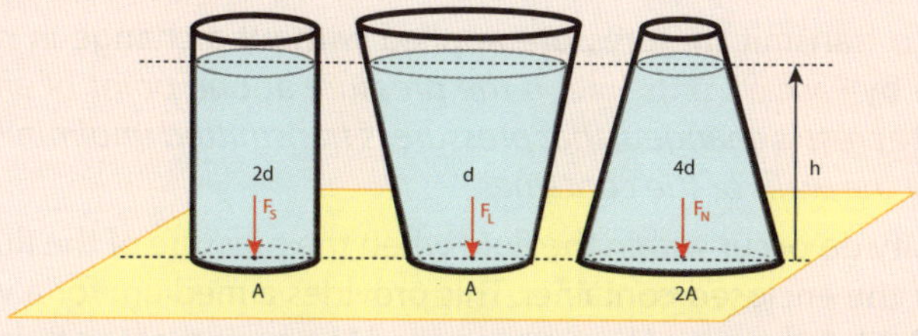

Containers S, L and N are filled with different liquids to the same height as shown in the figure.

Find the relationship between the forces of pressure F_S, F_L and F_N at the base of the containers due to these liquids.

Solution

$$F_S = h.2d.g.A = 2hdgA$$

$$F_L = h.d.g.A = hdgA$$

$$F_N = h.4d.g.2A = 8hdgA$$

$$F_L < F_S < F_N$$

We had mentioned that the force of pressure for solids is equal to the weight of the object. How will the force of pressure of liquids change with respect to the weight of the object?

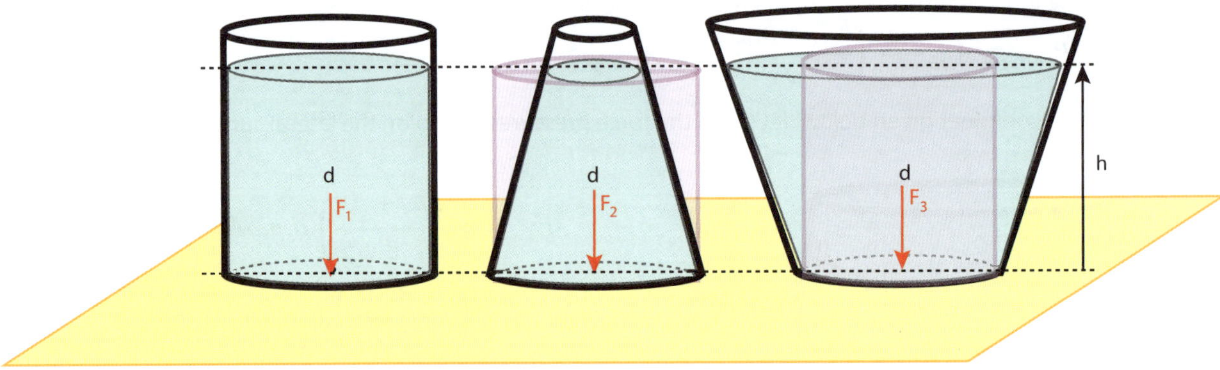

All of the above containers are filled with the same liquid up to the same height and they have the same base area. We will compare the force of pressures acting on the base of the containers to the weight of the liquids in the containers.

The first container is a cylindrical one where the lateral surfaces are perpendicular to the base. This also means that the liquid column above the base area is overlapping with the whole volume of liquid within the container. We can conclude that the whole weight of the liquid acting on the base thus producing the force of pressure on the base of the container.

$$W_1 = F_1$$

The second and third containers are conical containers which don't have perpendicular lateral surfaces. The liquid column above the base area (pink cylinders in the figure) are not overlapping with the volume of liquid within the containers. For the second container the weight of the liquid column is more when compared with the whole weight of the liquid in the container. Thus the force of pressure acting on the base of the container is going to be more than the weight of the liquid in the second container.

$$W_2 < F_2$$

For the third container the weight of the liquid column is less when compared with the whole weight of the liquid in the container. Thus the force of pressure acting on the base of the container is going to be less than the weight of the liquid in the third container.

$$W_3 > F_3$$

A plastic balloon filled with water, given a slight push

Fill a plastic balloon with water then give a slight push on one side of it. You will observe that slight push is transferred throughout the water within the balloon. The pressure of every other point within water will change with the same amount.

The ability of liquids to transmit the pressure applied without a change in magnitude is usually explained by Pascal's principle. *If the pressure at one point of an enclosed incompressible fluid at rest is changed, that pressure is transmitted undiminished throughout the fluid and to the walls of the container.*

Slight changes in density do occur within the fluid when the pressure of the fluid is changed from one point of the enclosed container. This provides a medium for a wave to propagate through the fluid. A change of pressure applied to one part of the fluid will propagate within the fluid as a wave with the speed of sound in that fluid. The reason for this will be explained further in unit 3, waves.

Hydraulic lifts is one of the first examples of practical devices that make use of Pascal's principle. With a hydraulic lift a small input force is used to exert a large output force by changing the surface areas of the lifts.

The pressure exerted by the input force on the input piston will be transferred to the liquid as the input pressure throughout the liquid. The output pressure produced by the heavy load on the output piston and the input pressure should be the same for equilibrium:

$$P_{output} = P_{input}$$

Physics Extra

For a hydraulic lift:

$$P_{output} = P_{input}$$

$$\frac{F_{output}}{A_{output}} = \frac{F_{input}}{A_{input}}$$

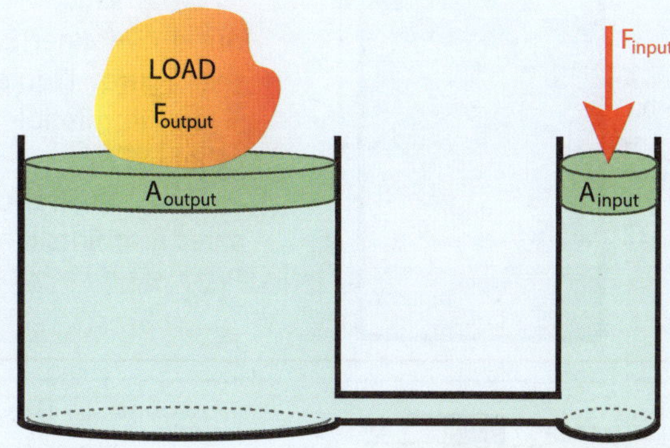

The ratio of output force to input force is called the mechanical advantage of the system.

$$MA = \frac{F_{output}}{F_{input}}$$

Example

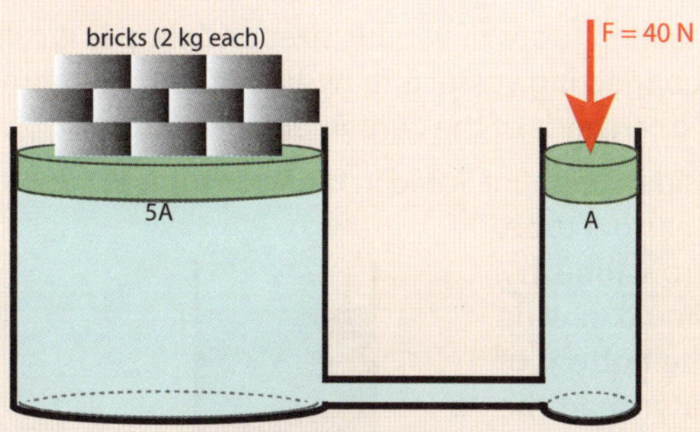

A worker wants to lift bricks by using the hydraulic lift given in the figure. The maximum force that can be applied by the worker is 40 N.

If mass of one brick is 2 kg, how many of the bricks can the worker lift in each time?

(g = 10 N/kg.)

Solution

$$P_{output} = P_{input}$$

$$\frac{F_{output}}{A_{output}} = \frac{F_{input}}{A_{input}}$$

$$\frac{F_{output}}{5A} = \frac{40}{A}$$

$$F_{output} = 200\ N$$

Weight of one brick;

$W = m.g = 2.\ 10 = 20\ N$

Number of bricks:

$$n = \frac{200}{20} = 10\ bricks$$

When you push the input piston downwards, work will be done by the input force in displacing the input piston by Δx_{input}. This work done, thus the energy input to the system is used by the output piston in lifting the load upwards (work against the force of gravity). This results as an increase in the potential energy of the load. Be careful about the rise of the load! Remember, the liquids were incompressible. When you push the input piston downwards, an equal volume of liquid will displace the output piston. The cross-section area of the output piston is larger, the rise of the liquid thus the rise of the output piston will be less than the input displacement. The increase in the force on the output piston is compensated by the decrease in the output displacement. Here there is not a violation of conservation of energy, the energy still is conserved.

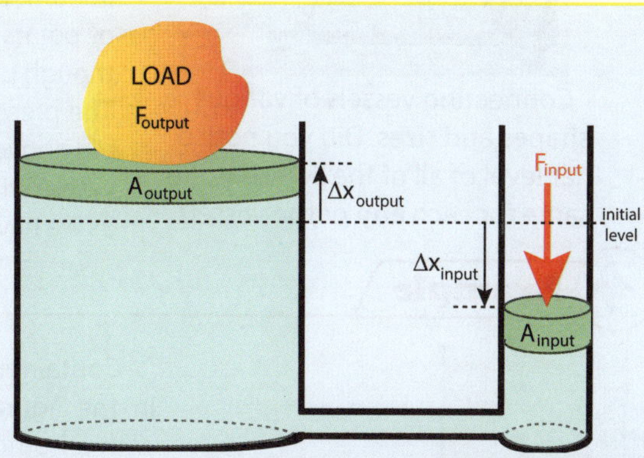

$$W_{input} = W_{output}$$

$$F_{input} \cdot \Delta x_{input} = F_{output} \cdot \Delta x_{output}$$

$$F_{input} < F_{output}\ thus\ \Delta x_{input} > \Delta x_{output}$$

Physics Extra

If two or more immiscible liquids are placed in the same container, the pressure caused by the weight of the above liquid is transmitted to the immiscible liquid below as a result of transmission of pressure by liquids (Pascal's principle).

Example

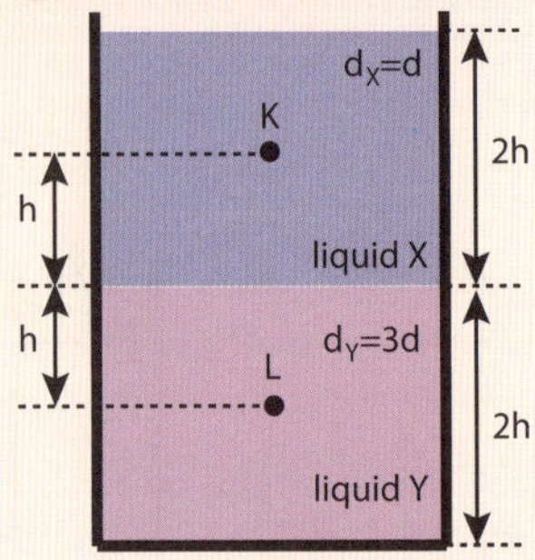

Cross section of a cylindrical container is given in the figure. Liquids X and Y are immiscible to each other.

Find the ratio of pressures at points K and L , P_K/P_L.

Solution

Pressure at point K is caused by the weight of liquid X above it:

$P_K = h.d.g$

Pressure at point L is caused by the weight of liquid X over liquid Y and the weight of liquid Y above it:

$P_L = 2h.d.g + h.3d.g = 5hdg$

$$\frac{P_K}{P_L} = \frac{h.d.g}{5h.d.g} = \frac{1}{5}$$

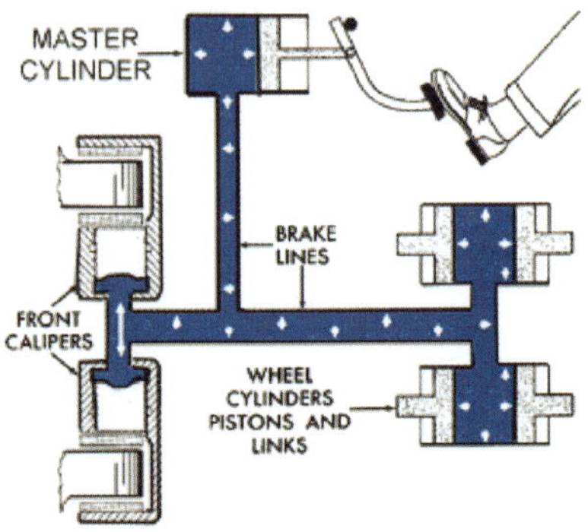

Schematic drawing of hydraulic brakes of a car

The brake systems of the cars also use advantage of Pascal's principle. Once the driver presses the brake pedal; the pressure in the master cylinder increases. This pressure increase is transmitted throughout the brake fluid. Thus the brake pads are pushed against the disk attached to the cars wheel.

A vessel made by interconnecting tubes of various sizes and shapes is called a connecting vessel. If the liquid in a connecting vessel is in hydrostatic equilibrium, then the liquid rises to same height in all open ends of the container.

For the connecting vessel aside, weight of liquid in the wide tube, W_1, is greater than the weight of the liquid in the narrow tube, W_2. If the height of liquid in narrow tube, h_2, was greater then the height of liquid in wide tube, h_1, then the pressure at the bottom of the narrow tube would be greater than the pressure of liquid at the bottom of the wide tube. This pressure difference would cause the liquid to flow from the narrow tube to the wide tube until the heights of liquids are equal on both of the tubes.

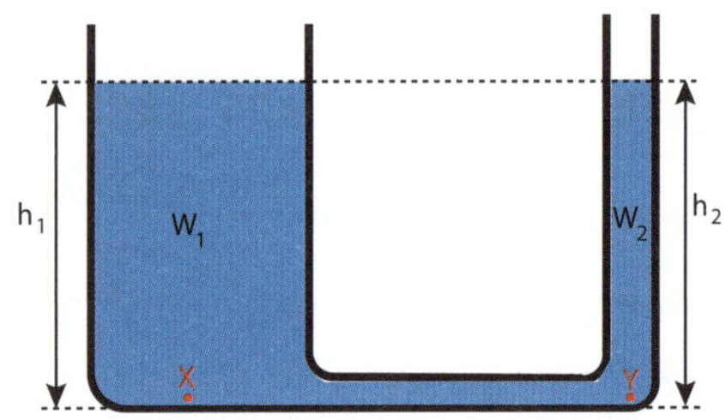

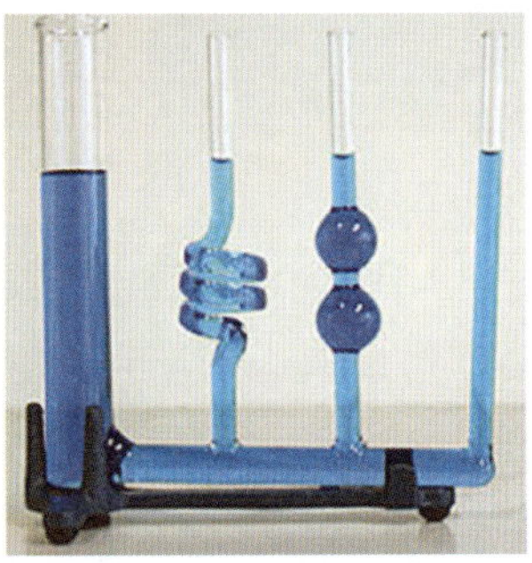

Connecting vessels of various shapes and sizes. Did you notice that level of all of the liquids is the same for each arm of the vessel?

As a conclusion the pressure of all points on a horizontal line through a connected vessel is the same, if the liquid is in hydrostatic equilibrium. For the container aside, the pressure of points K, L, M and N are all the same even though L and M seem to be below that height. The pressure caused by the weight of the liquid above these points is transmitted to everywhere within the liquid, including these points as we have explained in Pascal's principle.

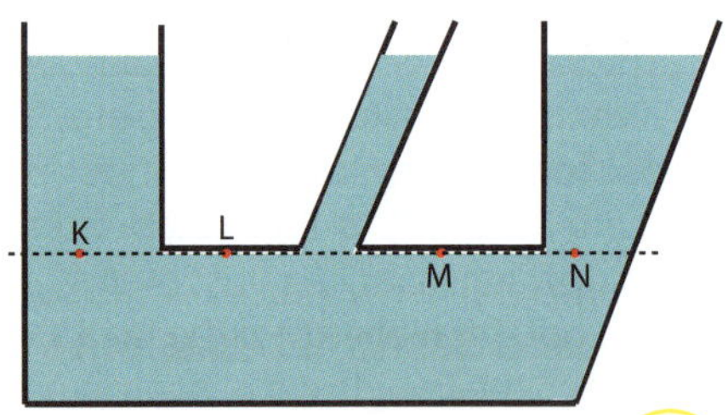

Example

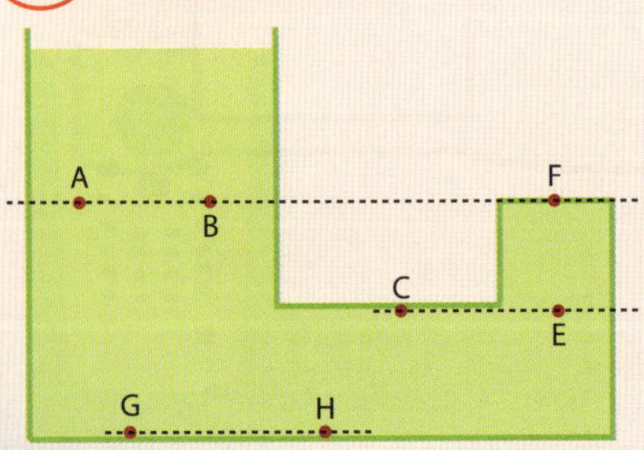

Container given in the figure is filled with a homogeneous liquid. Compare the pressure at the given points.

Solution

Points A, B and F are at the same level: $P_A = P_B = P_F$

Points C and E are at the same level: $P_C = P_E$

Points G and H are at the same level: $P_G = P_H$

The weight of liquid above the deepest points is the highest, so the pressure at the deepest points is going to be the highest:

$P_A = P_B = P_F < P_C = P_E < P_G = P_H$

The Earth is surrounded by an atmosphere. The gases in the atmosphere are pulled by the gravitational force of the Earth so that they cannot escape to outer space. They are energized by the sun; they have kinetic energy to move continuously and just don't *sink* on the Earth. The atmosphere that surrounds us extends up to many kilometers above the surface of the Earth. As you get away from the surface, the air above you gets less dense with the increasing height. Although there isn't a well defined top point for the atmosphere, 97% of the air in the atmosphere is below about 30 km.

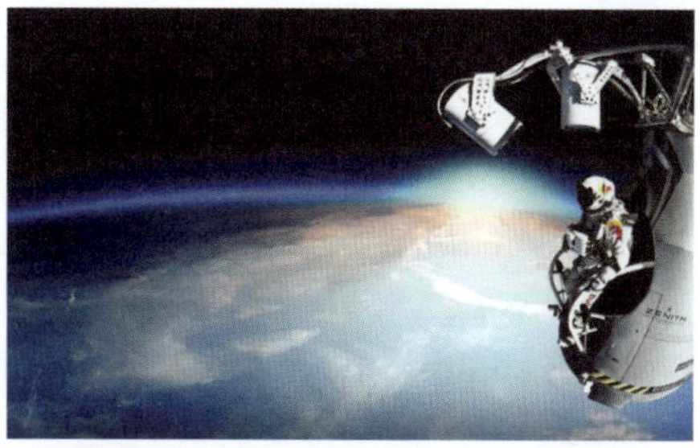

Felix Baumgartner just before his record breaking jump from 39 km above the surface of the Earth. He wears a special suit. Else his blood would boil at his body temperature. A tiny scratch of the suit would cause his death.

We have mentioned earlier that as you get closer to the bottom of a liquid, the pressure increases with the weight of the liquid above that point. This holds true for the atmosphere also. But air is compressible, so the air above that layer will compress the layer below it. Thus the density of air decreases as you get to higher elevations. At high altitudes, there is less air to push down; so the density is less. There will be variations of density depending on the weather conditions also. This concludes that we cannot use the equation $P = h.d.g$ for finding the pressure of air at any location. Remember what we had mentioned before; the equation is valid for a constant density.

Even though we are not much aware of it in our daily life, we are swimming in an ocean of air. The atmosphere exerts a pressure due to its weight on every object immersed in it, including our bodies. The pressure exerted by the atmosphere on average at sea level is:

$$101325\ N/m^2 = 10.1325\ N/cm^2$$

This means that 10.1325 N of force is applied on every cm^2 of our body by the atmosphere on average. Our body stands this huge pressure by the internal pressure maintained by the living cells of our body. This pressure is closely equal to the atmospheric pressure. A child's balloon will stand to the atmospheric pressure as long as its internal pressure is equal to the external (atmospheric) pressure. When the internal pressure is greater than the atmospheric pressure, the balloon explodes. The astronauts should wear special suits when they are in outer space to balance the pressure difference between their body and outer medium.

Magdeburg hemisphere experiment which was firstly conducted by Otto von Gueicke is one of the most famous experiments performed to show the effects of air pressure on objects. He placed two copper hemispheres of diameter 50 cm together, isolated the connection point of the hemispheres so that no air can get in or out. Then evacuated the air inside it. Then they tried to pull the hemispheres from each other by two teams of eight horses, but they couldn't. Once they pumped the air inside again; the hemispheres separated easily by one man. The experiment shows the force applied by the atmospheric pressure on the objects. Vacuum cleaners, suction cups make use of the unbalanced force produced by the atmospheric pressure.

Examples of suction cups

Air surrounding our body parts, let's say your arm, cancels out the forces acting on opposite sides of your arm. This is because the fluid (atmosphere) exerts a pressure in all directions. The net force on opposite sides of your arm is going to be zero thus you can move your arm easily.

The atmospheric pressure is firstly measured by Evangelista Torricelli, a student of Galileo Galilei. In his experiment, he filled a 1 meter long narrow tube completely with mercury and then he inverted the open end of the tube in a container which has mercury inside. The level of mercury dropped to a certain level leaving a *vacuum* at top of the tube. The atmospheric pressure can support this column of mercury so that it can stay at that height. This simple set up used to measure the atmospheric pressure is called a barometer. The level of mercury at sea level is equal to 76 cm.

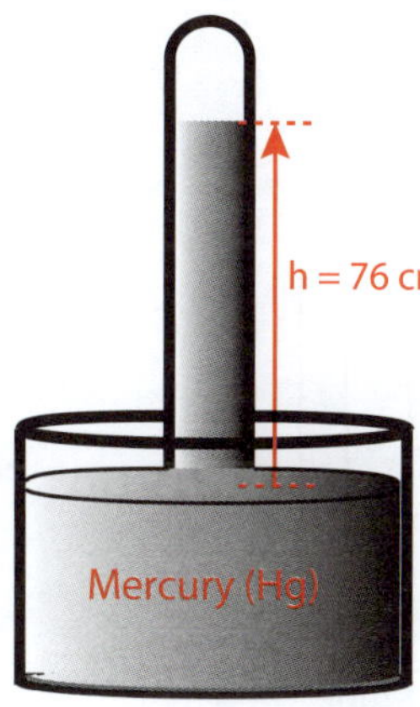

UNIT - 2 PRESSURE AND BUOYANCY

Physics Extra

$P_{Hg} = h.d.g = 76 \times 10^{-2}\ m \times 13.6 \times 10^3\ kg/m^3 \times 9.80\ m/s^2 = 1.013 \times 10^5\ N/m^2 = 1\ atm$

The unit *atm* is another unit of pressure which is most commonly used in expressing the atmospheric pressure.

$1\ atm = 1.013 \times 10^5\ Pa$

Bar is another unit of pressure. It is most often used in weather maps.

$1\ bar = 1.00 \times 10^5\ Pa$

However the pressure measured by a barometer is most often expressed in terms of cm-Hg or mm-Hg.

$\Delta P_{Hg} = h.d.g = 1 \times 10^{-3}\ m \times 13.6 \times 10^3\ kg/m^3 \times 9.80\ m/s^2 = 1.33\ N/m^2 = 1\ mm\text{-}Hg = 1\ torr$

Torr is the name given to mm-Hg in the honor of Evangelista Torricelli.

A barometer can be thought like a see-saw. If the pressure of the atmosphere increases, the air pushes the liquid in beaker down. This forces the liquid up the tube until the pressures at points A and B are equal. If the pressure falls, liquid has to flow out of the tube to keep the pressures equal at these two points.

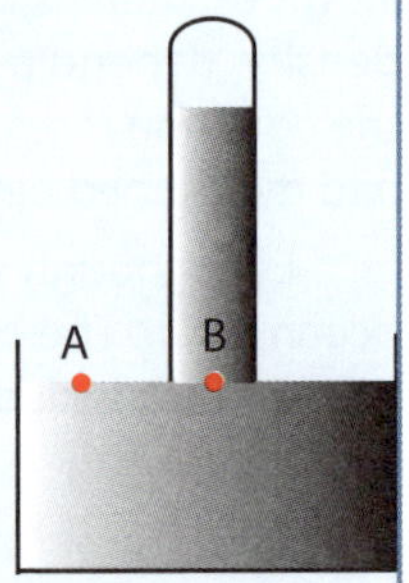

$$P_{atmosphere} = h_{Hg} d_{Hg} g$$
$$P_A = P_{atmosphere}$$
$$P_B = h_{Hg} d_{Hg} g$$
$$P_A = P_B$$

What if Torricelli had used water instead of mercury? How long should the length of the tube be to maintain the atmospheric pressure with water inside? If we perform a similar calculation by using the density of water $1.10^3\ kg/m^3$;

$$P_{atm} = P_{water}$$
$$1.013 \times 10^5\ N/m^2 = h \times 1.00 \times 10^3\ kg/m^3 \times 9.80\ m/s^2$$
$$h = 10.3\ meters$$

This results also concludes that, no matter how good a vacuum pump is; it cannot lift water more than about 10.3 m. What a pump really does is reducing the pressure at top of the tube. As the air pressure is decreased within the tube, atmospheric pressure pushes the water up the tube. The vacuum pump is not sucking the water. This is just like sucking a beverage with a straw. We are not sucking the beverage. We are lowering the pressure at top part of the straw so the atmosphere pushes the beverage up the straw.

A squirrel drinking his smoothie from a straw.

Torricelli's experiment shouldn't be mixed with capillary action. Remember from 9th grade that the rise or fall of a liquid in a fine hollow, narrow tube was called *capillarity*. As the tube gets narrower, the rise of the liquid increases more. The rise or the fall of the liquid depends on the relative strengths of *adhesive* and *cohesive forces* between the molecules. The amount of rise in the capillary depends on the *surface tension* which holds the liquid surface together. In Torricelli's experiment the amount of fall in the liquid level depends on the atmospheric pressure caused by the air surrounding the tube. The tube used in Torricelli's experiment is closed at top whereas a capillary tube is open at both ends. In capillarity liquid usually moves in opposition to external forces such as gravity depending on the difference of the adhesive and cohesive forces. But the liquid in Toricelli's experiment moves in the direction of gravity.

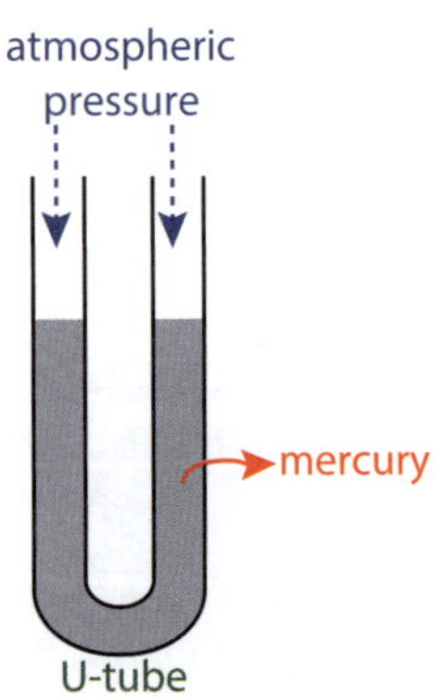

The pressure exerted by the weight of the gas is usually ignored if we are considering a mass of gas in a laboratory size closed container. In this case the pressure exerted by the gas is mainly because of the collisions of the gas molecules with the walls of the container. For measuring the pressure of the gas in an enclosed container a simple device called a manometer is used. A U-tube is filled with a liquid like mercury then it is connected to an enclosed container. Thus a simple manometer is obtained.

Physics Extra

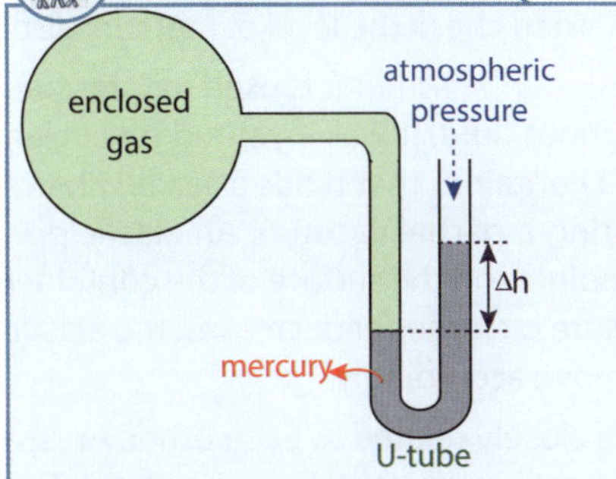

In the figure, the pressure of the gas in the enclosed container is balanced by the atmospheric pressure and the liquid pressure on the right hand side of the U-tube. The pressure of the gas is:

$$P_{gas} = P_{atmosphere} + \Delta h.d.g$$

With a barometer, the atmospheric pressure can be read in terms of cm-Hg. At sea level it is equal to 76 cm-Hg. The liquid in the U-tube is usually mercury. So giving the liquid pressure is in terms of cm-Hg will not cause a problem in terms of units. The pressure of the gas is read as cm-Hg with the manometer and given by:

$$P_{gas} = P_{atmosphere} + \Delta h \quad \text{(in cm-Hg)}$$

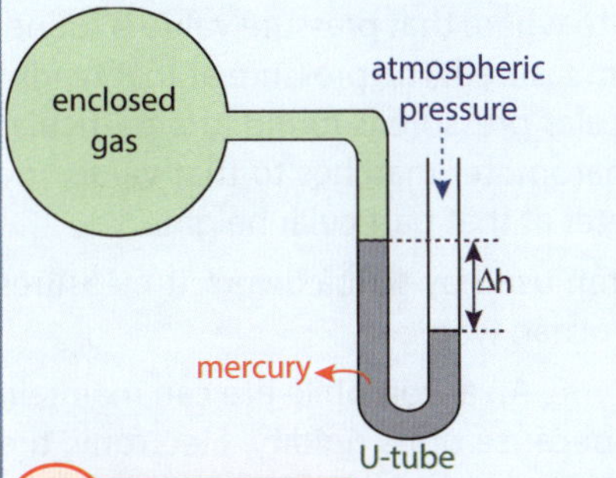

In the figure, the atmospheric pressure is balanced by the pressure of the gas in the enclosed container and the liquid pressure on the left hand side of the U-tube. The atmospheric pressure is:

$$P_{gas} + \Delta h.d.g = P_{atmosphere}$$

$$P_{gas} + \Delta h = P_{atmosphere} \quad \text{(in cm-Hg)}$$

Example

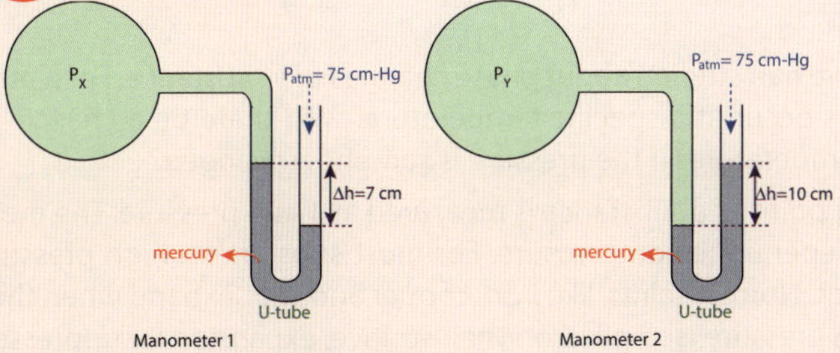

The atmospheric pressure is equal to 75 cm-Hg at a certain place. P_X is the pressure of gas in manometer 1 and P_Y is the pressure of the gas in manometer 2.

Find the ratio of the pressures of the gases P_X/P_Y.

Solution

In manometer 1 atmospheric pressure caused a rise of 7 cm on the other side of the U-tube.

$$P_{atmosphere} = P_{gas} + \Delta h$$
$$75 = P_X + 7$$
$$P_X = 68 \text{ cm-Hg}$$

In manometer 2 pressure of the gas Y caused a rise of 10 cm on the other side of the U-tube.

$$P_{gas} = P_{atmosphere} + \Delta h$$
$$P_Y = 75 + 10$$
$$P_Y = 85 \text{ cm-Hg}$$

$$\frac{P_X}{P_Y} = \frac{68}{85} = \frac{4}{5}$$

Example

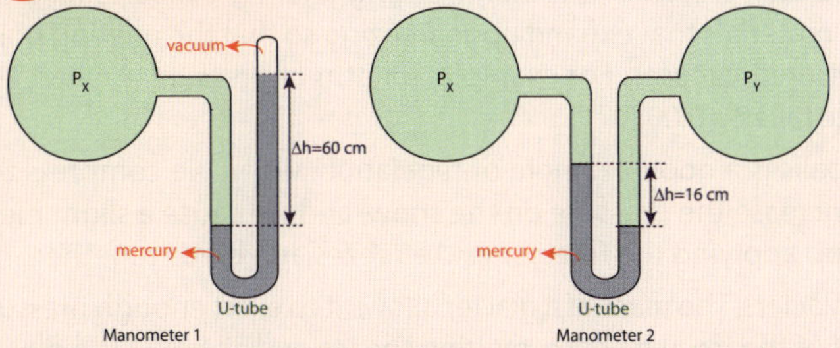

One end of the U-tube of manometer 1 is closed and its other end is connected to a glass balloon containing some amount of gas X. The gas X causes a rise of 60 cm on the other side of the U-tube. When another glass balloon is connected to the U-tube, the level of mercury rises by 16 cm on the left hand side of the U-tube.

Find the ratio of the pressures of the gases P_X/P_Y.

Solution

In manometer 1 pressure of the gas X caused a rise of 60 cm on the other side of the U-tube.

$$P_{gas} = \Delta h$$
$$P_X = 60 \text{ cm-Hg}$$

In manometer 2 pressure of the gas Y caused a rise of 16 cm on the X side of the U-tube.

$$P_Y = P_X + \Delta h$$
$$P_Y = 60 + 16$$
$$P_Y = 76 \text{ cm-Hg}$$

$$\frac{P_X}{P_Y} = \frac{60}{76} = \frac{15}{19}$$

An aneroid barometer is used to make weather forecast

An aneroid barograph can record the pressure changes until new paper is needed.

An electronic tire gauge

Although barometers and manometers are useful devices for measuring the pressure of gases, they are expensive, bulky, require careful handling and are not very portable. If you are travelling with a ship, it is usually so hard to check the level of mercury also.

So instead, a small portable device called an aneroid barometer is used for pressure measurements of atmosphere. Aneroid means without fluid. It was invented by Lucien Vidie, a French physicist. An aneroid barometer is a container that holds a sealed chamber from which some air has been removed, creating a partial vacuum. An elastic disc covering the chamber is connected to a needle or pointer on the surface of the container by a chain, lever, and springs. As atmospheric pressure increases or decreases, the elastic disc contracts or expands, causing the pointer to move accordingly.

The atmospheric pressure and the altitude are closely related to each other. An aneroid barometer calibrated to meters or feet for airplanes is called an altimeter. The calibration is made such that altitude corresponds to where that pressure value is found. With an altimeter the height of the airplane is not measured. The pressure at that height is measured. In the standard atmosphere a particular pressure is found at a particular height. When the measurement of the aneroid barometer matches to that value, the pilot can read the altitude on the calibrated altimeter at that particular height.

A depth gauge is another pressure gauge often used by scuba divers. It measures the equivalent depth in water, especially depth of ocean water.

The pressure of a tire is measured by a tire gauge. An automobile tire can maintain higher pressures than the atmospheric pressure because of its rigidity. Electronic tire gauges are used in nowadays measurements. The pressure is applied to a thin metal diaphragm. The resulting distortion is translated into an electrical signal by a transducer.

At 9th grade physics we had studied about phase changes of a substance. For a pure substance, phase changes occur at a specific temperature. Will there be a change of value for phase change temperatures if the pressure is somehow changed?

Normally the melting point of a substance is measured at 1 atm pressure. The melting point has a modest dependence on pressure. For most substances, high pressure means a high melting point. Normal solids, like iron, copper and wax, expand when they change into liquids. When pressure is applied on their surface, expansion is suppressed and melting is delayed. This results a rise in the melting point of a normal solid.

Abnormal solids, like ice and bismuth, contract as they are melting. When pressure is applied on the surface of such a solid, the change into the liquid is supported. As a result, the melting point of ice is lowered by increasing pressure.

In the case of water; we can also state that, its molecules form open structures in the solid state. When pressure is applied on these crystals they crush into the liquid state. When the added pressure is removed, molecules crystallize and refreezing occurs. This phenomenon of melting under pressure and then freezing again is called regelation. Regelation only occurs for materials that expand upon freezing so that the melting point decreases as external pressure increases. For example, for atm of pressure applied, the melting point of (water) ice falls by 0.0072 °C.

The making of snowballs is a good example of regelation. When we compress the snow with our hands we increase the pressure on the snow, and we cause a slight melting of the ice. When we stop applying the force, snow refreezes and binds together.

Regelation occurs in glaciers. The mass of a glacier allows it to exert enough pressure to lower the melting point of the ice at its base, melting the ice and allowing the glacier to slide over the liquid.

All liquids exert a certain vapor pressure at any certain temperature. The vapor pressure is the degree of the liquid molecules escaping into the vapor phase. At higher temperature the molecules move faster. They are more able to overcome the attractive intermolecular forces that tend to bind them together and the vapor pressure increases with temperature. If the pressure of that vapor phase is increased, the particles need more kinetic energy to escape because they must overcome the extra forces pushing them into the liquid phase.

A liquid boils when the vapor pressure of the liquid equals the atmospheric pressure of its surroundings. As the boiling point is reached tiny bubbles appear throughout the liquid. If the vapor pressure is less than the atmospheric pressure, the bubbles are crushed. If the vapor pressure is equal to the atmospheric pressure,then the bubble will increase its size as it rises to the surface of the liquid.

When the external pressure is:

a. less than 1 atm, the boiling point of the liquid is *lower* than its normal boiling point.

b. equal to 1 atm, the boiling point of a liquid is called the normal boiling point.

c. greater than 1 atm, the boiling point of the liquid is *greater* than its normal boiling point.

Water's vapor pressure curve has a great importance. It determines the relationship between temperature and pressure in any process where water is going from a liquid to a vapor state or vice-versa. These processes are important both in industry for example in steam power generation where water is made into steam in boilers and eventually condensed after running through turbines and in nature for example, in rainfall and evaporation from bodies of water.

Pressure cooker

The pressure cooker is a good application of effect of pressure on boiling point. It was invented by Denys Papin, a Frenchman, in 1681. By means of a valve arrangement, the steam formed is not allowed to escape freely. The pressure over the water surface consequently builds up, resulting in the boiling point of water being raised to about 120 °C. The food cooks faster due to the rise in temperature.

2.1.2. The relationship between the velocity and pressure of fluids

If the fluid you are working on is not at rest, if it is moving then there are going to be changes of pressure within the fluid. These changes also cause these fluids to accelerate. How airplanes stay in the air, how the heart pumps the blood in our circulatory system are just a few of the questions that fluid dynamics is seeking the answers for. Particularly if the working fluid is water, than it is called hydrodynamics.

There are two main types of fluid flow:

1. If the flow of the fluid is smooth, the neighboring layers of the fluid are sliding by each other smoothly then the flow is a laminar flow or a streamline. Each particle of the fluid follows a smooth path. These paths are called streamlines and they do not cross one another. The velocity of the liquid particle is always tangent to the streamlines.

2. If the velocity of the flow exceeds a certain value small, whirlpool-like circles (eddies) occur within the fluid. This kind of flow is called a turbulent flow. Water flowing in a rock-filled stream or river, the water in the wake of a speed boat, smoke rising from the chimney on a windy day, the air wakes left by air planes or cars are all examples of turbulent flow. The turbulent flow causes an increase in the internal friction of the fluid. Resistance of a fluid to flow is called viscosity.

To understand the motion of the fluids some assumptions are made and an ideal fluid is defined. These assumptions are:

1. An ideal fluid is *incompressible*, the density of the fluid remains constant.

2. The flow is *steady*, the velocity of the fluid at each point remains constant in time; its a laminar flow.

3. The flow is *irrotational*.

4. The fluid is *non viscous*, there are no internal friction or no retarding forces within the fluid.

Real fluids have viscosity and internal friction causes different layers of the fluid to flow at different speeds.

Let's consider the flow of the fluid within the tube given in the figure:

Mass flow rate through a cross-section area is: $\frac{\Delta m}{\Delta t}$

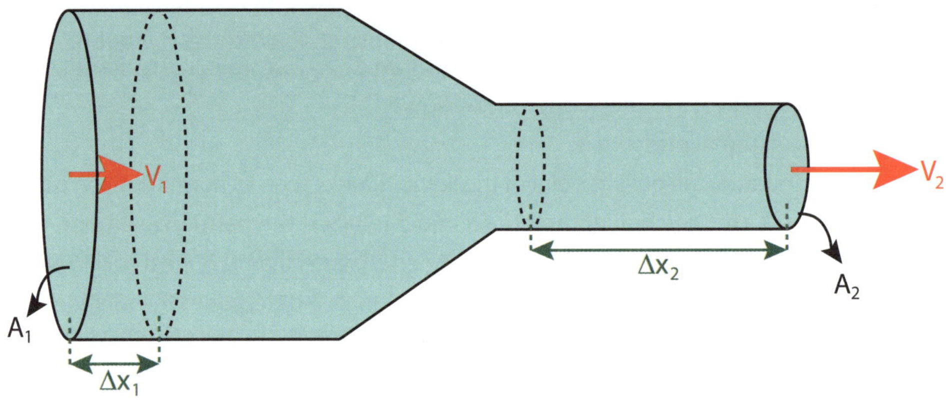

No fluid flows in or out of the sides of the tube, thus the flow rates through A_1 and A_2 should be equal for equal time intervals:

$$m = d.V \Rightarrow \Delta m = d.\Delta V = d.A.\Delta x$$

$$\frac{\Delta m}{\Delta t} = \frac{d.A.\Delta x}{\Delta t} = d.A.\frac{\Delta x}{\Delta t} = d.A.v$$

For an ideal fluid, we assumed that it is incompressible. This means that the density of the liquid is constant everywhere within the flow.

$$\frac{\Delta m_1}{\Delta t} = \frac{\Delta m_2}{\Delta t} \Rightarrow d_1A_1v_1 = d_2A_2v_2$$

$$A_1v_1 = A_2v_2$$

Symbol	Quantity	Unit
A	cross-section area	m^2
v	velocity	m/s

This equation is known as the equation of continuity. The volume of an incompressible fluid entering one part of a tube or pipe must be equal to the volume of fluid leaving the tube. According to equation of continuity; a fluid flows slower when the cross-section area is large and vice-versa. For a fluid in a flow, to have the product of A.v remaining constant along the tube of flow, in the narrow part of the tube the streamlines must come close together than the wide part. As the distance between the streamlines decreases, the speed must increase.

Rivers flow faster in narrow gorges

A river flows slowly through a meadow where it is broad. But when it is passing through a narrow gorge it reaches to very high speeds. If you put a nozzle to the end of a garden hose, water squirts farther when compared the previous case. The narrow opening of a nozzle gives the water a higher exit speed.

The product of $A.v$ is equal to volume of fluid flowing in unit time; $\Delta V/\Delta t$ in m^3/s. This ratio is often called volume flow rate. The volume flow rate is constant at all points of a tube.

A small unit of a fluid will contain many particles of fluid. We will consider that small unit of fluid and call is the fluid element.

Nozzles help water to squirt farther

In the above tube, we have mentioned that the fluid element is moving with velocity v_1 at cross-section A_1 and velocity v_2 at cross-section A_2. With respect to equation of continuity; since $A_1 > A_2$, then $v_1 < v_2$. So far we have learned about the laws of motion at 9th grade. There should be a reason for a change of velocity when an object is moving. That reason is usually a force applied on the object. For the fluid element that we are considering within the tube, there are no external forces. The fluid is moving in the horizontal plane so there is no effect of weight of the object on its motion. So what causes the force for the change of velocity?

Remember that the force of pressure applied on a surface was equal to pressure times area; $F = P \times A$. Let's say that the pressure on the left side of the fluid element in the figure has more pressure than its right side. As a result the force on the left side is going to be larger than the force on the right side. This force difference resulting from the pressure difference gives rise to a net force on the fluid element.

$$F_{net} = F_{left} - F_{right} = P_{left}.A - P_{right}.A = \Delta P.A$$

We can conclude that, if and only if there is a pressure difference between the two faces, there will be a net force on the fluid element causing it to accelerate, or to change its velocity.

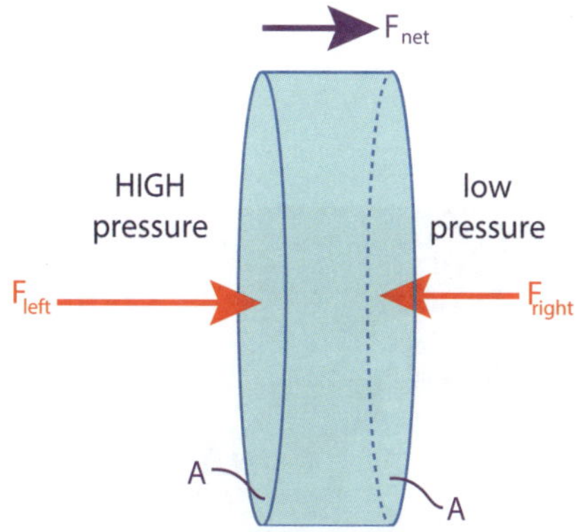

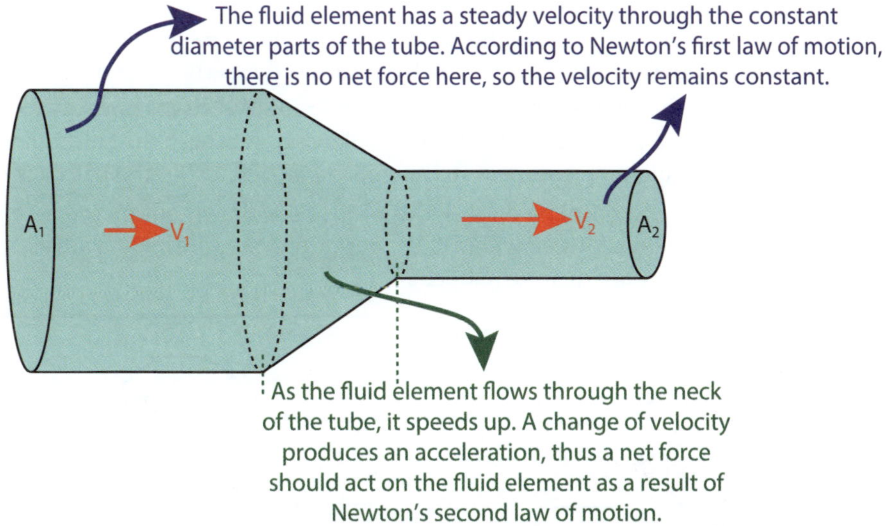

For the tube in the above figure; in order to accelerate the fluid elements through the neck, the pressure in the wider section, P_1, should be higher than the pressure at the narrow section, P_2. If there is a pressure difference from one point to the other; the fluid accelerates.

As a result; a fluid moves slower at a wider cross-section area and its pressure is higher when compared to the fast moving fluid in the narrow cross-section area. This effect of the fluids is first discovered by Swiss scientist, Daniel Bernoulli and called the Bernoulli effect.

Bernoulli has derived an equation for the pressure of moving liquids by using the equations of motion. His equation is actually an expression of the law of conservation of energy. Now, let's use the equation of continuity and Bernoulli's effect to explain some daily phenomenons:

If you place a sheet of paper on a table and blow across its top surface (or wind may be blowing across its top), you will observe that top surface of the paper rises. When you blow the air over the paper, you cause the air above it to move faster. Faster moving air causes a reduction in air pressure above the paper. Thus a net force is produced and the paper rises.

The walls of our houses are under the influence of the huge force exerted by the atmospheric pressure. The walls of the house do not collapse because the same atmospheric pressure is present inside the house. In the case of strong winds, hurricanes or tornadoes; the atmospheric pressure decreases. If a building in not well-vented the air pressure inside the house may be much more greater than the external air pressure. Thus the roof of the building can easily be blown off.

An approaching tornado

The pressure inside a tornado is almost 10% lower than the atmospheric pressure. When a tornado comes over a house, there will be a pressure difference between the inner and the outer walls. Inside the house is at normal atmospheric pressure whereas the outer pressure is lower. This pressure difference produces a net force from the inside of the house to outwards which will be enough to explode the walls of the house outwards. In the case of a tornado, opening the doors and the windows will help a great deal of the air inside the house to flow out through them. Hence the pressure difference between the inside and outside of the walls of the house will be minimized.

Many victims of tornadoes do not follow this procedure because mostly tornadoes are spawn out of severe thunder storms. These storms are usually accompanied with torrential rain. Usually the first thing the victims does in the house to close the windows.

The perfume atomizers that we are using in our daily life work according to the Bernoulli's effect. Air blown at high speed across the top of a vertical tube of perfume atomizer reduces the pressure above the tube. This pressure is less than the normal air pressure inside the bowl. This pressure in the bowl causes the perfume to rise upwards to the air stream. The perfume disperses into a fine spray of droplets.

Floating orchestra of ping pong balls.

A ping-pong ball can be made to float above a blowing jet of air. If the ball begins to leave the jet of air, the higher pressure in still air outside the jet pushes the ball back in.

In the carburetor of a gasoline engine, the air inside is drawn out by the air filter thus a low pressure region is formed within it. The vaporized gasoline mixes with air, and enters the cylinder of the engine for combustion.

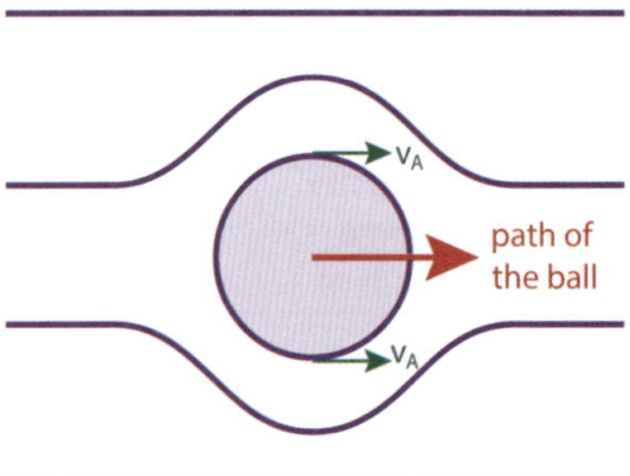

non-spinning ball

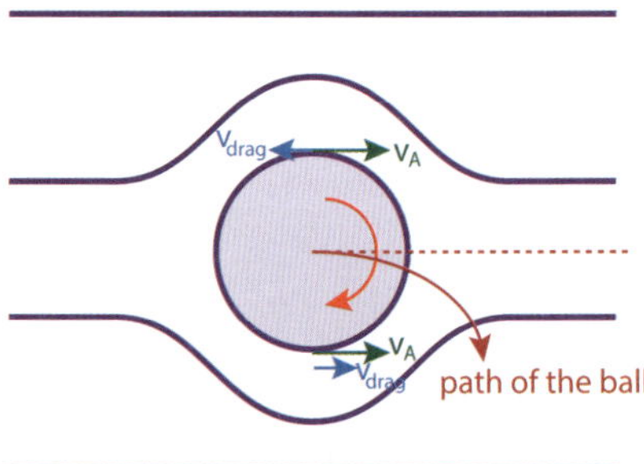

spinning ball

A non-spinning ball thrown through the air, follows a straight line path. The air at top and bottom of the ball moves with the same speed. If the ball is released with a downward spin, the spinning ball drags some air around with it. This produces a velocity difference at top and bottom of the ball. A velocity difference at top and bottom causes a pressure difference.

$$v_{top} < v_{bottom} \text{ so } P_{top} > P_{bottom} \text{ thus } F_{top} > F_{bottom}$$

Therefore the ball curves downward or sinks as it approaches to the other player. By spinning the ball to the right, the ball curves toward the right or vice versa.

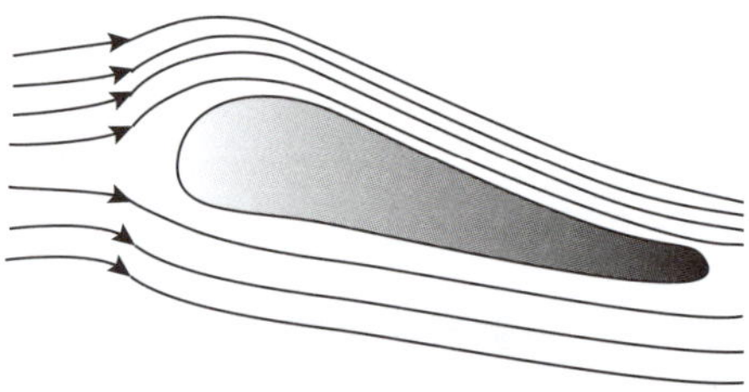

Streamlines on top and bottom surface of an airplane's wing. Notice that the streamlines are not intersecting with each other, and closer to each other at top surface of the wing

Now let's examine the conditions for an airplane to fly. The air flowing over the top of the wing has a longer distance to travel than the air flowing under the wing. Remember that the streamlines were never intersecting with each other. So, the air at the leading edge should reach to the trailing edge at the same time either for top and bottom surfaces of the wing. To have this, the velocity of the air at the top surface should be larger than the velocity of the air at the bottom surface. If the velocity at top is higher than the velocity at the bottom of the wing, the pressure at the top of the wing is smaller than the pressure at the bottom of the wing. If a vehicle is moving at sufficiently high speeds, the pressure difference between the top and bottom of the vehicle may become so large. The force produced by this pressure difference may be large enough to overcome the force of gravity. In the case of an air plane, if it is moving at a high speed with its wings tilted upward at a small angle which is called the attack angle, it experiences a lift force. The upward tilt and the rounded upper surface of the wing causes streamlines to be forced upward and become closer to each other above the wing. As the streamlines get closer to each other, the speed of air above the wing increases. Since the speed is greater above the wing than the below surface, the air pressure above the wing is smaller than the air pressure below the wing. This net upward force produced, called a dynamic lift.

The friction between the air and the wing exerts a drag force which the engines of the airplane must overcome.

The lift force is not dependent on the shape of the wing; a flat wing or a wing with a symmetric cross-section area wouldn't matter. As long as the front wing is tilted upwards the airplane flies.

The wing shown in the figure can experience the dynamic lift even if the attack angle is zero. The rounded upper surface deflects the air up and thus the streamlines are squeezed together. If the attack angle is about 15°, turbulence sets. Turbulence causes a greater drag and less lift and plane to drop.

A sailboat can move against the wind by setting the sails at an angle. When the wind fills in the sails, the wind travels rapidly over the bulging surface of the sails. The relatively still air behind the sail exerts a greater pressure on the sail. As a result, the wind exerts a net force on the sail. This force will cause the boat to move sideways. The keel, extending vertically downward beneath the water is under the influence of the force applied by water. This force is almost perpendicular to the keel. The resultant of the force applied by the wind and the water is almost directly forward.

Sailing with the wind

The smoke of a stove goes up to the chimney partly because hot air rises. The wind blowing across the chimney lowers the pressure. The pressure inside the house is more than the air outside the house. Hence the air and the smoke are pushed up the chimney by the higher indoor pressure.

Animals like Prairie dogs, rabbits living under ground must circulate the air in their burrows to avoid suffocation. For circulating the air, they always have at least two openings with slight different heights for their burrows. The flow of air across different holes is usually slightly different. This results a slight pressure difference. This difference forces a flow of air through the burrow.

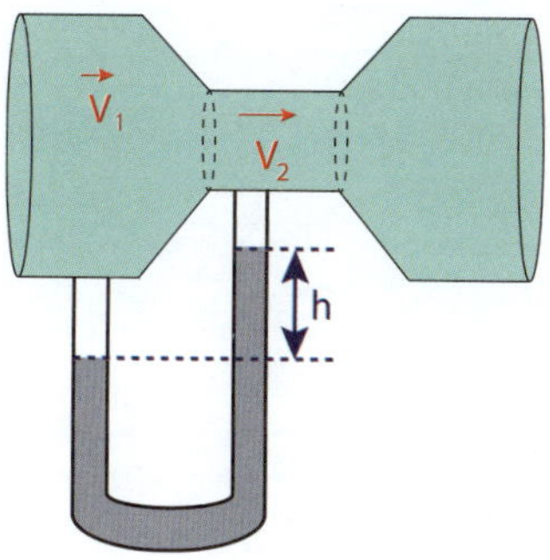

A venturi tube is used to measure the flow speeds of liquids. The velocity of the blood in the arteries is measured by a venturi tube. A liquid with a density of d flows through the tube with velocity v_1. At the throat the cross-section area is reduced to a smaller value, so the velocity increases to v_2 and the pressure decreases. The level difference of mercury at the attached U-tube gives the pressure difference between the larger and the narrower parts of the pipe. The velocity is then can be calculated or using the pressure gauges and speedometers the values can be read as in the figure.

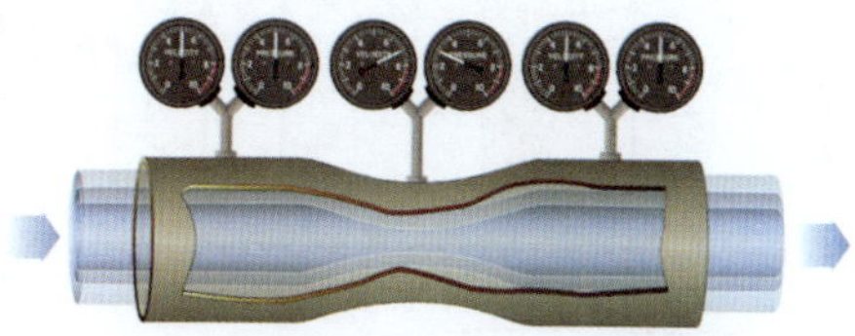

Venturi tube with speedometers and pressure gauges

If the speed of a gas is measured then it is called a Pitot tube.

2.2. Buoyant Force

2.2.1. Buoyant Force

Depending on your everyday observations, you can easily say that a rock thrown into water will sink, a piece of wood will float on water. Although a coin sinks in water a huge ship made of steel can float. A child's ball made of plastic can easily float on water. But if you ever try to push the ball under water, you will experience a force which doesn't allow you to push it under water. Now we will seek for the answer of the questions "*why*" and "*how*" for these observations.

First let's weigh an object in vacuum with a dynamometer. Then submerge it slowly into a container filled by a fluid. We will observe that as the volume of object in the fluid increases (immersed part of its volume) the value read on the dynamometer decreases until the object is totally submerged. Once the object is totally submerged, the dynamometer displays no change in the final reading.

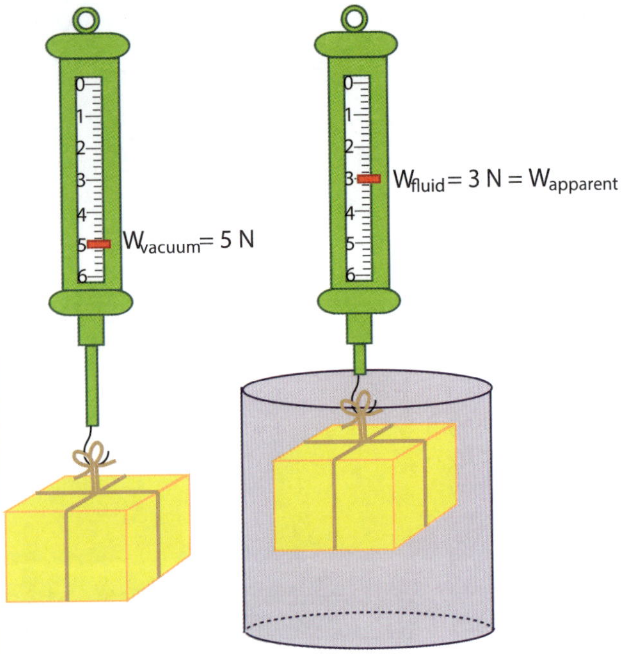

In the figure the object weighs 5 N in vacuum, which means the real weight of the object is 5 N. Its weight is 3 N in the fluid. This value is less than the real weight of the object. So we will call it the apparent weight of the object. The apparent weight of the object is 3 N in the figure.

For the object to weigh less, there should be a force acting in the opposite direction of weight. This force acting on the object within the liquid is called the buoyant force. Any object when immersed in a fluid, experiences the effects of the buoyant force.

The buoyant force is:

$$F_B = W_{real} - W_{apparent}$$

According to our figure:

$$F_B = 5 - 3 = 2\ N$$

The main reason for the buoyant force to occur is the pressure difference between the top and bottom surfaces of the object within the fluid. The pressure of a fluid increases with depth, so does the force of pressure. The difference of force of pressures between the bottom and the top surfaces is equal to the buoyant force.

Physics Extra

Let's consider a cylindrical object within a liquid. The upward force of pressure acting on the bottom surface of the cylinder is larger than the downward force of pressure acting on the top surface of the cylinder. Lateral force of pressures cancel out each other since average heights for both is the same.

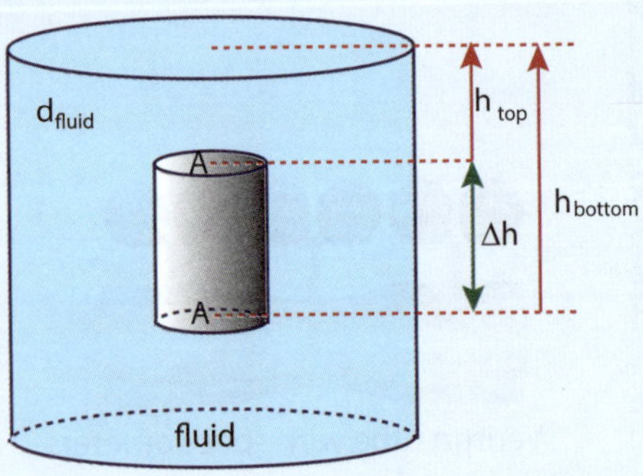

$$\left.\begin{aligned} F_{bottom} &= h_{bottom} \cdot d_{liquid} \cdot g.A \\ F_{top} &= h_{top} \cdot d_{liquid} \cdot g.A \end{aligned}\right\} \Delta F = F_{bottom} - F_{top}$$

$$\Delta F = F_{bottom} - F_{top}$$

$$\Delta F = (h_{bottom} - h_{top}) \cdot d_{fluid} \cdot g \cdot A$$

$$\Delta F = \Delta h \cdot d_{fluid} \cdot g \cdot A$$

$$V = h \cdot A \quad \text{so} \quad \Delta F = d_{fluid} \cdot \Delta V \cdot g$$

$$m = d \cdot V \quad \text{so} \quad \Delta F = \Delta m_{fluid} \cdot g$$

$$W = m \cdot g \quad \text{so} \quad \Delta F = \Delta W_{fluid}$$

We can conclude that the difference of the force of pressures is equal to the weight of the fluid displaced by the cylinder.

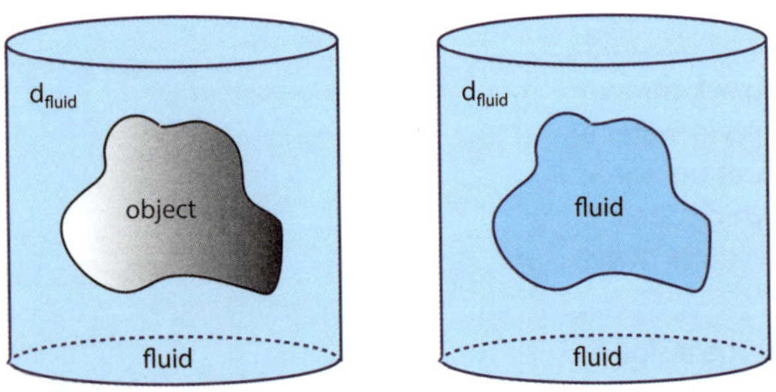

What if the object has an irregular shape? Will it experience the buoyant force as well? To explain this, let's consider an irregular shape object within a fluid as in the figure. We will replace the object with another identical shape but this time the shape is composed of the fluid itself. The mass of the fluid is in equilibrium with its surroundings. Thus the resultant force due to the pressure of the surrounding liquid should be an upward force and its magnitude should be equal to the weight of the liquid contained within the irregular boundary. The object was also in the same position and any other object of the same shape and position will be under the influence of the same magnitude of buoyant force. Thus we can conclude that the pressure exerted by the surrounding fluid does not depend on the composition of the material within the body.

Archimedes, an ancient Greek Philosopher, is famous about a story about him and the crown of King of Syracuse. According to the story; the King of Syracuse wanted to have himself a crown made of pure gold, so he gave his goldsmith some gold. When he got his new crown he felt a suspicion about the goldsmith. Did he use the whole gold for the crown or did he keep some of it for himself? He ordered Archimedes to solve the problem. As the days were passing Archimedes was hopeless about the solution until he got into a bath tube completely filled with water. As he stepped inside, some of the water had overflown. He got out of the bath shouting "Eureka / I found it!" He knew that the crown had the same mass with the gold given to the goldsmith. Then he thought about the volumes, were they the same? Usually the story tellers give the answer as a "no".

Years after the times of Archimedes, a physicist who is working on the buoyancy of fluids should know about his principle.

"EUREKA" shouted Archimedes in his bath tube.

Archimede's Principle is: *The buoyant force on an object immersed in a fluid is equal to the weight of the fluid displaced by that object.*

In this sentence, fluid displaced means *a volume of fluid equal to the volume of the submerged object, or the part of the object submerged in the fluid in the case of floating.* Shortly, the fluid that used to be where the object is.

So far we have been using the word fluid, to emphasise that air is also a fluid. Objects in air weigh less than they weigh in vacuum because of the buoyant force of air applied on the object. The density of air is so small, that's why the effect of buoyancy of air on solids is usually ignored.

Suppose you are holding an object within a fluid, let's take it as a liquid. When you release the object:

Symbol	Quantity	Unit
F_B	buoyant force	N
d	density	kg/m^3
V	volume	m^3
W	weight	N

1. The object may move up and come to rest as some portion of its volume within the liquid (floating)

In the free body diagram, notice that the buoyant force is drawn from the center of the immersed part and its weight is drawn from the center of the object assuming that the object is a homogeneous one. With respect to the free body diagram; the forces acting on the object are its weight and the buoyant force of the liquid. The object does not move up or down, so the net force acting on the object should be equal to zero according to first law of motion; inertia.

$$W = F_B$$
$$m_{object} \cdot g = m_{liquid\ displaced} \cdot g$$
$$d_{object} \cdot V_{object} \cdot g = d_{liquid} \cdot V_{immersed} g$$
$$V_{object} > V_{immersed}$$
$$d_{object} < d_{liquid}$$

If the density of the object is smaller than the density of the fluid, then the object floats as some part of its volume within the fluid.

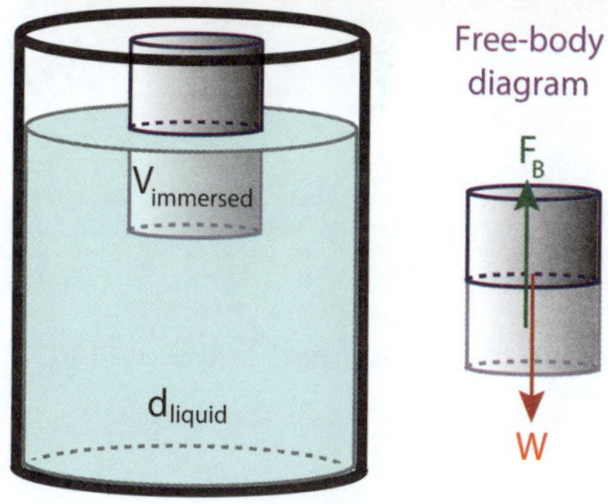

Floating

2. The object may remain in the position that you had released the object.

With respect to the free body diagram; the forces acting on the object are its weight and the buoyant force of the liquid. The object does not move up or down, so the net force acting on the object should be equal to zero according to first law of motion; inertia.

$$W = F_B$$
$$m_{object} \cdot g = m_{liquid\ displaced} \cdot g$$
$$d_{object} \cdot V_{object} \cdot g = d_{liquid} \cdot V_{immersed} g$$
$$V_{object} = V_{immersed}$$
$$d_{object} = d_{liquid}$$

If the density of the object is equal to the density of the fluid, then the object remains at where it was released.

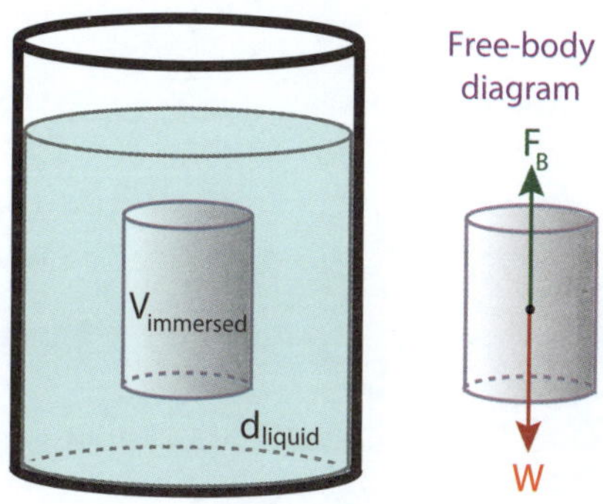

Suspension

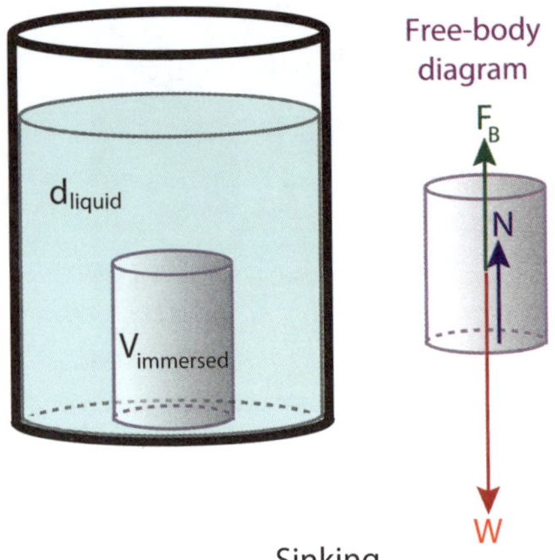

Sinking

3. The object may move down towards the base of container and stop moving as it reaches to the bottom (sinking).

With respect to the free body diagram; the forces acting on the object are its weight, the buoyant force of the liquid and the reaction (normal) force of the surface. The object would continue to move down if the container was deeper.

$$W = F_B + N$$
$$W > F_B$$
$$m_{object} \cdot g > m_{liquid\ displaced} \cdot g$$
$$d_{object} \cdot V_{object} \cdot g > d_{liquid} \cdot V_{immersed} \cdot g$$
$$V_{object} = V_{immersed}$$
$$d_{object} > d_{liquid}$$

If the density of the object is larger than the density of the fluid, then the object sinks in the fluid.

Example

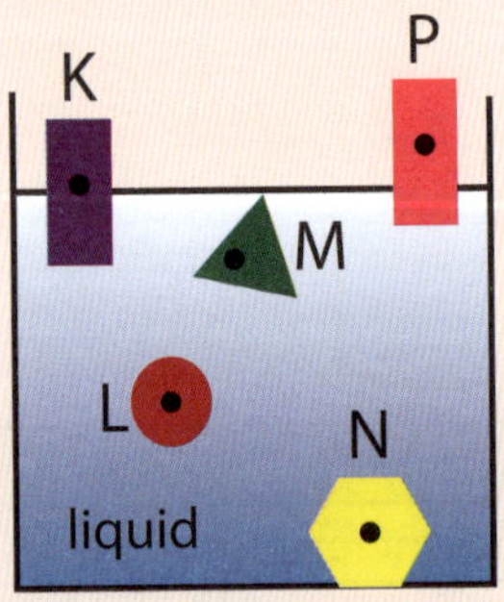

Final positions of the objects K, L, M, N and P in a liquid are as shown in the figure. K and P have the same volume. The black dots on the objects represent their center of mass.

Compare the densities of these objects and the density of the liquid.

Solution

Object N sinks in the liquid, so: $d_{liquid} < d_N$

Objects L and M are under water and remain at their positions, so: $d_L = d_M = d_{liquid}$

Objects K and P are floating in water, so: $d_K < d_{liquid}$ and $d_P < d_{liquid}$

But the immersed volume of K within the liquid is greater than the immersed volume of P so: $d_P < d_K < d_{liquid}$

As a result:

$d_P < d_K < d_{liquid} = d_L = d_M < d_N$

2.2.2. Applications of Buoyant Force

Although a block of iron sinks in water, a ship made of iron floats in water. We can have the same weight of iron made into thin slabs and then weld these slabs to each other to form the shape of a ship. By increasing the total size hence the volume, a greater volume of immersed part can be obtained. More water will be displaced and the magnitude of the buoyant force will be increased. If the magnitude of the buoyant force can be made equal to the weight of the iron ship, then the ship can be made to float in water.

Fish fill their swim bladders with air to arrange their average density to the density of water. When a fish needs to rise or descend, it changes the volume of its swim bladder by filling it with gas. As a result the average density of the fish changes. Likewise, a submarine uses ballast for descending and ascending underwater. In order to descend the submarine, the craft is allowed to take water into its ballast tanks. When the captain decides to rise toward the surface, compressed air is pumped into the tanks by a valve. Thus the water is pushed out and the submarine ascends to the surface.

For an object to float in air, it should weight less than the weight of air it displaces. If you fill a regular balloon with air, the weight of air inside the balloon will be equal to the weight of air displaced by the balloon. There won't be an upward net force. The balloon itself will have a weight so it will sink in air. For a balloon to float, we must fill it with a gas which has a lower density than that of air such as hydrogen, helium or hot air. Even though hydrogen is the least dense one and cheap to produce it is extremely flammable. The use of hydrogen for airships ended when the airship Hindenburg crashed in 1937. Helium is perfectly safe and only slightly less dense than hydrogen but it is very expensive. It is ideal for the party balloons but its price makes it impractical to be used for large air balloons. Hence air heated up to 299 °C is the only viable option. Balloon also have a role in aiding space exploration. Telescopes mounted on huge, unmanned balloons provide high-resolution images at elevations as high as 35 km above the surface of the Earth. Such balloons are not only used for the images of the Earth but also from Venus and Mars.

Solved Questions

Question

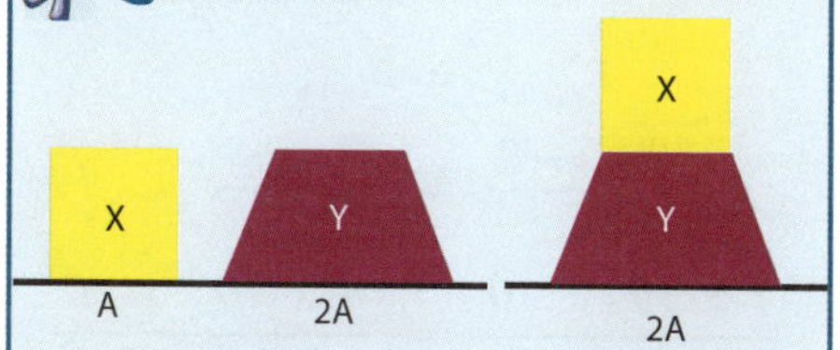

Figure 1 Figure 2

Base areas of X and Y are A and 2A respectively. The pressures exerted by both of the objects are P in figure 1.

a. What are the force of pressures exerted on the ground by X and Y?

b. What would the pressure exerted by Y on the ground be if X was placed over it as in figure 2?

c. What is the force of pressure acting on the ground in figure 2.

Solution

a.

$F = P \,.\, A$ *for solids*

$F_X = P \,.\, A = PA$

$F_Y = P \,.\, 2A = 2PA$

b.

$$P = \frac{W_{total}}{A} = \frac{PA + 2PA}{2A} = \frac{3}{2}P$$

c.

$$F_{final} = \frac{3}{2}P.2A = 3PA$$

Question

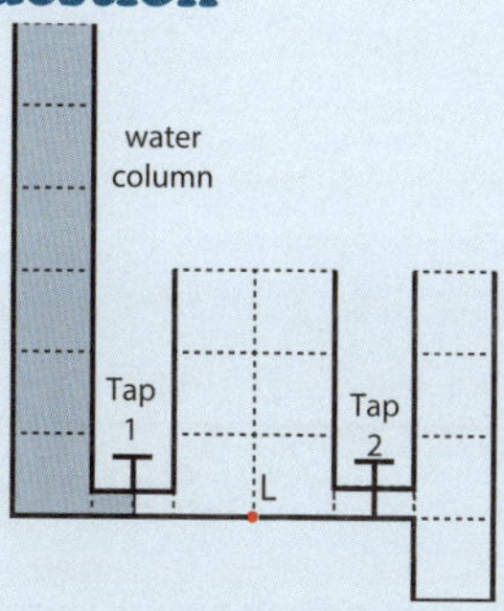

Cross-section of a connected vessel is given in the figure.

a. Find the pressure at point L when only tap I is opened.

b. Find the pressure at point L when both of the taps are opened.

(Distances between the broken lines are equal, ignore the volume of liquid in the connection pipes)

Solution

The level of liquid should be the same for all arms of a connecting vessel. When the tap is opened, the volume of water will spread over the base, so that it has the same height on each arm.

a. The height of water is going to be 2 units above point L when equally distributed:

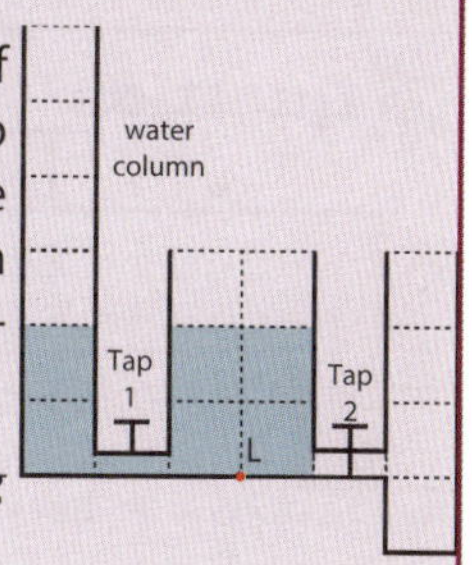

$P_L = 2h.\, d_{water} g$

b. When the water in the container is distributed, 1 extra unit will spread over 4 units of base area for this container.

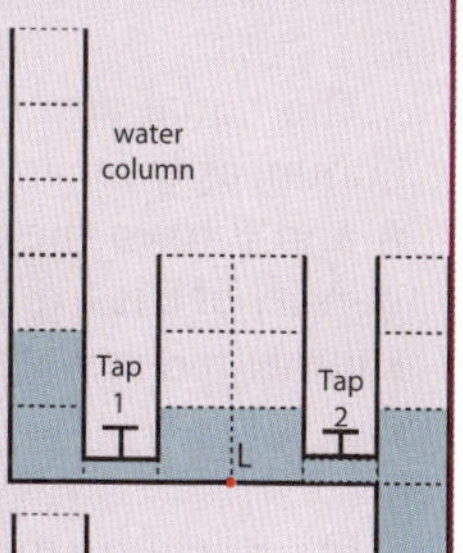

The height of water is going to be 1.25 units above point L. Thus:

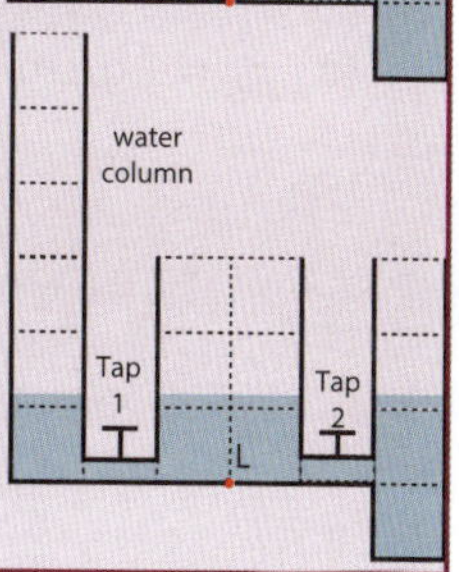

$P_L = 1.25h.\, d_{water} g$

HOW BLOOD PRESSURE IS MEASURED?

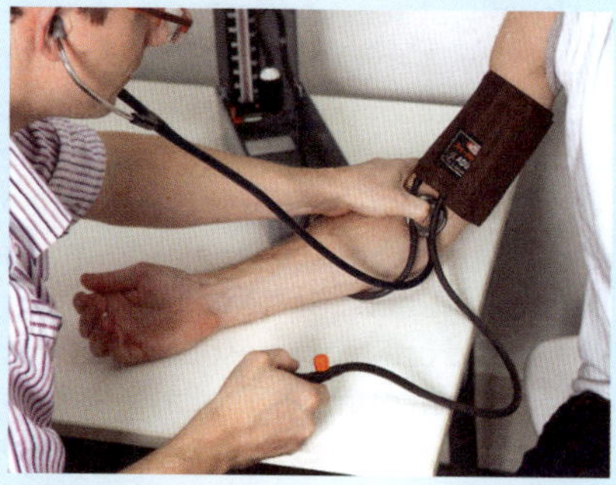

Blood pressure is measured with an instrument called sphygmomanometer . The device includes a cuff, a pump with a valve, and a gauge. To measure your blood pressure:

1. Cuff is wrapped around your upper arm and made sure that the cuff is the right size for your arm.
2. The cuff is inflated by closing the valve and squeezing a small pump. This stops the flow of blood to your arm for a few seconds.
3. The end of a stethoscope is placed on the inner surface of your elbow.
4. The valve is opened to slowly release air from the cuff. As your blood begins to flow back through your arm it makes a pulsing sound. This sound is listened with the stethoscope.
5. When the first sound of the pulse is heard, the number from the pressure gauge is recorded. This is your systolic blood pressure. It represents the pressure in your arteries when your heart beats or contracts.
6. When the sound fades or disappears, that number from the pressure gauge is again recorded. This is your diastolic blood pressure. It represents the pressure of the blood in your arteries when your heart rests between beats.

Your blood pressure is noted as a fraction. The top number is your systolic blood pressure. The bottom number is your diastolic blood pressure. For instance, a systolic pressure of 120 and a diastolic pressure of 80 is written as 120/80 and read as 120 over 80. Blood pressure is measured in millimeters of mercury, shown as mm Hg.

Usually at least two readings separated by 2 minutes are taken, then your doctor will average the readings. The top number, or systolic pressure, is the measurement of the pressure of the blood in your arteries when your heart contracts, or beats, and is forcing blood out to the rest of your body. The bottom number, or diastolic pressure, is the measurement of pressure of the blood in your arteries when your heart is relaxing between beats. This is when your heart is filling for the next beat.

What your blood pressure test results mean:

Systolic Blood Pressure (your top number)	Diastolic Blood Pressure (your bottom number)	Your Blood Pressure
Less than 120	Less than 80	Optimal
Less than 130	Less than 85	Normal
130 to 139	85 to 89	High to normal
140 to 159	90 to 99	Stage 1 hypertension
160 to 179	100 to 109	Stage 2 hypertension
180 or greater	110 or greater	Stage 3 hypertension

Normal blood pressure ranges. Normal blood pressure is a range of values. Don't be alarmed if your blood pressure is a bit higher or lower than normal at times. Blood pressure readings can fluctuate by as much as 15 to 20 mm Hg at various times of the day. However, if your readings are consistently high, you have high blood pressure.

Question

Average weight of an elephant is 4 tonnes and the average weight of a mouse is 40 g.

What should the ratio of cross-sections of a hydraulic lift be to lift an elephant by using a mouse?

($g = 10\ m/s^2$)

Solution

$4\ tonnes = 4 \times 10^3\ kg$

$40\ g = 4 \times 10^{-2}\ kg$

$$P_{elephant} = P_{mouse}$$

$$\frac{m_{elephant} \cdot g}{A_{elephant}} = \frac{m_{mouse} \cdot g}{A_{mouse}}$$

$$\frac{4 \times 10^3 \times 10}{A_{elephant}} = \frac{4 \times 10^{-2} \times 10}{A_{mouse}}$$

$$\frac{A_{elephant}}{A_{mouse}} = 1.10^5$$

Question

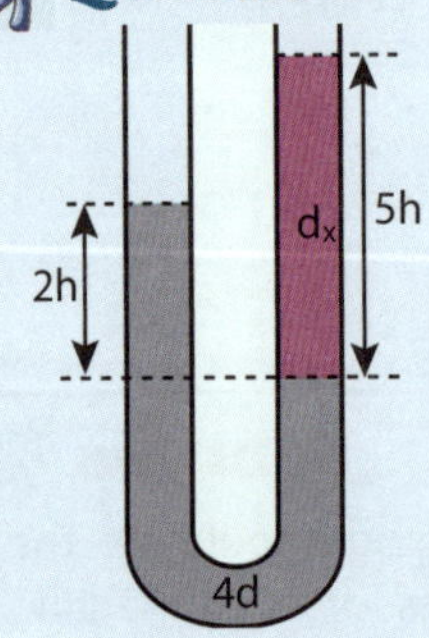

A U-tube is filled with a liquid of density 4d and another liquid of unknown density. The height of this liquid on the right arm of the U-tube is 5h.

Find the density of the liquid.

Solution

Pressures on both arms should be equal.

$P_{left} = P_{right}$

$2h \cdot 4d \cdot g = 5h \cdot d_X \cdot g$

$d_X = 1.6d$

Question

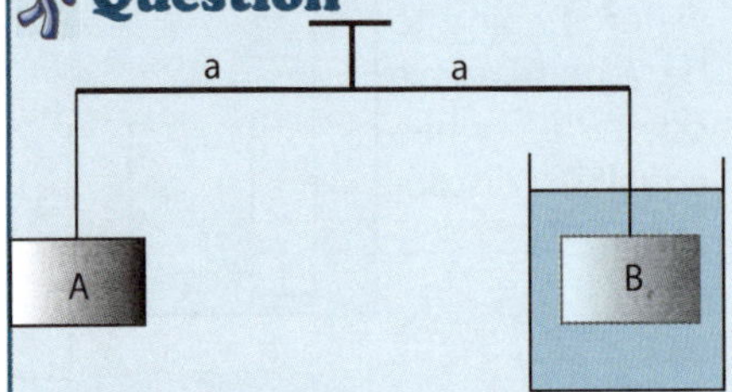

The equal arm balance is in equilibrium as given in the figure. Objects A and B have the same volume. The density of B is 4 g/cm^3 and it is totally submerged in water.

Find the density of A g/cm^3.

($d_{water} = 1\ g/cm^3$)

Solution

Since the equal arm balance is in equilibrium, weights on both sides should be equal. But B is submerged in water so the apparent weight of B is equal to the weight of A.

$W_{real} = W_{apparent} + F_B$

$W_{apparent} = W_{real} - F_B$

$W_{apparent} = d_B \cdot V_B \cdot g - d_{water} V_{immersed} g$

$W_{apparent} = 4.V.g - 1.V.g = 3Vg$

For the equal arm balance:

$W_A = W_{apparent}$

$d_A \cdot V g = 3Vg$

$d_A = 3\ g/cm^3$

Conceptual Questions

1. Why would a sharp knife cut better than a dull knife?

2. Would the suction cups work for an astronaut who is working on the exterior of the space craft? Give your reasoning.

3. A large block of concrete is thrown into a deep lake. As it sinks towards the bottom of the lake, will the magnitude of the buoyant force acting on it increase?

4. A vacuum pump does not operate for a well that is more than 10.3 m deep. Explain why?

5. How will the internal pressure of a fluid flowing through a pipe change when it is speeding up?

True - False

1. 1. Pressure is a vector quantity. (___)

2. If holes at different heights are punched on one side of a plastic bottle, the water spurting from the highest hole will have the largest horizontal displacement. (___)

3. Solids transmit the force whereas the fluids transmit the pressure without a change in magnitude. (___)

4. Astronauts should wear special suits for balancing the internal pressure produced by their cells with the external pressure. (___)

5. A fluid would move faster at the wider cross-section of a pipe. (___)

6. If you open your windows of your house in the case of an approaching tornado, it will cause less damage. (___)

7. Planes can fly because they are hollow inside. (___)

8. The melting point of ice increases with the increasing pressure. (___)

9. If the external pressure is greater than 1 atm, the liquid boils at a temperature higher than its boiling point. (___)

10. A balloon filled with ordinary air can also float in air because it has the same density. (___)

Fill in the Blanks

1. The unit of pressure in SI unit system is ______________.

2. Solids exert a pressure because of their ______________ on their standing surface.

3. Atmospheric pressure is measured by a ______________.

4. For measuring the pressure of a gas in an enclosed container ______________ are used.

5. Smooth paths followed by the fluid molecules in motion are called ______________.

6. The flow speeds of liquids is measured by ______________ / ______________.

7. Melting under pressure and then freezing again is called ______________.

8. An object submerged in a fluid will weigh less than it is in vacuum. This weight is usually referred as the ______________ weight.

9. If the density of an object is smaller than the density of a fluid, it ______________ within the fluid.

10. A fluid moves slower at a ______________ cross-section area and it has a ______________ pressure.

Conceptual Questions

1. Why would a sharp knife cut better than a dull knife?

A sharp knife cuts better because it has a smaller cross-section area than a dull knife. The same force applied produces a greater pressure on the smaller cross-section area of the knife.

2. Would the suction cups work for an astronaut who is working on the exterior of the space craft? Give your reasoning.

No, because the suction cups work by pushing out the air underneath the cup. The air pressure outside of the cup holds it in its space. But in outer space there is no air pressure to hold the suction cup in its position.

3. A large block of concrete is thrown into a deep lake. As it sinks towards the bottom of the lake, will the magnitude of the buoyant force acting on it increase?

There won't be a change in the magnitude of the buoyant force because there won't be a change in the displaced volume of water.

4. A vacuum pump does not operate for a well that is more than 10.3 m deep. Explain why?

A vacuum pump operates by producing a partial vacuum in the pipe that reaches to the water down the surface. The atmospheric pressure exerted on the surface of the water pushes the water up the pipe where the pressure is reduced. At sea level water cannot be pushed up by the atmosphere more than 10.3 m.

5. How will the internal pressure of a fluid flowing through a pipe change when it is speeding up?

According to Bernoulli's effect, the pressure of a fluid is inversely proportional to its speed. Thus the pressure decreases as the fluid speeds up.

True - False

1. Pressure is a vector quantity. (_F_)

2. If holes at different heights are punched on one side of a plastic bottle, the water spurting from the highest hole will have the largest horizontal displacement. (_F_)

3. Solids transmit the force whereas the fluids transmit the pressure without a change in magnitude. (_T_)

4. Astronauts should wear special suits for balancing the internal pressure produced by their cells with the external pressure. (_T_)

5. A fluid would move faster at the wider cross-section of a pipe. (_F_)

6. If you open your windows of your house in the case of an approaching tornado, it will cause less damage. (_T_)

7. Planes can fly because they are hollow inside. (_F_)

8. The melting point of ice increases with the increasing pressure. (_F_)

9. If the external pressure is greater than 1 atm, the liquid boils at a temperature higher than its boiling point. (_T_)

10. A balloon filled with ordinary air can also float in air because it has the same density. (_F_)

Fill in the Blanks

1. The unit of pressure in SI unit system is pascal.

2. Solids exert a pressure because of their weight on their standing surface.

3. Atmospheric pressure is measured by a barometer.

4. For measuring the pressure of a gas in an enclosed container manometers are used.

5. Smooth paths followed by the fluid molecules in motion are called streamlines.

6. The flow speeds of liquids is measured by venturi tube.

7. Melting under pressure and then freezing again is called regelation.

8. An object submerged in a fluid will weigh less than it is in vacuum. This weight is usually referred as the apparent weight.

9. If the density of an object is smaller than the density of a fluid, it floats within the fluid.

10. A fluid moves slower at a wider cross-section area and it has a higher pressure.

Problems

1.

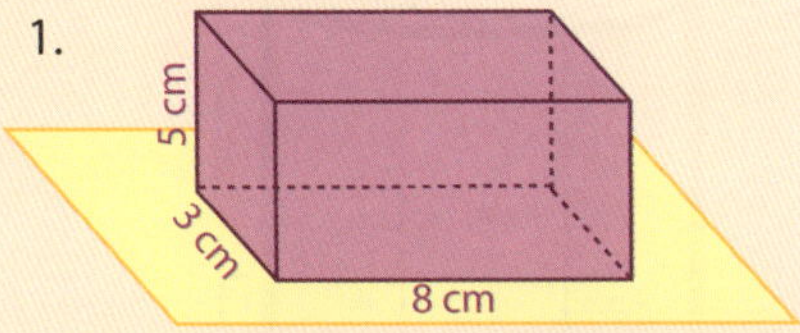

Dimensions of a rectangular box are 8 cm, 3 cm and 5 cm respectively. The pressure exerted by the block on the given horizontal plane is P, when it is placed as in the figure.

Find the pressure of the block on the horizontal plane in terms of P, if it were standing on the smallest base area.

2.

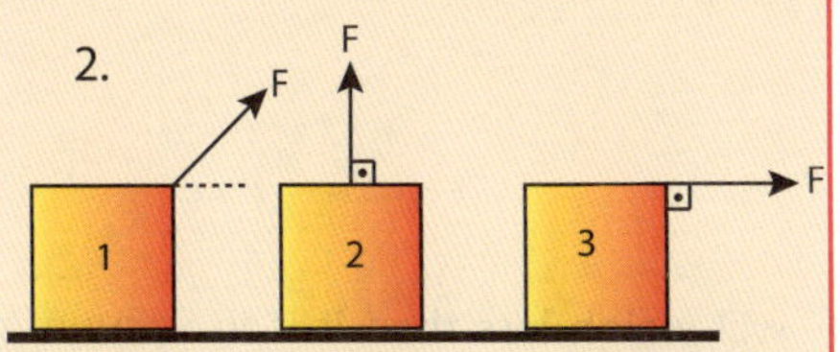

The same magnitude of force at different action of lines are applied on the same box as in the figure. Force applied is not large enough to move or to lift the object.

What is the relationship between the pressures exerted on the ground P_1, P_2, P_3?

3.

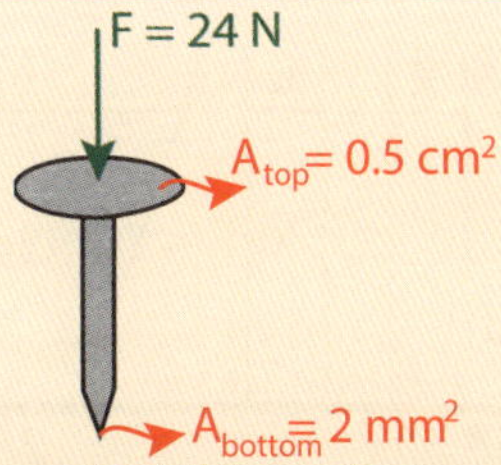

A force of 24 N is applied on the top of a nail with negligible weight.

Find the ratio of pressures at top and bottom surfaces, P_{top}/P_{bottom}.

4.

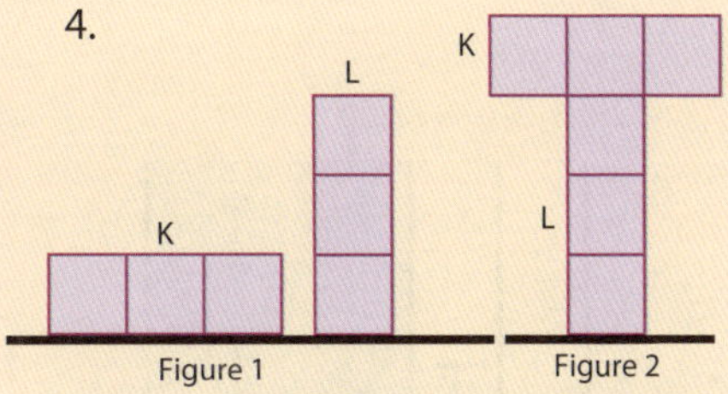

K and L are made of three identical cubes. The pressure exerted by K on the ground is P.

What would the pressure of L on the ground be in terms of P if K was placed over it as in figure 2?

5.

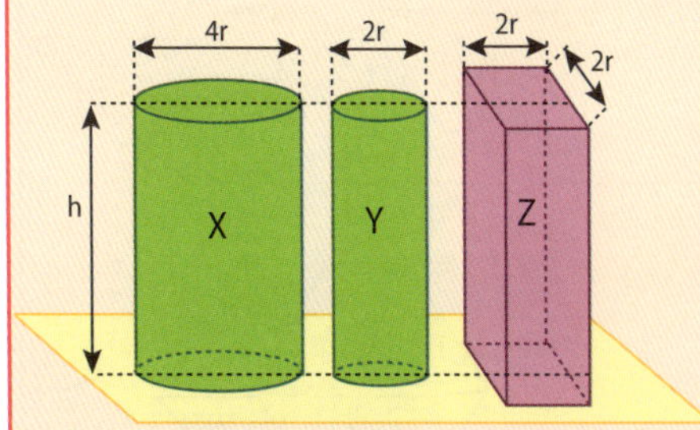

Pressures exerted on the ground by the solid objects X, Y, Z are the same.

What is the relationship between their densities; d_X, d_Y, d_Z?

1. 1.6P / 2. $P_3 > P_1 > P_2$ / 3. 1/25 / 4. 6P / 5. $d_X = d_Y = d_Z$

6.

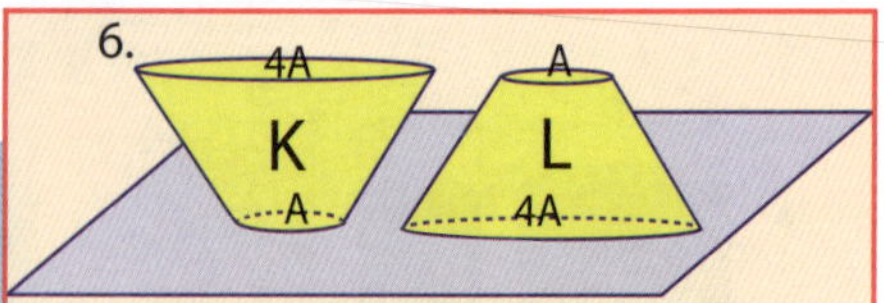

Figure 1

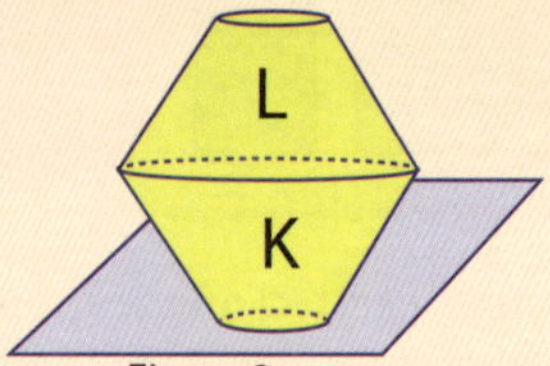

Figure 2

Base area of K is A and the pressure exerted by K on the ground is 4P. Base area of L is 4A and the pressure exerted by L on the ground is P in figure 1.

Find the pressure exerted on the ground when L is placed on K as in figure 2.

7.

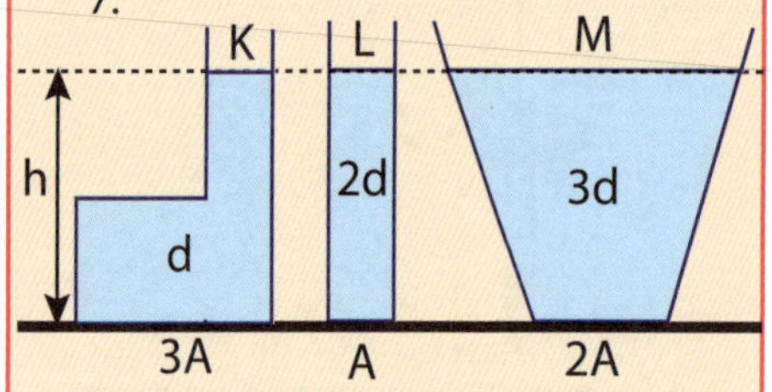

Containers K, L, and M are filled with liquids of densities d, 2d and 3d as shown in the figure above.

a. Compare the liquid pressures acting on to the base of the containers P_K, P_L and P_M.

b. Compare the forces of pressure F_K, F_L and F_M acting on to the base of each container.

8.

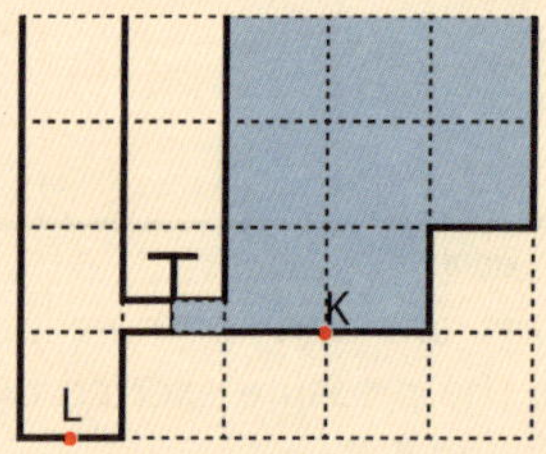

The liquid pressure at point K is P.

Find the pressure of point L when the tap in between is opened.

(Distances between the broken lines are equal, ignore the volume of liquid in the connection pipe)

9.

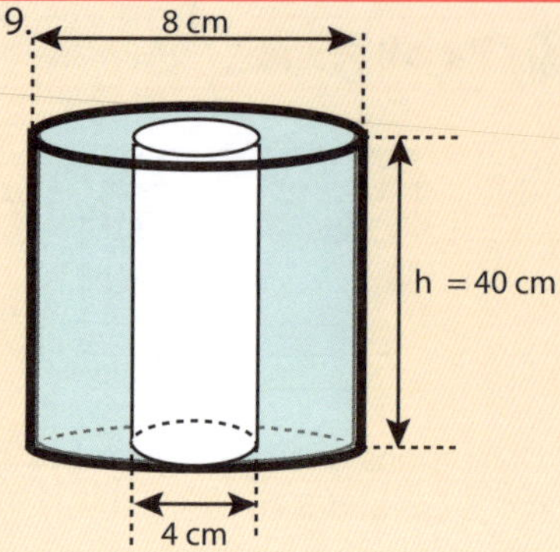

The diameter of the outer cylinder is 8 cm and that of inner cylinder is 4 cm. Both cylinders have a height of 40 cm. The outer cylinder is full of water and the inner cylinder is empty.

What will be the height of water in the cylinders if a hole is punched at the bottom of the inner cylinder?

(Neglect the thickness of the inner cylinder, $\pi = 3$.)

6. 8P /7. a. $P_K < P_L < P_M$; b. $F_L < F_K < F_M$ /8. P /9. 30 cm

10.

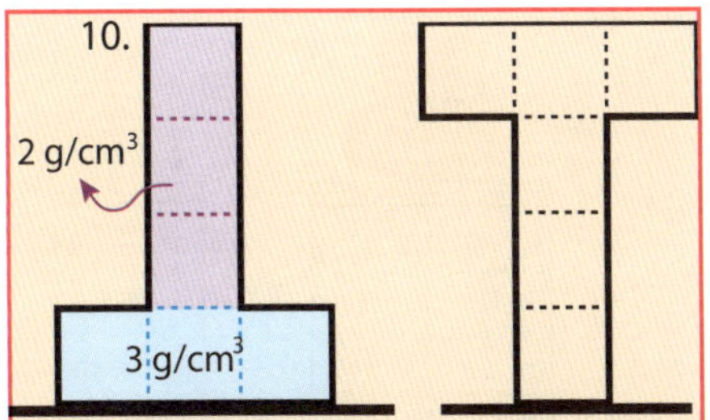

Figure 1 Figure 2

Two immiscible liquids of densities 2 g/cm^3 and 3 g/cm^3 exert a total pressure of P_1 at the base of the container in figure 1. When the container is turned upside down as in figure 2, total pressure at the base becomes P_2.

a. Find the ratio P_1/P_2.

b. If F_1 and F_2 are the forces of pressure at the base of the container in figures 1 and 2, find their ratio, F_1/F_2.

(distances between the broken lines are equal)

11.

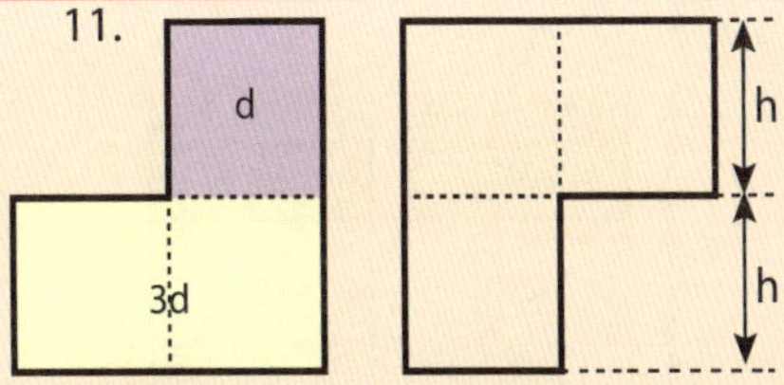

Figure 1 Figure 2

A container is filled by immiscible liquids of densities d and 3d as shown in figure 1. Liquid pressure on the base of the container is P and the force of pressure is F.

a. Find the height of the liquids in terms of h when the container is turned upside down as in figure 2.

b. Find the new pressure acting on the base of the container in terms of P in figure 2.

c. Find the new force of pressure acting on the base of the container in terms of F in figure 2.

12.

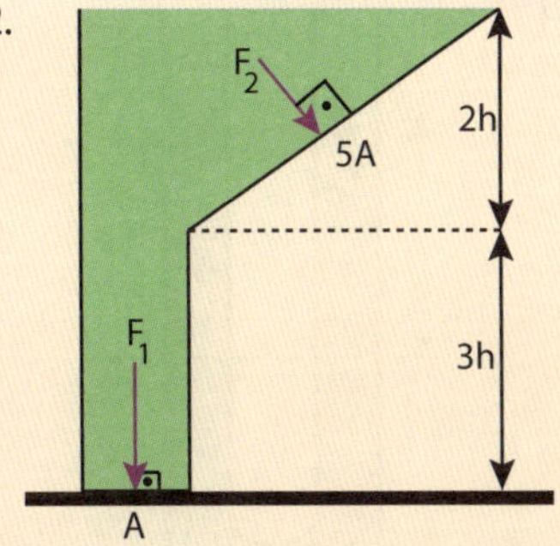

A container is filled with a liquid of density d. Its cross-section is given in the figure.

Find the ratio of forces of pressure applied on the given surfaces, F_1/F_2.

10. a. 9/11; b. 27/11 /11. a. 3h/2 for 3d, h/2 for d; b. 5P/4 c. 5F/8 /12. 1

13.

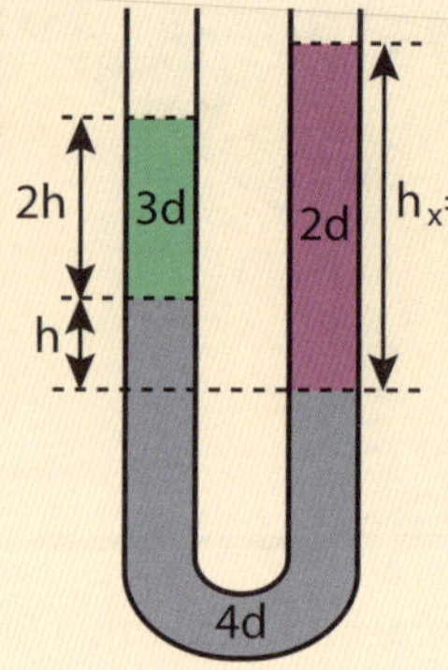

Immiscible liquids with densities 2d, 3d and 4d are poured in a U-tube. The system is in equilibrium as given in the figure.

Find the height of liquid with density 2d.

14.

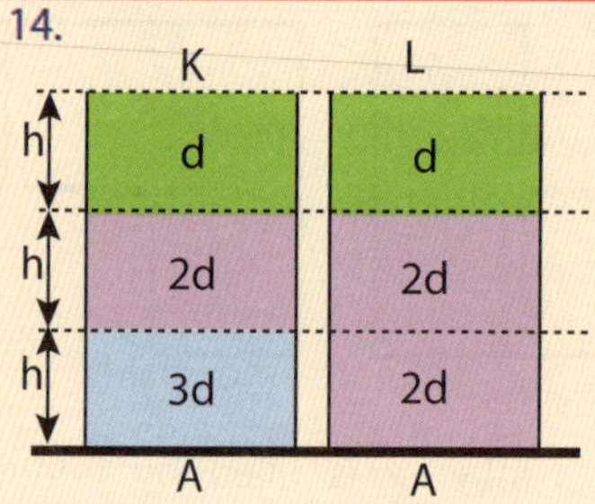

Two identical containers, K and L, are filled with immiscible liquids of densities d, 2d and 3d as shown in the figure. Pressures at the base of the containers are P_K and P_L where force of liquid pressure acting at the base of the containers are F_K and F_L.

a. Find the ratio of P_K/P_L.

b. Find the the ratio F_K/F_L.

15.

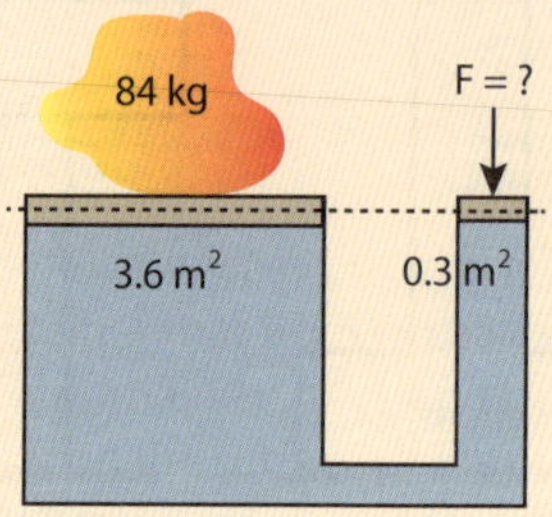

Cross-section areas of the pistons are 3.6 m^2 and 0.3 m^2.

What is the minimum force needed to lift the object?

16.

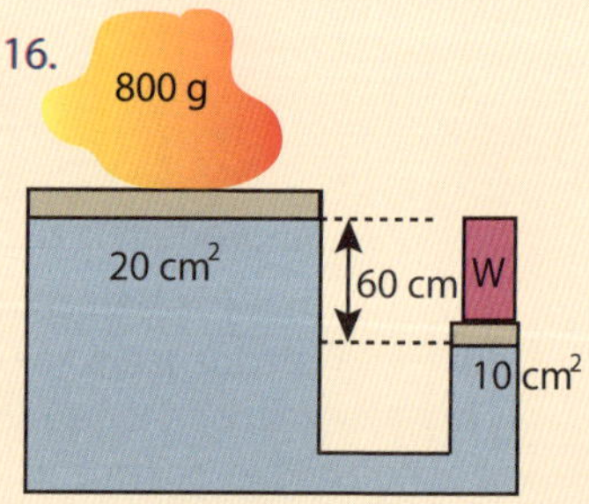

The hydraulic lift system in the figure, uses water as the fluid. An object with a mass of 800 g is placed on the larger piston. An object with a mass of W is placed on the smaller piston. The system comes to equilibrium as given in the figure.

Find the mass of W in terms of kg.

(d_{water}=1000kg/m^3, g=10N/kg.)

13. 5h /14. a. 6/5; b. 6/5 /15. 70 N /16. 1 kg

17.

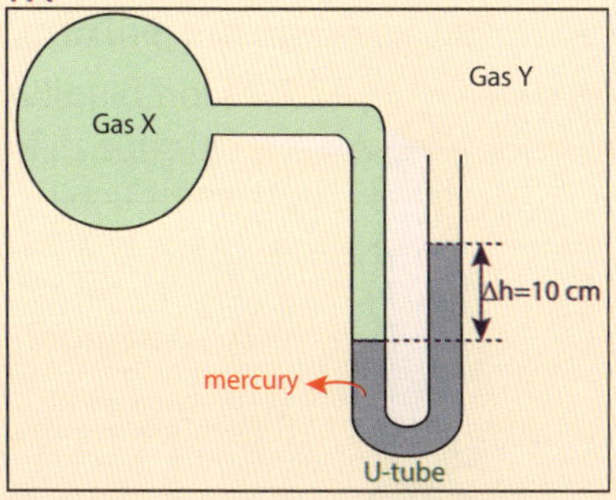

The pressure of gas X 50 cm-Hg.

Find the pressure of gas Y in cm-Hg.

18.

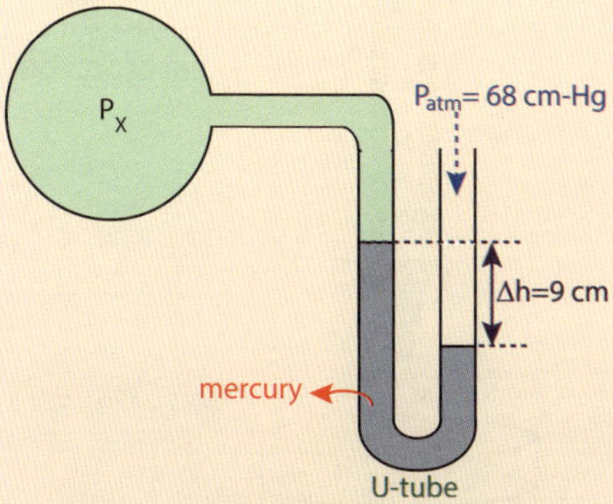

The manometer given in the figure is filled with gas X. The measurement is done at an elevation which has an atmospheric pressure of 68 cm-Hg. The difference in the levels of mercury is 9 cm at this elevation.

What will the height of mercury column be if the same measurement is repeated at sea level?

(P_{atm} = 76 cm-Hg at sea level)

19.

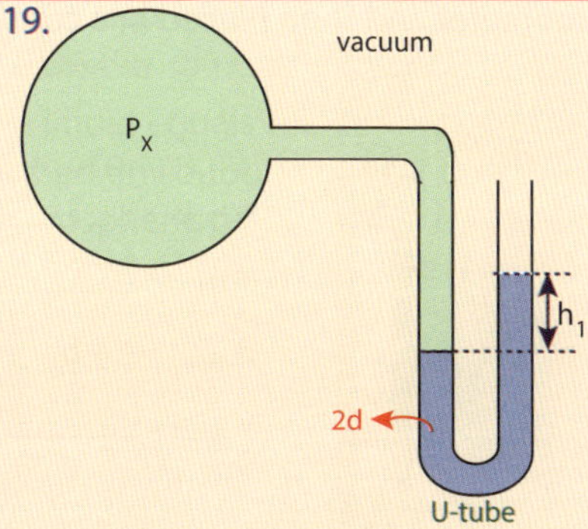

Manometer 1

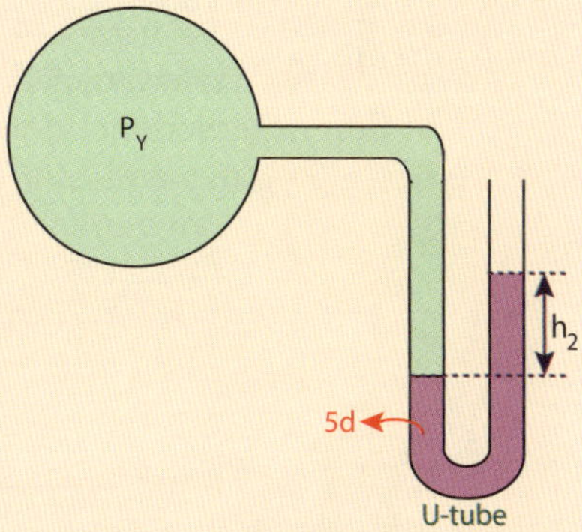

Manometer 2

Manometers given in the figure are placed in a vacuum. Density of the liquid in manometer 1 is 2d and that of manometer 2 is 5d.

If $3P_X = 2P_Y$ then find the ratio of h_1/h_2.

20. Two buckets are filled with water up to the brim. First bucket has water only whereas a piece of wooden block floats in water in the second bucket.

Which bucket weighs more? Why?

21. Styrofoam, which is most often used in the production of disposable coffee cups, plates, etc. has a very low density of 0.3 g/cm^3. A ship is loaded with a cargo of styrofoam.

Will the ship float more or less in water when compared to the empty ship?

17. 40 cm-Hg /18. 17cm /19. 5/3 /20. Both same /21. More

22. An object is weighed in air by a dynamometer and then placed in a container or 'eureka can' filled with water. Complete the missing parts in the figures. (d_{water} = 1 g/cm^3; g = 10 m/s^2)

a.

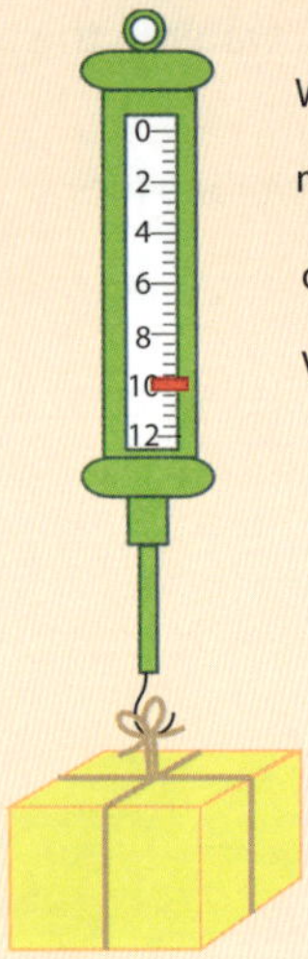

W_{object} = ___

m_{object} = ___

d_{object} = 2 g/cm^3

V_{object} = ___

b.

$W_{apparent}$ = ___

F_B = ___

c.

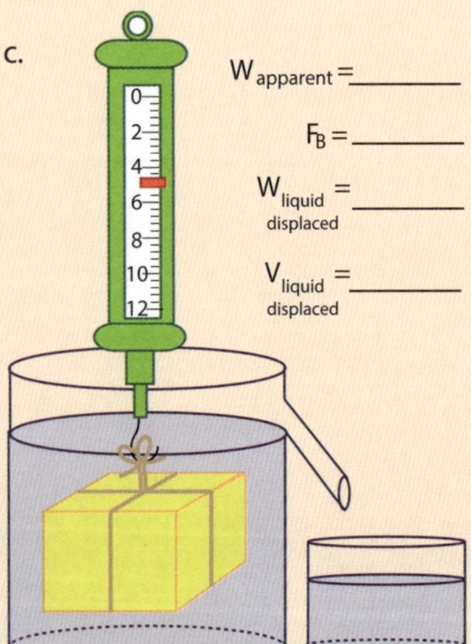

$W_{apparent}$ = ___

F_B = ___

$W_{liquid\ displaced}$ = ___

$V_{liquid\ displaced}$ = ___

d.

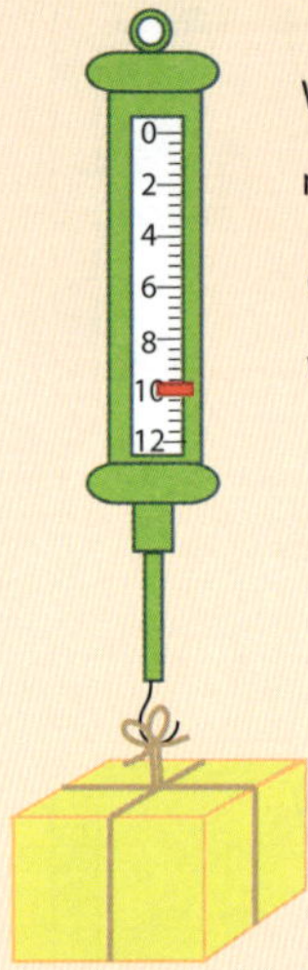

W_{object} = ___

m_{object} = ___

d_{object} = 1 g/cm^3

V_{object} = ___

e.

$W_{apparent}$ = ___

F_B = ___

f.

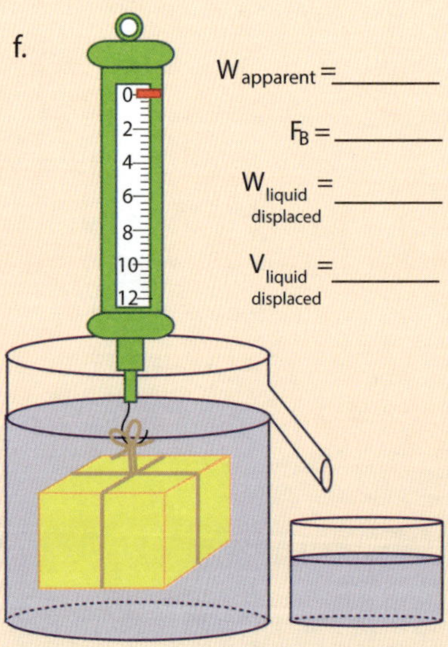

$W_{apparent}$ = ___

F_B = ___

$W_{liquid\ displaced}$ = ___

$V_{liquid\ displaced}$ = ___

g.

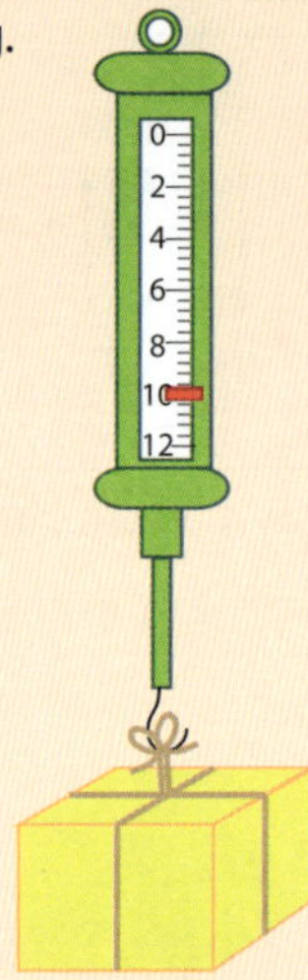

W_{object} = ___

m_{object} = ___

d_{object} = 0.5 g/cm^3

V_{object} = ___

h.

$W_{apparent}$ = ___

F_B = ___

i.

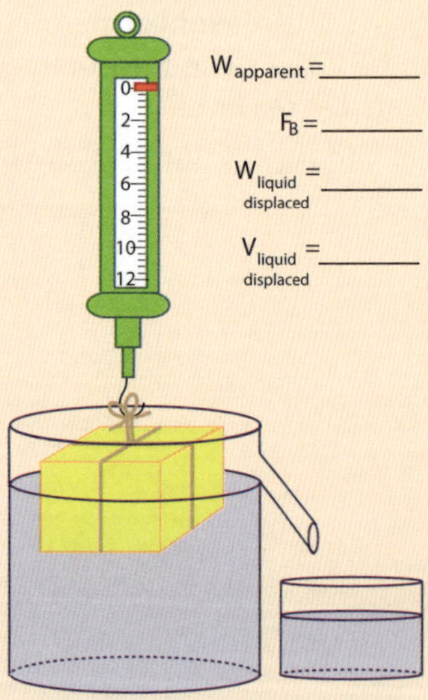

$W_{apparent}$ = ___

F_B = ___

$W_{liquid\ displaced}$ = ___

$V_{liquid\ displaced}$ = ___

23. Objects A, B and C are placed in a Eureka can filled with water. Complete the missing parts in the figures. Draw free body diagram of the object when it is in water.

(Spring scales measure the weight in newtons, $d_{water} = 1\ g/cm^3$; $g = 10\ m/s^2$)

a.

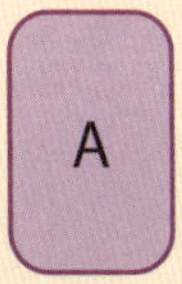

$d = 0.5\ g/cm^3$
$m = 1\ kg$
$V =$ ____________
$W =$ ____________

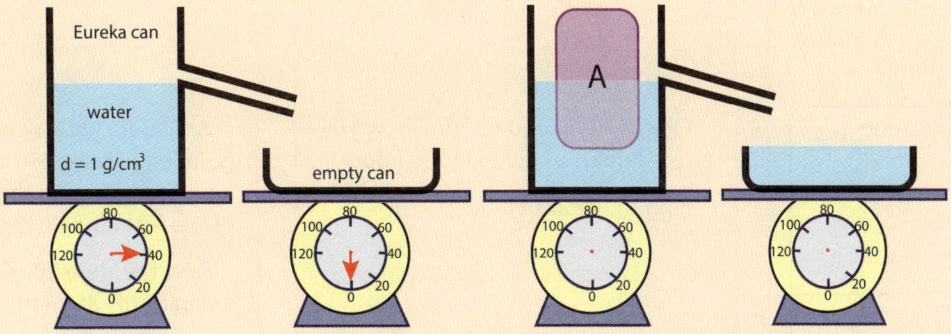

$W_{new} =$ ____________ $W_{new} =$ ____________
$m_{new} =$ ____________ $m_{new} =$ ____________

$W_{apparent} =$ ____________
$F_B =$ ____________
$W_{disp.\ liq.} =$ ____________
$m_{disp.\ liq.} =$ ____________
$V_{disp.\ liq.} =$ ____________
$V_{immersed.} =$ ____________

b.

$d = 1\ g/cm^3$
$m = 1\ kg$
$V =$ ____________
$W =$ ____________

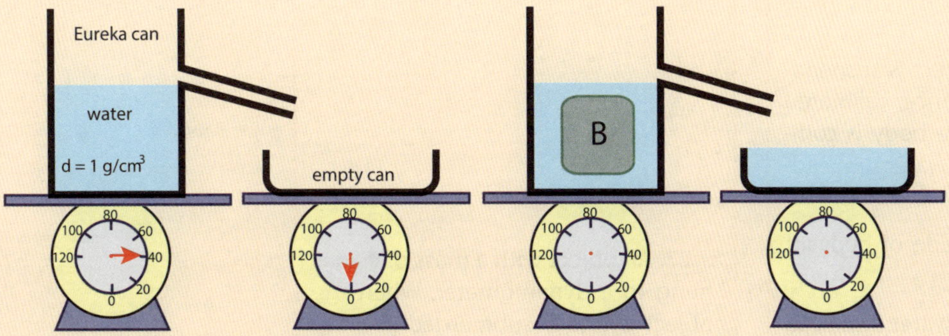

$W_{new} =$ ____________ $W_{new} =$ ____________
$m_{new} =$ ____________ $m_{new} =$ ____________

$W_{apparent} =$ ____________
$F_B =$ ____________
$W_{disp.\ liq.} =$ ____________
$m_{disp.\ liq.} =$ ____________
$V_{disp.\ liq.} =$ ____________
$V_{immersed.} =$ ____________

c.

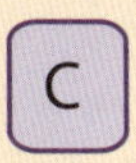

$d = 2\ g/cm^3$
$m = 1\ kg$
$V =$ ____________
$W =$ ____________

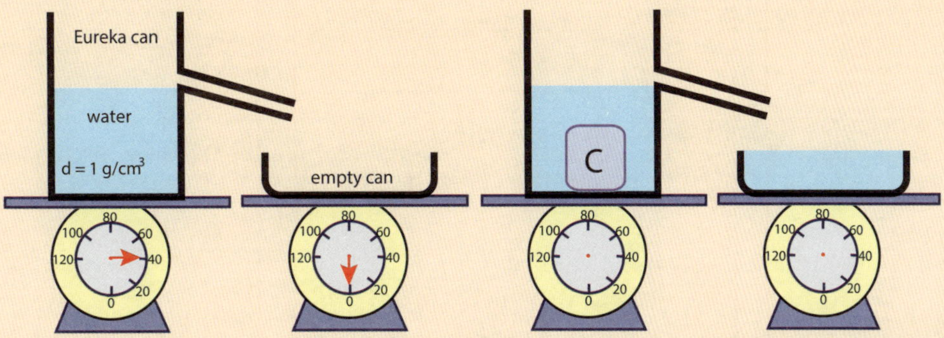

$W_{new} =$ ____________ $W_{new} =$ ____________
$m_{new} =$ ____________ $m_{new} =$ ____________

$W_{apparent} =$ ____________
$F_B =$ ____________
$W_{disp.\ liq.} =$ ____________
$m_{disp.\ liq.} =$ ____________
$V_{disp.\ liq.} =$ ____________
$V_{immersed.} =$ ____________

UNIT - 2 PRESSURE AND BUOYANCY

24. A metal block is weighed in air, half submerged in water, fully submerged in water and fully submerged in brine; a strong salt solution. The readings obtained, though not necessarily in the correct order, were 5 N, 8 N, 10 N and 6 N. Which reading was obtained for each weighing?

Medium	Reading
Air	
Half submerged in water	
Completely submerged in water	
Completely submerged in brine	

25. A 4 kg mass object is suspended from a spring balance, calibrated in newtons. When the body is completely submerged in water the reading of the spring balance is 34 N.

a. Find the magnitude of buoyant force in newtons.

b. Find the mass of water displaced by the object in kg?

c. The density of water is 1000 kg/m^3. Find the volume of water displaced in cm^3.

d. Find the density of the object in kg/m^3?

(g = 10 m/s^2)

26.

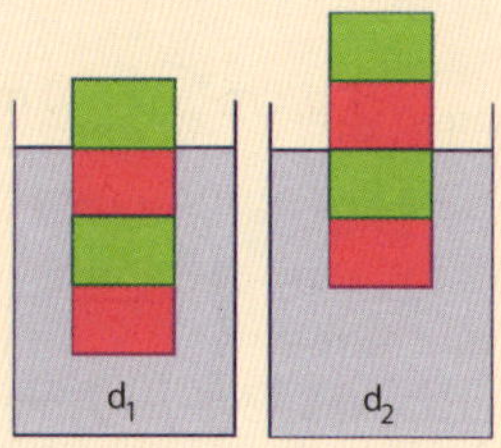

Object K is in equilibrium in two different liquids as given in the figure.

Find the ratio of d_1/d_2.

27. An object with a mass of 4 kg is hung to a dynamometer. When the object is totally submerged in a liquid of density d_1 the dynamometer shows 10 N. When the object is submerged in a liquid of density d_2 the tension becomes 20 N.

Find the ratio of d_1/d_2 ?

28.

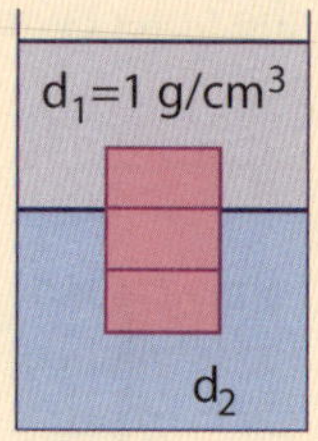

An object with a density of 3 g/cm^3 is in equilibrium in two immiscible liquids as given in the figure. Density of the first liquid is 1 g/cm^3.

Find the density of the second liquid in terms of g/cm^3.

24. 10, 8, 6, 5 /25. a. 6 N; b. 0.6 kg; c. 600 cm^3; d. 2000/3 kg/m^3 /26. 2/3 /27. 3/2 /28. 4 g/cm^3

Multiple Choice Questions

1.

Figure 1 Figure 2

In figure 1 the pressure exerted by the object is P and the force of pressure is F.

How would pressure and force of pressure change if the object was placed as in figure 2?

	Pressure	Force
A.	Increases	Decreases
B.	Remains same	Increases
C.	Decreases	Remains same
D.	Decreases	Increases
E.	Increases	Remains same

2.

Identical blocks are placed as given in the figure.

What is the relationship between the pressure that they exert on the ground P_1, P_2, P_3?

A. $P_1 = P_2 < P_3$

B. $P_3 = P_1 < P_2$

C. $P_2 < P_3 = P_1$

D. $P_3 < P_1 < P_2$

E. $P_3 < P_1 = P_2$

3. For the identical blocks given in question 2, what is the relationship between the force of pressures F_1, F_2, F_3?

A. $F_1 = F_2 = F_3$

B. $F_3 = F_1 < F_2$

C. $F_2 < F_3 = F_1$

D. $F_3 < F_1 < F_2$

E. $F_3 < F_1 = F_2$

4.

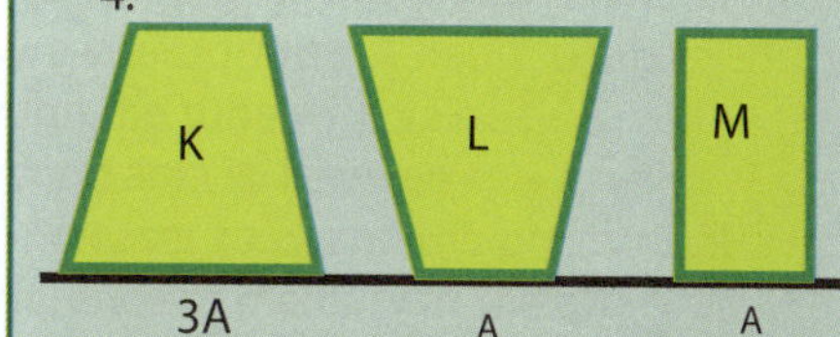

Pressures exerted on the ground by the solid objects K, L, M are the same.

What is the relationship between their weights; W_K, W_L, W_M?

A. $W_K = W_L < W_M$

B. $W_L = W_M < W_K$

C. $W_L < W_M < W_K$

D. $W_M < W_K < W_L$

E. $W_M < W_K = W_L$

5.

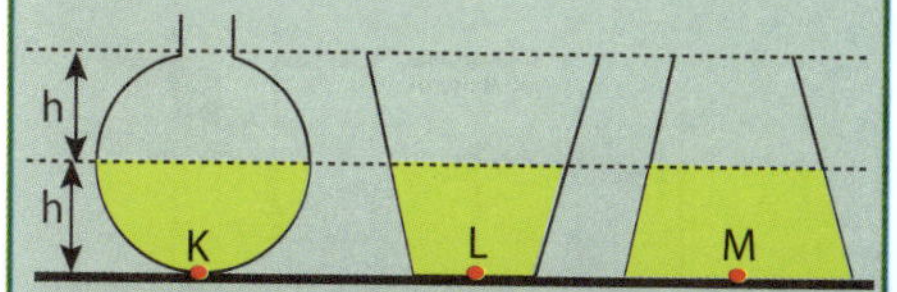

Half height of the containers in the figure are filled with water. The rest of the containers are filled by another miscible liquid whose density is smaller than the density of water.

What is the relationship between the liquid pressures acting on points K, L, and M when the containers are completely full?

A. $P_K < P_L < P_M$

B. $P_L = P_M = P_K$

C. $P_L < P_M < P_K$

D. $P_L < P_K < P_M$

E. $P_M < P_K = P_L$

6.

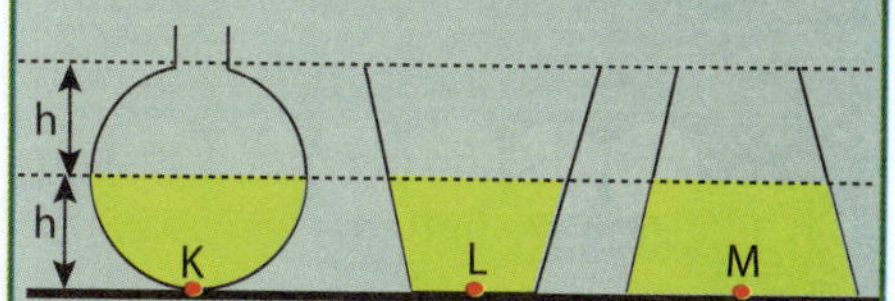

Half height of the containers given in the figure are filled by water. The rest of the containers are filled by another miscible liquid whose density is larger than the density of water.

What is the relationship between the liquid pressures acting on points K, L, and M when the containers are completely full?

A. $P_K < P_L < P_M$

B. $P_L = P_M < P_K$

C. $P_M < P_K < P_L$

D. $P_L = P_K < P_M$

E. $P_M = P_K = P_L$

7.

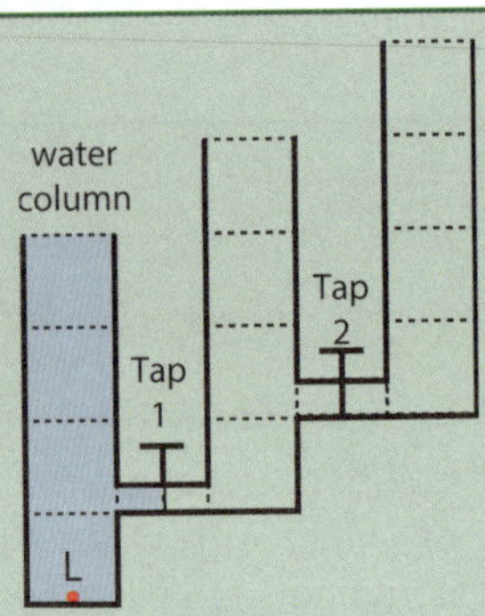

Cross-section of a connecting vessel is given in the figure. The first arm of the container is filled with water. The water pressure at point L is P.

Find the pressure of point L in terms of P when tap 1 is opened.

(distances between the broken lines are equal, ignore the volume of liquid in the connection pipe)

A. $\frac{3}{4}$ B. $\frac{1}{2}$ C. $\frac{2}{3}$

D. $\frac{3}{5}$ E. $\frac{5}{8}$

8. For the connecting vessel given in question 7, find the pressure of point L in terms of P when both taps are opened.

A. $\frac{9}{12}$ B. $\frac{7}{12}$ C. $\frac{5}{12}$

D. $\frac{3}{12}$ E. $\frac{1}{12}$

9. Which of the followings is not an assumption made in defining an ideal fluid?

A. The fluid is incompressible

B. The flow is turbulent

C. The fluid is non-vicous

D. The flow is irrotational

E. The flow is laminar

10.

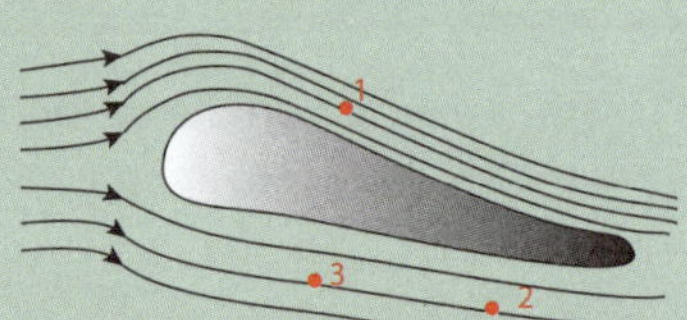

The figure is for the streamlines of an airplane wing.

What is the relationship between the velocity of the air molecules at points 1, 2, 3.

A. $v_1 = v_2 < v_3$

B. $v_3 = v_1 < v_2$

C. $v_2 = v_3 < v_1$

D. $v_3 < v_1 < v_2$

E. $v_3 < v_1 = v_2$

11.

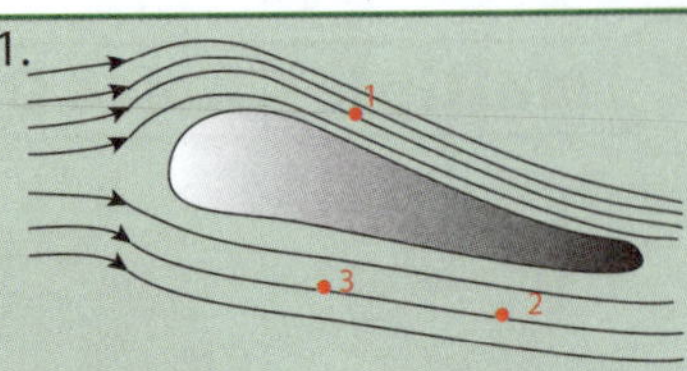

The figure is for the streamlines of an airplane wing.

What is the relationship between the pressure of the air molecules at points A, B, C.

A. $P_1 < P_2 = P_3$

B. $P_3 < P_1 < P_2$

C. $P_2 = P_3 = P_1$

D. $P_3 < P_1 = P_2$

E. $P_3 = P_2 < P_1$

12.

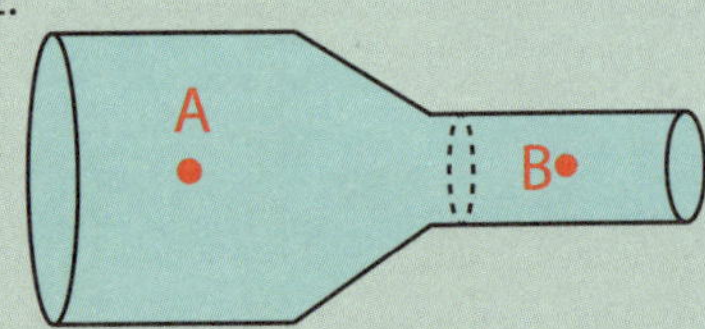

Water flows through the pipe shown in the figure.

What is the relationship between the pressure and the velocity of points A and B?

A. $v_A > v_B$; $P_A < P_B$

B. $v_A < v_B$; $P_A = P_B$

C. $v_A = v_B$; $P_A < P_B$

D. $v_A < v_B$; $P_A > P_B$

E. $v_A < v_B$; $P_A < P_B$

7. E 8. B 9. B 10. C 11. A 12. D

13.

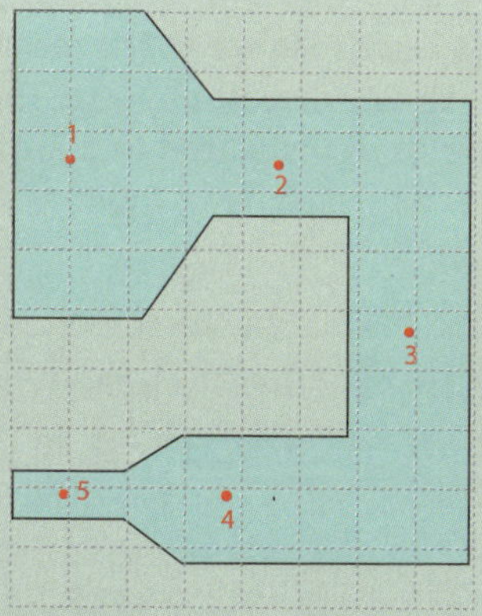

Overhead view of a pipe is given in the figure. A fluid with a negligible viscosity is flowing through it.

At which point fluid flow has the highest velocity?

A. 1 B. 2 C. 3 D. 4 E. 5

14.

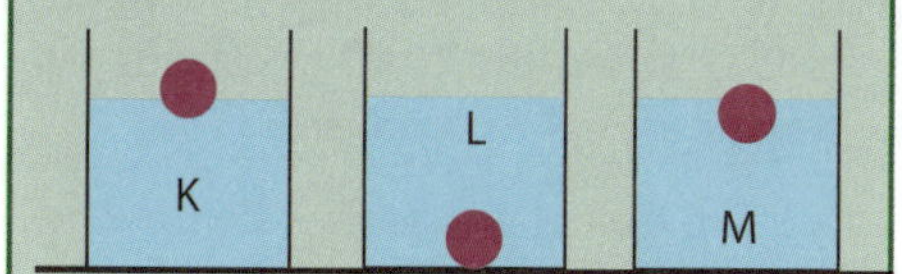

Final positions of an object in different liquids are given in the figures.

What is the relationship between the density of the liquids d_K, d_L, d_M and density of the object d_O.

A. $d_O = d_K < d_L < d_M$

B. $d_L < d_O < d_M < d_K$

C. $d_M < d_K < d_L < d_O$

D. $d_K < d_M < d_O < d_L$

E. $d_M = d_K = d_L = d_O$

15.

Three objects with equal masses are in equilibrium in water as shown in the figure.

Which of the following statements is definitely true for these objects?

A. The objects have equal volumes.

B. The objects have equal densities

C. Volumes of the immersed part of each object is the same

D. Objects have the same densities with water

E. Objects have larger densities than water

16. An object is hung to a dynamometer and then submerged in a liquid gradually.

Which of the followings is correct for the weight of the object read on the dynamometer?

A. It remains the same.

B. First increases, then decreases.

C. No change occurs until it is totally submerged in liquid.

D. First decreases, then remains the same.

E. First decreases then increases.

17.

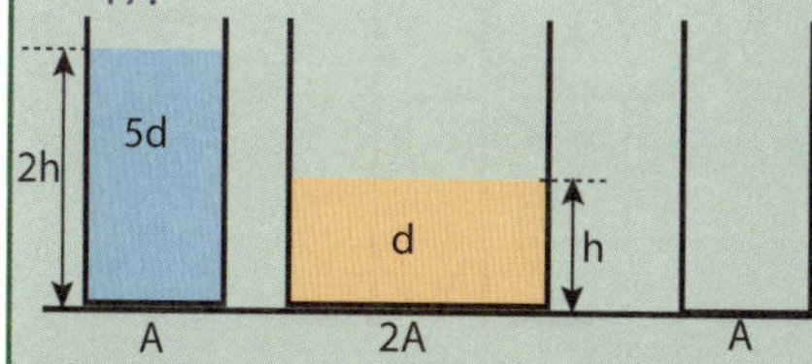

Liquid pressure acting on the base of the first container is P. The liquids are miscible with each other. They are poured in the third container which is large enough to take the mixture.

What is the pressure of the mixture acting on the base of the third container in terms of P?

A. $\frac{3}{5}$ B. $\frac{7}{4}$ C. $\frac{5}{6}$

D. $\frac{4}{7}$ E. $\frac{6}{5}$

18.

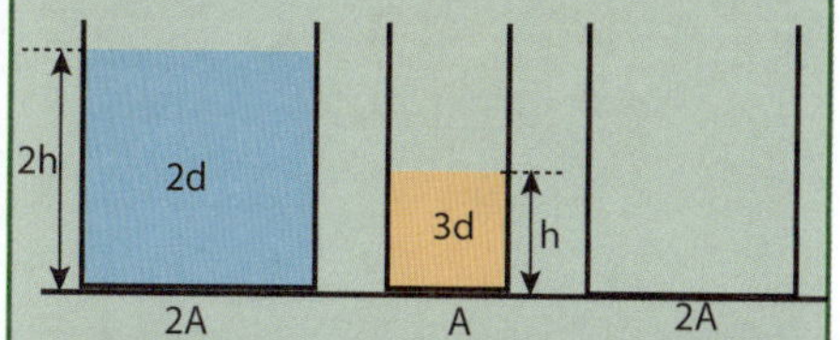

The force of pressure at the base of the first container is F. The liquids are miscible to each other. They are poured in the third container which is large enough to take the mixture.

What is the force of pressure of the mixture acting on the base of the third container in terms of F?

A. $\frac{11}{8}$ B. $\frac{7}{9}$ C. $\frac{5}{7}$

D. $\frac{11}{16}$ E. $\frac{3}{5}$

19.

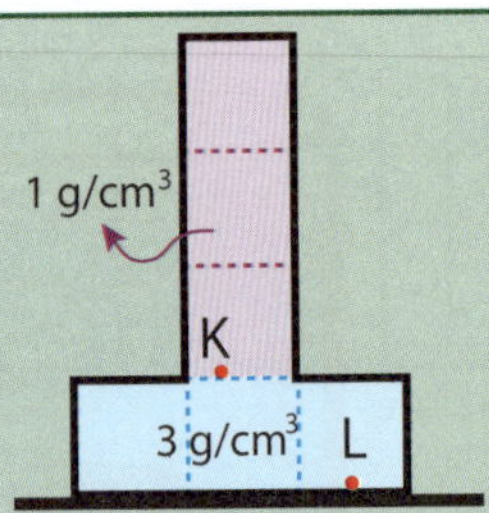

Cross section of a container is given in the figure. Liquids are immiscible to each other.

Find the ratio of pressures at points K and L, P_K/P_L.

A. $\frac{1}{3}$ B. $\frac{1}{2}$ C. $\frac{2}{3}$

D. 2 E. $\frac{3}{4}$

20.

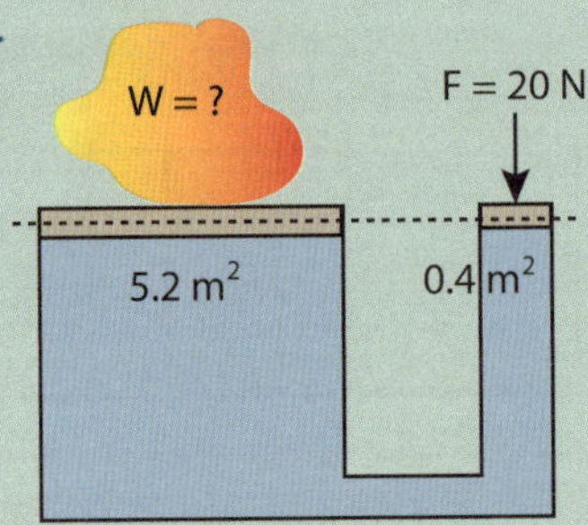

For the hydraulic lift is given in the figure, cross-section area of the larger piston is 5.2 m^2 and that of smaller piston is 0.4 m^2. A force of 20 N is applied on the smaller piston.

Find the weight that can be lifted on the larger piston in terms of newtons.

A. 300 B. 280 C. 260 D. 240 E. 220

21.

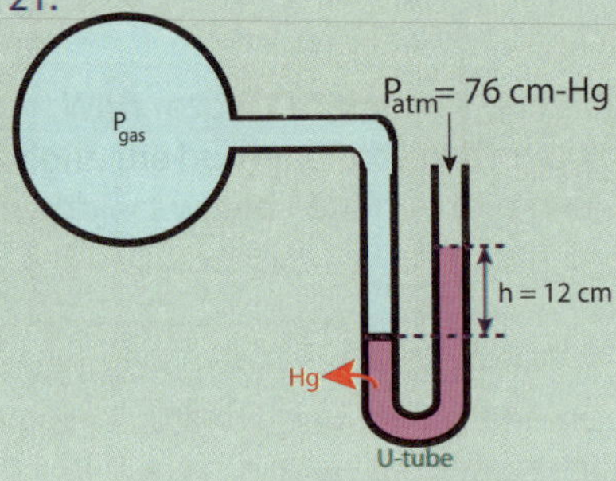

The pressure of a gas is measured by using a mercury manometer. The gas rises the level of mercury by 12 cm on the right side of the U-tube. The atmospheric pressure is 76 cm-Hg.

Find the pressure of the gas in cm-Hg.

A. 64 B. 72 C. 80 D. 84 E. 88

22. Weight of an object is 16 N when measured in air. It weighs 4 N when completely submerged in liquid K, and it weighs 6 N when completely submerged in liquid L.

Find the ratio of densities, d_K/d_L.

A. $\frac{3}{5}$ B. $\frac{7}{4}$ C. $\frac{5}{6}$

D. $\frac{4}{7}$ E. $\frac{6}{5}$

23. An object with a volume 10 cm^3 is submerged in water. The mass of the eureka can increases by 60 g. Then the same object is placed in another eureka can which has a liquid of density 3 g/cm^3.

Find the mass increase of the second eureka can in grams. ($d_{water} = 1$ g/cm^3)

A. 10 B. 20 C. 30 D. 40 E. 50

24. A child holds an object which has a smaller density than that of water is placed at the bottom of a container.

Which of the following statements is true when the child takes out his hand out of the container?

I. The object will accelerate upward and float.

II. The upthrust on the object will decrease and become equal to the weight of the object.

III. The object remains at the bottom of the container.

A. Only I B. Only II

C. I and II D. II and III

E. I, II and III

19. B 20. C 21. E 22. E 23. D 24. C

UNIT 3 WAVES

The concepts that will be covered in this unit are:
Wave, vibration, wave motion, wavelength, period, frequency, velocity, amplitude, pulse, resonance, loudness, sound intensity, resonance, echo, seismic waves.

CONTENTS PAGE

An anechoic chamber is the quietest room in the world. It is so quiet so that you can hear your blood running through your veins, heart pumping the blood or even the sound made by the tiny bones in your ear. The silence is maddening, it can cause hallucinations. The longest time that anyone has been able to bear the room was 45 minutes

WAVES

3.1. Wave and Fundamentals of Wave Motion

Water waves in a pond

Whether we recognize it or not, the wave motion is everywhere around us in our daily life. The water waves, the light waves, the earthquake waves, microwaves, radio waves are just a few examples of wave motion from our daily life. Even when you are saying hello or goodbye and moving your hand, you do a wave-like motion there.

The first image that comes to peoples mind is a wave moving on the surface of a sea, lake or pond. Dropping a pebble into the water will cause circular pattern to spread out from the point of impact.

As such a wave moves forward, the water does not move with the wave. A piece of wood moves up and down as the wave passes but it does not travel along with the wave. Here we can give the example of the "stadium waves". Think about the fans standing up and sitting down successively in a football game. Their rise and falls create a wave motion. The people at the stadium are just moving "up" and "down" in their seats but they are not changing their seats.

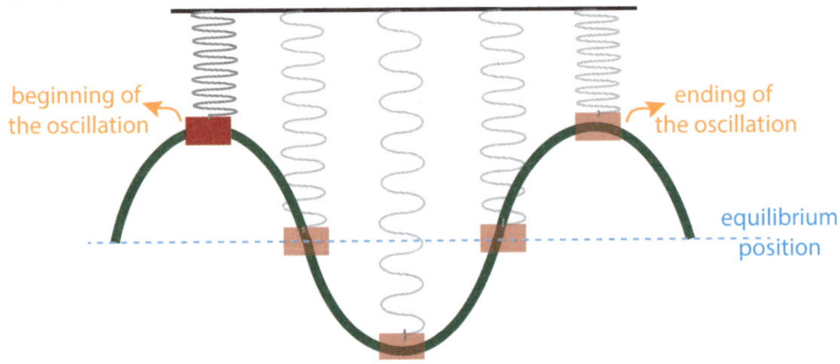

Bees and dragon flies vibrate their wings continuously

The up and down motion of the fans is analogous to the oscillatory motion. An object tied to the end of a spring moving back and forth (or up and down) about an equilibrium position is an example of the oscillatory motion. Such a system is usually called a simple harmonic oscillator. One complete oscillation is the two successive passes of the object from the same position in the same direction. In the above illustration, successive snapshots of the motion of the object is drawn. The object is not moving away from its original position. It just moves up and down about the equilibrium position.

Let's think about the motion of the pendulum of a clock. The pendulum of the clock moves back and forth over the same path which is called a vibration or oscillation. The vibrational motion can find examples from the animal kingdom such as the bees, the dragonflies, etc… The humming sound of a bee or the buzzes of a fly are caused by the vibration of their wings.

The vibration of the pendulum takes the same amount of time. Such a motion repeating itself in time is called a periodic motion. *When a disturbance is caused in a medium, the disturbance causes the particles of the medium to oscillate about an equilibrium point just like the motion of a pendulum of a clock.*

A wave disturbance is created by a source. When you drop a pebble in water pebble is the reason creating the disturbance in the medium. Hold your hand at your throat as you are speaking. The vibrations of the vocal cords that you feel are the cause so that the source of a sound wave travels from you to the ears of the people around you.

Water waves caused by the wind at open seas reach to the shore, the radio waves reach to your antenna, sound waves from a speaker reach to your ears… Waves can move over large distances. They can go from one place to another, travel through some medium. But the medium does not go along with the wave. The particles of the medium oscillates about an equilibrium position, they have a limited motion.

The particles of the medium do not travel away with the wave. They are only "disturbed" from their position and forced to move up and down or back and forth. The disturbance, that is the wave pattern moving away from the source. The waves carry the energy of the disturbance with them, not the matter.

3.1.1. Fundamental Concepts About Waves

Though there are many kinds of waves, when describing a wave, we have some fundamental concepts. Let's investigate the motion of a simple harmonic oscillator a bit more detailed, before we talk about these concepts.

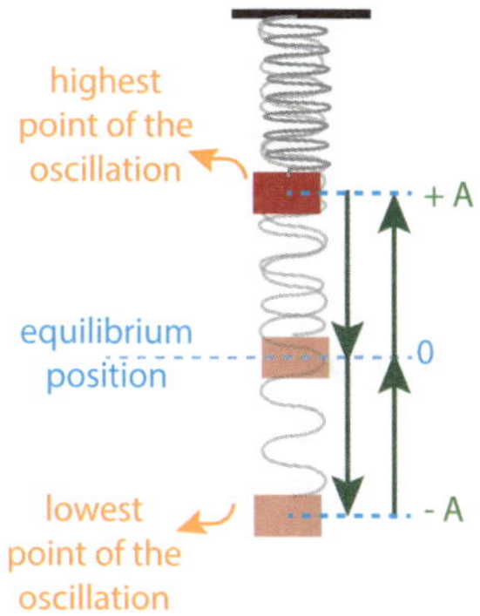

If we release the object at t = 0 from a compression of A, the object will begin to move down. It will try to reach its equilibrium position where it restores its original length. But because of its inertia it cannot stop there and it will continue to move down until all of its kinetic energy due to its motion is transferred to the system to stretch the spring. As the spring reaches to the lowest point of the oscillation once again it will try to reach to the equilibrium position. This back-forth and back again motion is one complete oscillation. This is the motion performed by the particles of the medium.

Now let's change the object with a piece of crayon, and draw its motion on a rolling paper. As the crayon moves up and down with the spring, a pattern will be drawn on the paper.

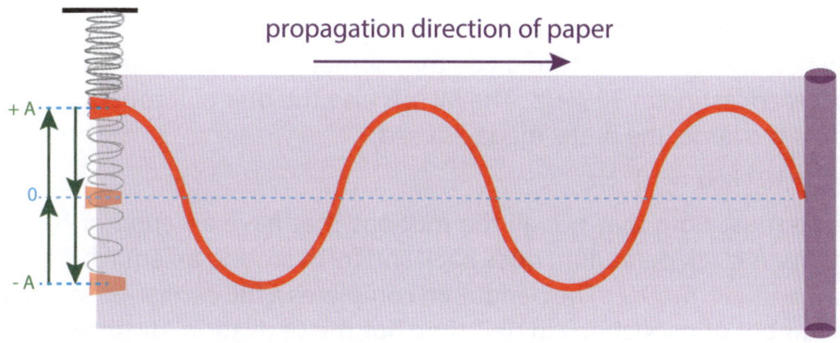

You can observe such a pattern if you stretch a string and move its one end continuously left and right or with the water waves travelling in a pond. Most often this pattern is used to represent a travelling wave in a medium.

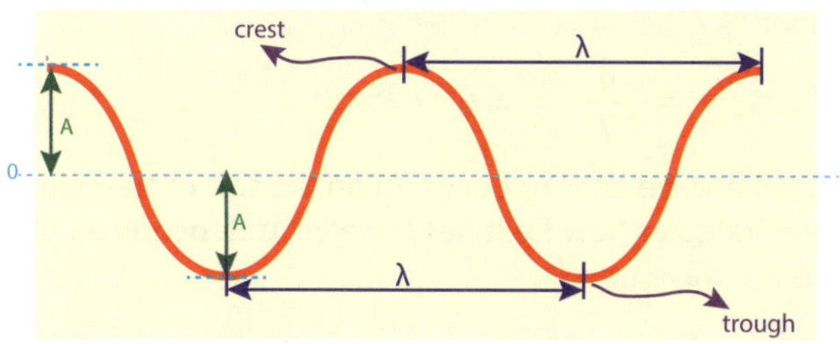

The highest points of the wave pattern are called crests and lowest points are called troughs. The maximum height of a crest or the depth of a trough relative to the equilibrium position is called the amplitude of the wave.

A wave is periodic in both space and time. The space period is the wavelength. The distance between two successive crests or two successive troughs (any two identical successive points on the wave) is the wavelength. It is denoted by λ (lambda) a letter from Greek alphabet. Since it is a length it is measured in terms of meters but for very small wavelengths such as the wavelength of visible light nanometers is used most commonly.

HUMMING BIRD

An example of vibrational motion used by animals is the hummingbirds. Hummingbirds are known for their ability to hover in mid-air by rapidly flapping their wings 15–80 times per second depending on the species. This means that they are vibrating their wings 15 -80 in each second.

Capable of sustained hovering, the hummingbird also has the ability to move backward through the air, being the only group of birds able to do so. Hummingbirds may also fly vertically or horizontally, and are capable of maintaining a position while drinking nectar or eating tiny arthropods from flower blossoms. Their English name derives from the characteristic hum made by their wings.

Example

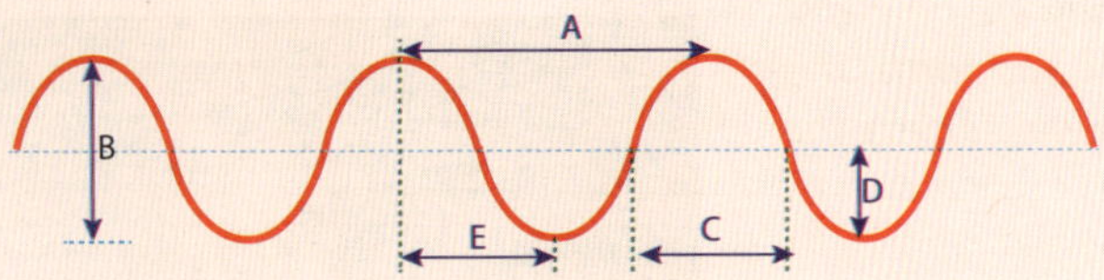

Answer the following questions for the given figure.
a. Show a crest and a trough on the wave.
b. Which letter represents the amplitude of the wave?
c. Which letter stands for one wave length?
d. What is depicted by letter B?
e. What is depicted by letter C?
f. What is depicted by letter E?

Solution

a.

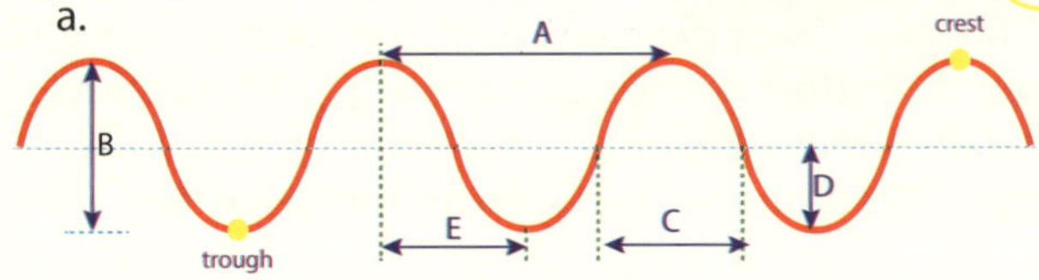

b. The amplitude is the maximum height of a crest or a trough, D is the amplitude of the wave.

c. The wavelength is the distance between two successive identical points such as crests, the wavelength is represented by A.

d. B represents the twice of the amplitude.

e. C is the distance between two successive equilibrium positions but the points on the wave are moving in opposite directions (such as one up the other down), it represents half of the wavelength.

f. E is the distance between a crest and a trough, so it represents half wave length.

The time elapsed between two successive crests passing by the same point in space is the period and the number of crest or complete cycles that pass a point per unit time is the frequency of the wave. As a result of their definition period and frequency are related to each other with:

$$T = \frac{1}{f} \Rightarrow T.f = 1$$

Since period is time it is measured in terms of seconds and frequency is the reciprocal of period so its unit is 1/s or s^{-1} or Hz.

When you are tuning your radio you may sometimes hear a sentence such as "this is …radio, broadcasting at 30 MHz". The MHz is also the unit of frequency but since the magnitude is too large, the prefix mega- is used.

1 MHz = 10^3 kHz = 10^6 Hz

"Similar frequencies flow in the same media"

Here we must note that not all the motions that have a period or frequency are wave motions. A clock pendulum goes back and forth around an equilibrium point and this is also a periodic motion. The pendulum completes a full oscillation in one period or does any number of oscillations in unit time but the oscillation (disturbance) does not propagate in space.

Symbol	Quantity	Unit
v	velocity	m/s
λ	wavelength	m
T	period	s
f	frequency	s^{-1} or Hz

The velocity at which the wave crests (or any other part of the wave) moves is the wave velocity. A crest travels a distance of one wavelength in one period so the velocity of the wave is given by the equation:

$$v = \frac{x}{t} \Rightarrow v = \frac{\lambda}{T} = \lambda.f$$

The magnitude of the velocity of a wave does not depend on the size or the shape of the wave, how the wave was generated or how far it has traveled. It depends on the properties of the medium in which the wave travels.

Example

A wave is formed on a thin wire held tight at each end. It has an amplitude of 4 cm, a frequency of 50 Hz and a distance from a crest to the neighboring trough of 12 cm.

a. What is the period of the wave?
b. What is the velocity of the wave?

Solution

a. $T = \frac{1}{f} = \frac{1}{50} = 0.02\ s$

b. The distance between a crest and a neighbouring trough is half of the wavelength. So λ = 2 x 12 = 24 cm.

$$v = \frac{\lambda}{T} = \lambda.f = 24 \times 50 = 1200\ cm/s = 12\ m/s$$

Check Your Understanding

Decide whether the followings are examples of oscillatory motion or wave motion.
Put a check sign in the related box.

	Oscillatory motion	Wave motion
A flag flapping at wind		
A person swinging on a rocking chair		
An earthquake		
A chandelier moving back and forth with the wind		
Voice from a singer reaching to your ear		
Waving good bye to a friend		
Sunlight reaching your eye		
The wings of a hovering humming bird		

3.1.2. Waves According to Their Direction of Vibration and Propagation

As we were defining the wave motion, we had emphasized that a wave does not carry matter but transport energy from one point of the medium to the other. We experience the energy transferred by waves too often in our daily life. Our skin is warmed by light waves transmitted by sun, a singing person emits energy in the form of sound waves, a microwave oven heats the meal by using the microwaves… Moreover the huge energy released by the seismic waves of an earthquake collapses the buildings and so on…

The amount of energy carried by a wave is related to its amplitude. A high energy wave is characterized by high amplitude; a low energy wave is characterized by low amplitude. The amplitude of a wave refers to the maximum amount of displacement of a particle in the medium from its equilibrium position.

Large amplitude = High energy wave

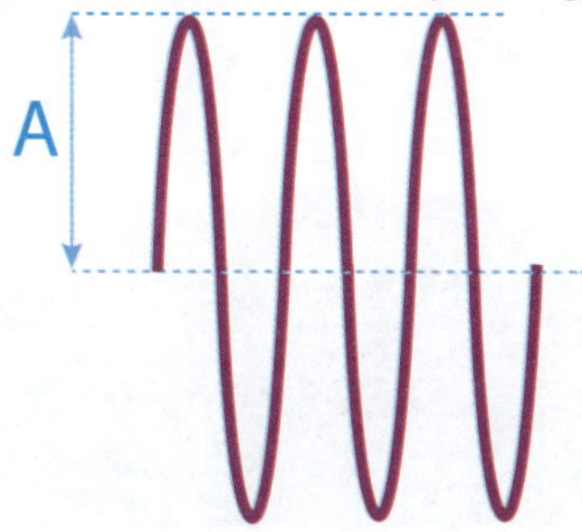

Small amplitude = Low energy wave

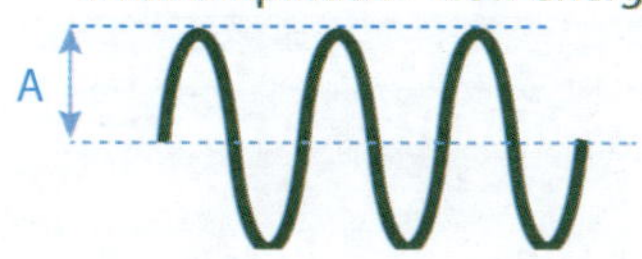

The energy carried by a wave also depends on the frequency, the rate at which the source oscillates. As an example, in the spectrum of the visible light, even though the velocity and the amplitude are the same, the energy of different colors of light are different. Violet has the highest frequency thus the highest energy whereas red light has the lowest frequency thus the lowest energy in the spectrum of the visible light.

We can categorize waves according to their ability to *transfer energy in vacuum*. These two categories contain *mechanical* and *electromagnetic* waves.

In order for a mechanical wave to exist there must be some source of disturbance, a medium that can be disturbed, and some physical connection through which adjacent particles in the medium can influence one another. As a mechanical wave propagates through a medium, the atoms or the molecules of the medium are displaced from their equilibrium position and perform vibrational motion about the equilibrium position. Mechanical waves are the disturbance propagating through the medium. They transfer their energy through the medium but not through a vacuum. Their velocity is dependent on the properties of the medium.

Water waves need water to travel

A sound wave is an example of a mechanical wave. Sound waves cannot travel through a vacuum. Spring waves require the coils of the spring, water waves require water, and stadium waves require fans in a stadium....

Electromagnetic waves are produced by the vibration of charged particles. The visible light waves, microwaves, radio waves are examples of electromagnetic waves. Electromagnetic waves are capable of transmitting their energy through a vacuum such as empty space. The electromagnetic waves which are produced on the sun travel to Earth through the vacuum of outer space. If they were not able to travel through the vacuum there would not be life on earth. Another important note about them is all kinds of electromagnetic waves travel with the same speed in vacuum which is the speed of light, 3 x 10^5 km/s. Electromagnetic waves travel much faster than mechanical waves.

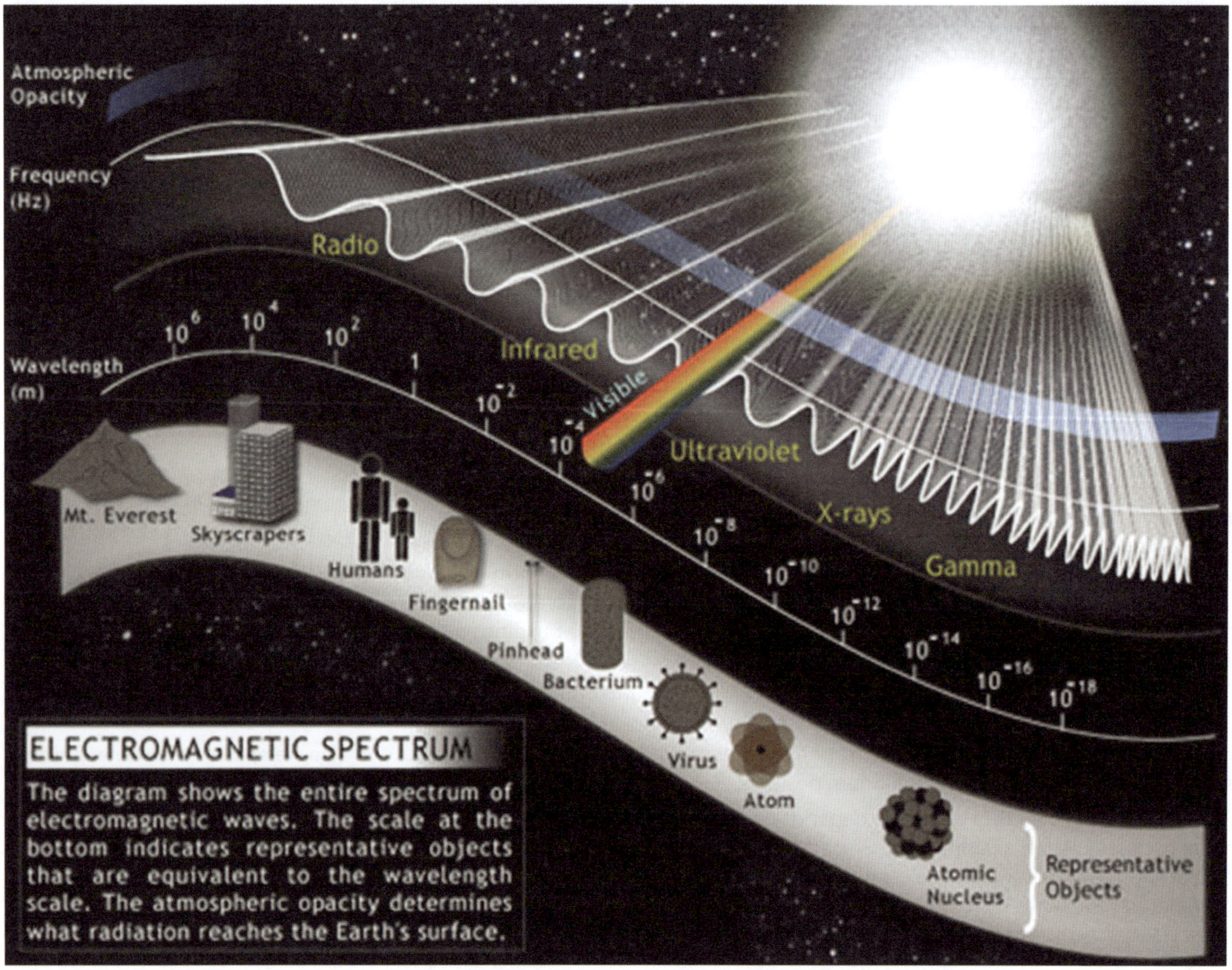

It should always be kept in mind that no matter the kind of the wave is, all forms of traveling waves transport energy. Energy is given to a water wave for example by a rock thrown into the water or by the wind far out at the sea. The energy is transported by the waves to the shore.

We can also categorize waves according to the *direction of movement of the individual particles of the medium relative to the direction which the waves travel*. This categorization leads us to three notable categories: *transverse, longitudinal,* and *surface waves.*

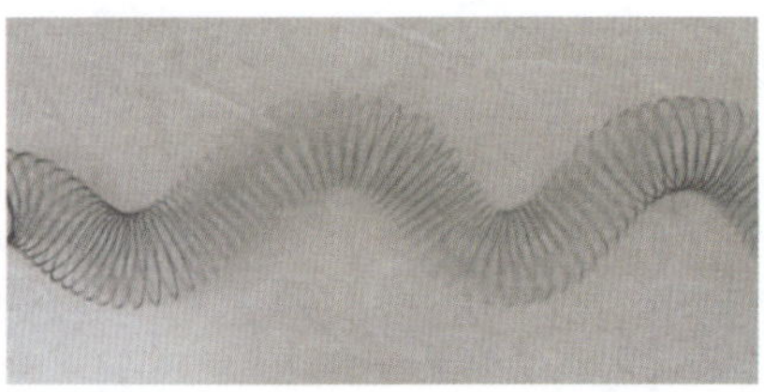

Transverse waves on a spring

Shaking the end of a stretched spring up and down is an easy way to produce transverse waves. The coils of the spring move perpendicular to the direction in which the wave travels. *Transverse waves are always characterized by particle motion being perpendicular to wave motion.* Electromagnetic waves are examples of transverse waves. High points of a transverse wave are called crests and low points are called troughs.

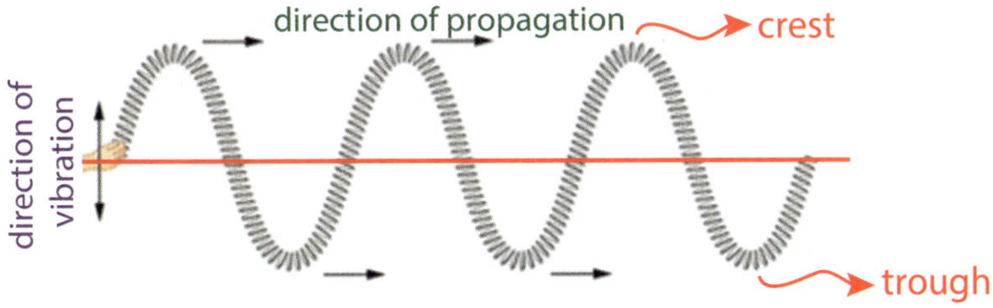

Giving a push to a spring is an easy way to produce a longitudinal wave. *The coils of the spring move parallel to the direction in which the wave travels.* Longitudinal waves are always characterized by particle motion being parallel to wave motion.

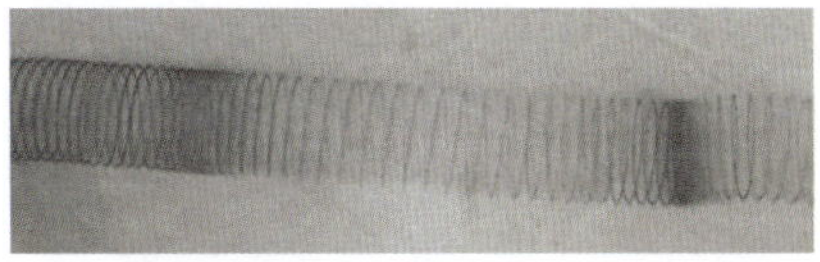

Longitudinal waves on a spring

When the coils of the spring are momentarily far apart we observe the rarefactions and when they are momentarily close we observe compressions of a longitudinal wave. Sound waves are examples of longitudinal waves.

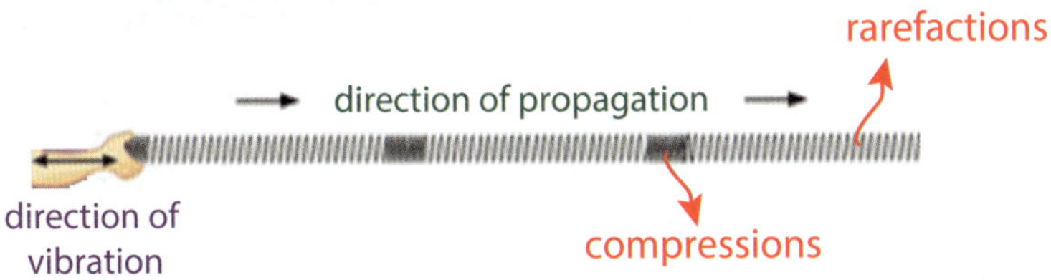

A loudspeaker cone moving with a periodic motion will create a sound wave. Each time the cone moves forward, it collides with air molecules and pushes them close to each other. This produces a region of higher pressure. Half cycle later the cone moves backward and the compressed air has space to expand. So the pressure decreases. The higher and lower pressure regions produce the compressions and the rarefactions of the longitudinal wave.

Speaker cones vibrate back and forth and produce the sound waves in air

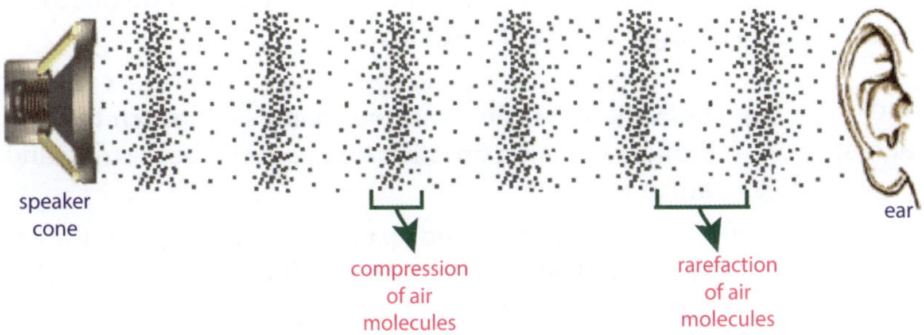

Sound waves created by a loudspeaker reaches to your ear

Both longitudinal and transverse waves can travel in solids since the atoms or the molecules can vibrate about their relatively fixed positions at any direction. In a fluid (liquids and gases) only longitudinal waves can travel because there is no restoring force for perpendicular motion of the wave. The fluids are readily deformable. Geophysicists used this fact to infer that a portion of the earth must be liquid. After an earthquake the longitudinal waves are detected diametrically across the earth but not the transverse waves.

The surface waves travel along the boundary between two materials. A water wave moves along the boundary between water and air and do not extend to downward to any appreciable depth. *The motion of each particle of water at the surface is circular or elliptical.*

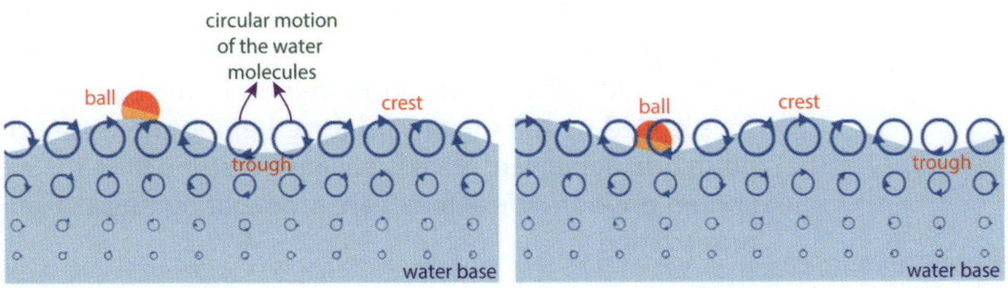

Surface waves are neither longitudinal nor transverse. In longitudinal waves all the particles of the medium move in a parallel direction and for the transverse waves all the particles move in a perpendicular direction relative to the direction of energy transport. In a surface wave, it is only the particles at the surface of the medium which undergo the circular motion. We may conclude that the surface waves exhibit both transverse and longitudinal waves at the same time.

The motion of particles tends to decrease as you go down from the surface. At the bottom the motion is only longitudinal. This is why a submarine travels unaffected from a storm while a ship faces huge waves on the surface of the sea. When a wave approaches to the shore, the water drags at the bottom and it is slowed down while the crests move at a higher speed. So it spills over the top.

3.2 Spring Waves

3.2.1. Differences Between Pulse and Periodic Waves

Water waves spill over the top when the water medium ends. What happens to other waves when their medium ends? What happens to the wave when it reaches to a boundary or it changes its propagation medium? What if two waves move towards each other?

To find the answer to these questions, instead of using the whole periodic wave pattern, we will investigate the motion of one portion of a wave. We will call this portion a pulse. A pulse is a single disturbance moving through a medium from one location to another location. Let's produce a pulse on a string now:

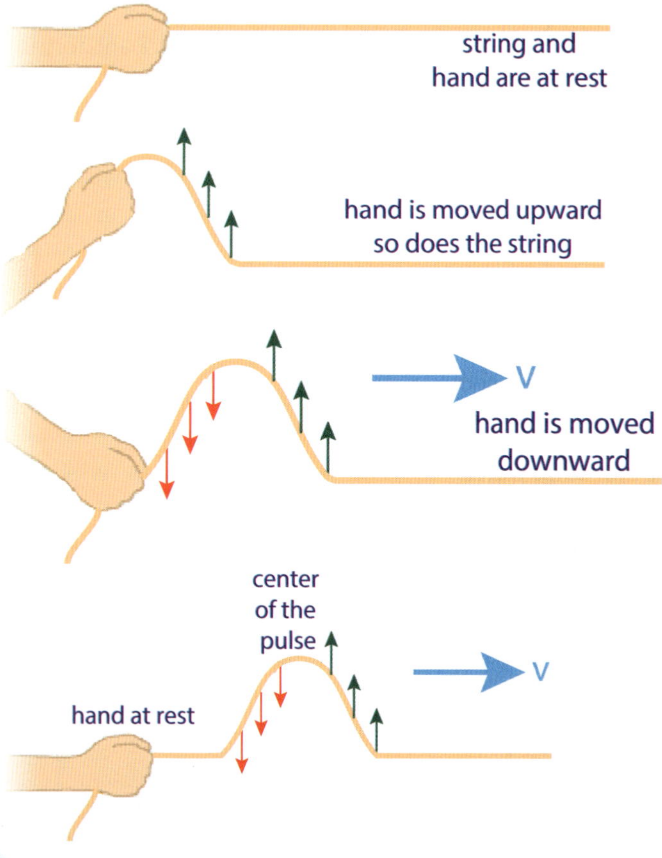

We take a long enough piece of a string, fix it at one end, hold the other end with our hand. We stretch the string for storing a tension on it. For the moment, both string and hand are at rest.

We move our hand up rapidly. The string near the hand will also be pulled up with our hand. The green arrows in the figure represent the upward force acting on the particles of the string. Each particle that moves upward will exert a force on the particle to its right by the tension of the string. The upward force that will be passed is from particle to particle on the string.

Now we pull our hand down rapidly so that the string near the hand is pulled down with our hand. The red arrows in the figure represent the downward force acting on the particles of the string. The upward force that will be passed is from particle to particle by the tension of the string. The upward arrows caused by the upward force applied by the hand are still upward and they are moving towards right now.

Our hand has returned back to the equilibrium position and it is at rest now. But its rapid up and down motion has created a pulse which is moving along the string. The upward arrows represent the portions of the rope pulled by the upward force and the downward arrows represent the portions of the rope pulled by the downward force behind the center of the pulse. The downward force is pulling the string back to its equilibrium position. These forces are propagating thus the pulse is propagating.

3.2.2. Reflection and Transmission of Spring Waves

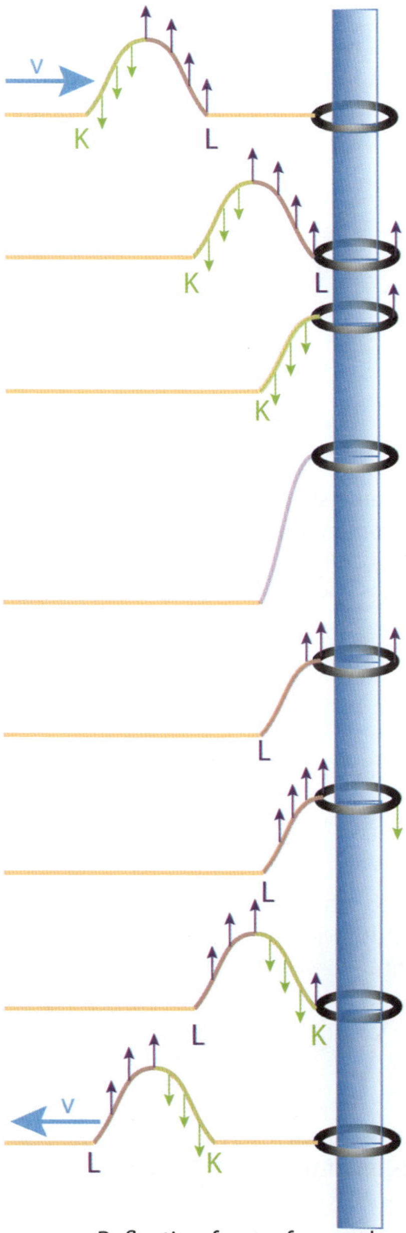

Reflection from a free end

The pulse we have formed on the string is a useful and simplifying method to analyze the behavior of the periodic waves. If you continuously move your hand up and down, then you have a continuous disturbance thus a periodic wave. A hand oscillating the string at one end produces a wave on a string, a vibrating drum membrane produces sound waves, oscillating charges produce electromagnetic waves and so on... Once created the disturbance travels outward through the medium at wave speed.

We had produced a pulse on a string. Now we will have an *end* for the string.

The end of the string is not fixed rigidly (free end): The end is attached to ring and allowed to move in the vertical direction on a frictionless rod. For a simpler approach we will ignore the gravity and just consider the forces on the string coming from the pulse as it is propagating. The pulse we had produced is moving towards the free end now.

The upward force has been propagating on the string and now it reaches to the ring at the free end. As the pulse propagates toward the rod, the upward force will cause the ring to move upward. The ring rises uses the height of the pulse to rise as the center of the ring reaches to the rod. As the ring rises up, it has a velocity thus a kinetic energy. It will use this energy and continue to rise until all of it is lost. The string is attached to the ring so it will rise with the ring. Eventually it will overcome the downward forces of the pulse at its back end and the ring will continue to rise. Because the string to the left side of the ring will have a net upward force on it. This upward force now propagates along the string to the left by pulling each adjacent particle to its left upward. The ring is pulled upward by the string. Newton's third law states that there is an equal but opposite direction of reaction force to every action force. So the ring pulls the string downward. Thus a downward force acts on the string. Now the upward and downward forces on the string are moving towards left.

As a result of interaction of the pulse moving to the right and the ring, we have a reflected pulse on the string. *The reflected pulse has the same size, shape and velocity as the incident pulse, but it is moving towards left now.*

The end of the string is fixed rigidly: It is fixed to a wall and not allowed to move.

The leading end of the pulse has upwards forces acting on it and arrives at the wall. An upward force is exerted on the wall. But the unlike the tiny ring that we had in the previous case, the wall has an enormous mass. The upward force cannot move the wall upwards. The end of the string remains fixed on the wall. The upward force on the wall is an action force. The wall causes a downward reaction force on the string and pulls it downward. The front portion of the pulse is pulled downward and a motion to the left of the wall is initiated. The incident pulse still propagates toward right on the string. The front portion of the pulse has been reflected and now moving downward toward left. The back portion of the pulse is still moving upward toward right. The force acting on each particle on the string is the vector sum of the up and down forces acting on them. Once the back portion of the pulse reaches to the wall, a downward force is applied on the wall. Again a reaction force will be exerted by the wall on the string. But this time the reaction force is upward. The pulse has been completely reflected by the wall now.

As a result of reaction forces applied by the wall on the pulse moving to the right, we have a inverted reflected pulse on the string. *The reflected pulse has the same size, shape and velocity as the incident pulse, but it is an inverted pulse moving towards left now.*

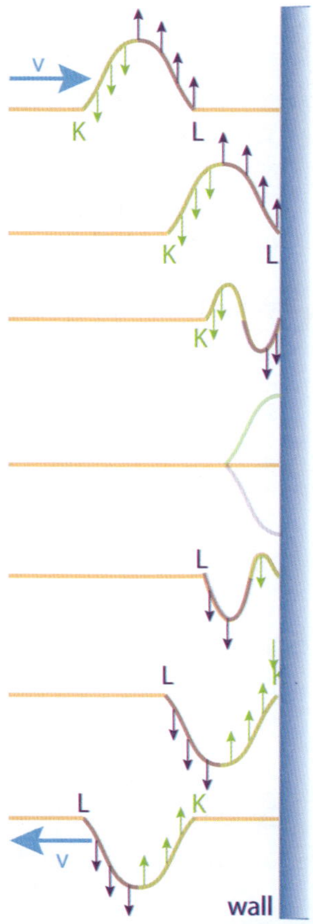

Reflection from a fixed end

Example

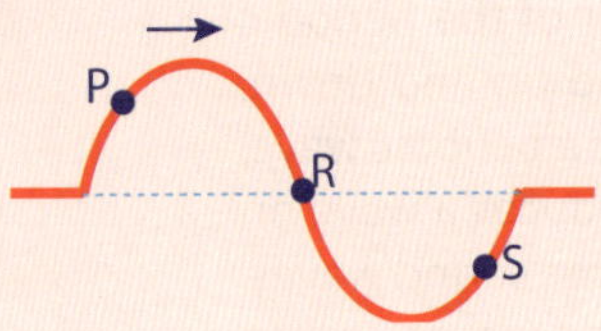

Show the direction of motion of the points P, R, S on the propagating wave given in the figure.

Solution

Below, the position of the wave just a moment later is drawn. The arrows drawn to the projectiles of each point shows the direction of motion of each point.

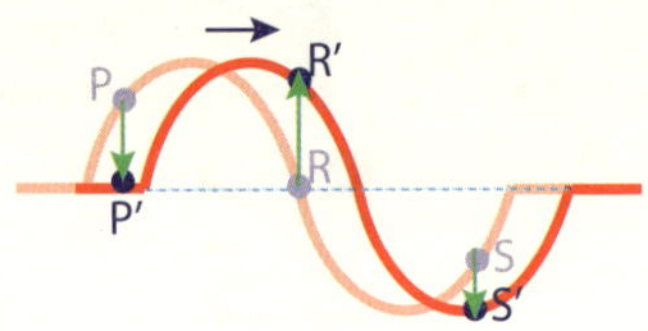

Example

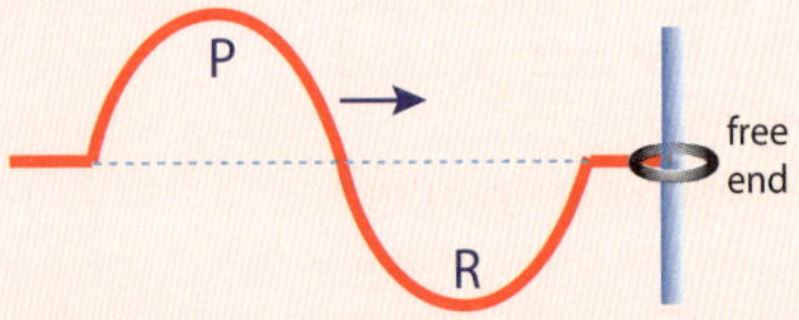

Pulses K and L are moving in the given direction towards a free end.

How will the reflected pulses look like after being reflected from the free end?

Solution

Pulses will have the same shape when they are reflected from a free end. First coming pulse will be the first to be reflected. So pulse R will be in front.

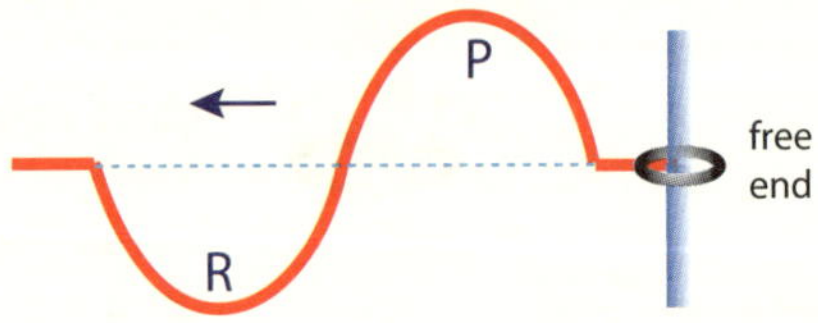

Example

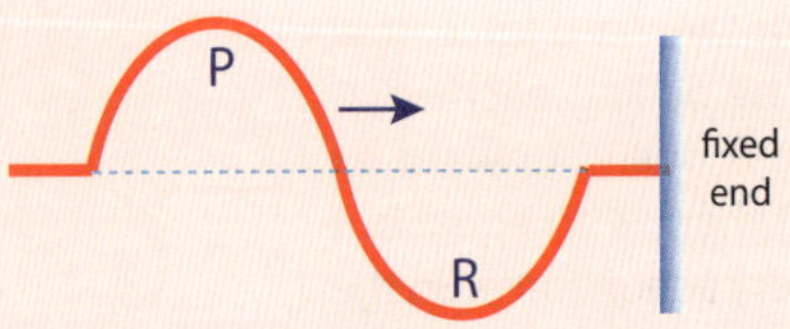

Pulses K and L are moving in the given direction towards a fixed end.

How will the reflected pulses look like after being reflected from the fixed end?

Solution

Pulses will be inverted when they are reflected from a fixed end. First coming pulse will be the first to be reflected. So pulse R will be in front.

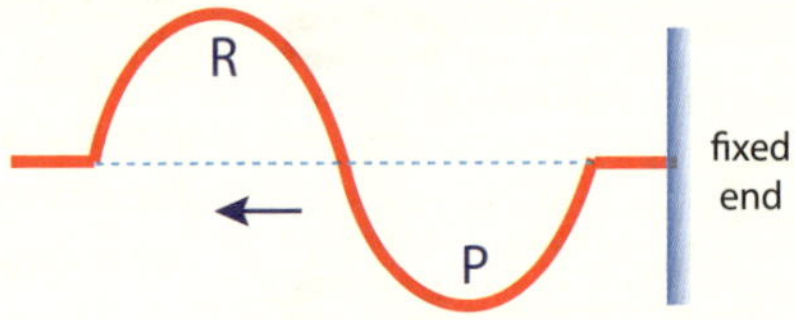

The velocity of a pulse on a string depends on the mass to length ratio which is called linear density. A string with a greater density has more inertia. It will respond less rapidly so the pulse will travel slower. *Pulse speed is inversely proportional to the linear density.* A string with a greater tension will exert larger forces on the segments of the spring. The string will respond more rapidly so the pulse will move faster. *Pulse speed is directly proportional to the tension.*

What if two strings with different linear densities are attached to each other and a pulse is produced on them? Can the pulse propagate from one string to the other? Will there be any changes of velocity or any other properties of the incident pulse? Experimentation is a good method to investigate the changes on the pulses.

We will have strings with different linear densities. The linear density is our independent variable. We will investigate the velocity and the amplitude of the pulse. They will be our dependent variables. We will apply the same tension on the strings that we had attached. So the tension of the strings will be our controlled variable.

As we have mentioned the speed of a wave on a string is inversely proportional to linear density. Waves in a heavier string are slower than the waves on a lighter string.

If a pulse is incident from a less dense medium (lighter string) to a more dense medium (heavier string):

In passing from the less dense medium to the more dense medium, an upward force will be applied on the boundary of the media by the incident pulse. More dense medium acts like a fixed end (wall) at the boundary. Thus the reaction force applied on the lighter medium will invert the reflected pulse. The reflected pulse is still in the same medium, so it will move with the same velocity. But it is now inverted and moving backwards.

Although the more dense medium acts like a wall as in reflection from a fixed end, it is not as massive as a wall. A transmitted pulse will be formed on the more dense medium caused by the upward force applied by the incident pulse. The upwards forces are initiating the pulse in the more dense medium, so it will have the orientation and direction of motion as the incident pulse. But now the linear density is greater so its velocity will be smaller than the incident pulse.

The width of the reflected pulse will be the same as the incident pulse since they have equal velocities. Equal velocities will result equal displacements in the same time interval. But the width of the transmitted pulse will be smaller than the incident pulse, because it is moving slower in the medium. Smaller velocity will result a smaller displacement in the same time interval for the transmitted pulse.

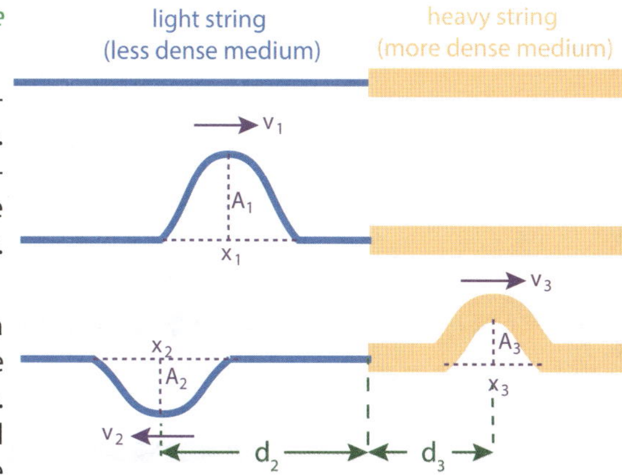

Transmission from a less dense medium to a more dense medium:	
Velocities	$v_1 = v_2 > v_3$
Widths	$x_1 = x_2 > x_3$
Amplitudes	$A_1 > A_2; A_1 > A_3$
Frequencies	$f_1 = f_2 = f_3$
Wavelengths	$\lambda_1 = \lambda_2 > \lambda_3$
Distances travelled	$d_2 > d_3$

If a pulse is incident from a more dense medium (heavier string) to a less dense medium (lighter string);

In passing from the more dense medium to the less dense medium, an upward force will be applied on the boundary of the mediums by the incident pulse. Less dense medium acts like a free end (ring on the rod) at the boundary. The reflected pulse will have the same orientation as the incident wave, but now it is moving backwards. The reflected pulse is still in the same medium, so it will move with the same velocity.

A transmitted pulse will be formed on the less dense medium caused by the upward force applied by the incident pulse. The upwards forces are initiating the pulse in the less dense medium, so it will have the same direction as the incident pulse. But now the linear density is smaller so its velocity will be larger than the incident pulse.

The width of the reflected pulse will be the same as the incident pulse since they have equal velocities. But the width of the transmitted pulse will be larger than the incident pulse, because it is moving faster in the medium. Larger velocity will result a larger displacement in the same time interval for the transmitted pulse.

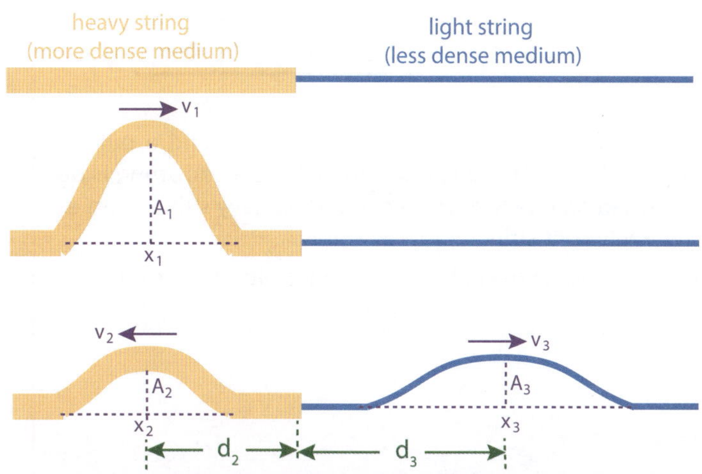

Transmission from a more dense medium to a less dense medium:	
Velocities	$v_1 = v_2 < v_3$
Widths	$x_1 = x_2 < x_3$
Amplitudes	$A_1 > A_2; A_1 > A_3$ or $A_3 > A_1$
Frequencies	$f_1 = f_2 = f_3$
Wavelengths	$\lambda_1 = \lambda_2 < \lambda_3$
Distances travelled	$d_2 < d_3$

Some energy of the incident pulse will be transferred to the transmitted pulse in both cases. The reflected pulse will usually have less energy thus a smaller amplitude than the incident pulse. The transmitted pulse may have a larger amplitude in the case of transmission from a heavy spring to a lighter one. But there is no distinct relationship between the amplitudes of the reflected and transmitted waves. One can be smaller than the other or sometimes be equal to each other depending on the ratio of their linear densities.

If a wave is incident on a boundary, either for transmission from a less dense medium to a more dense one or vice versa, the frequency of the wave remains constant for both sides of the boundary. Because *the frequency is dependent on the source creating the wave, not the properties of the media.* Since the frequency is constant for both sides of the boundary, the wavelength will change depending on the velocity of the wave in the media.

Example

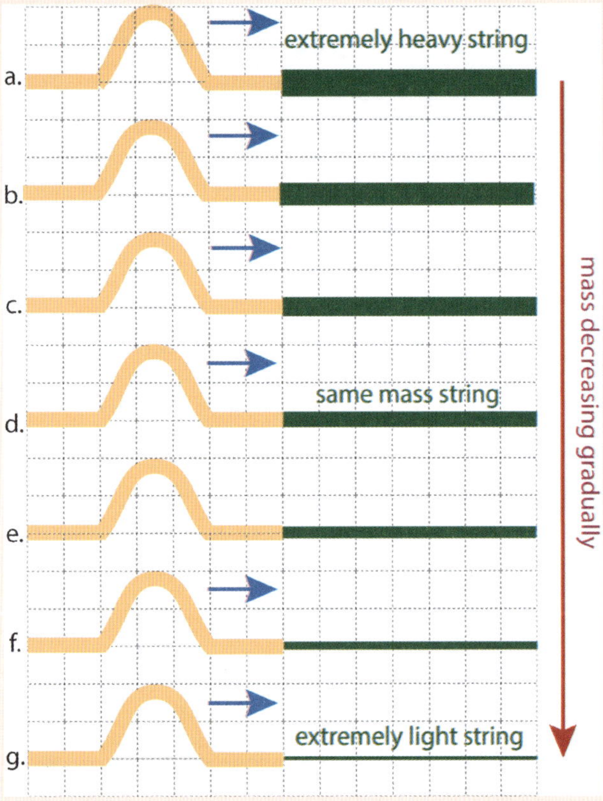

A string is attached to another string and a pulse is formed on it. The linear density of the second string is decreased gradually from an extremely heavy string to an extremely light string.

Investigate the reflected and transmitted pulses.

Solution

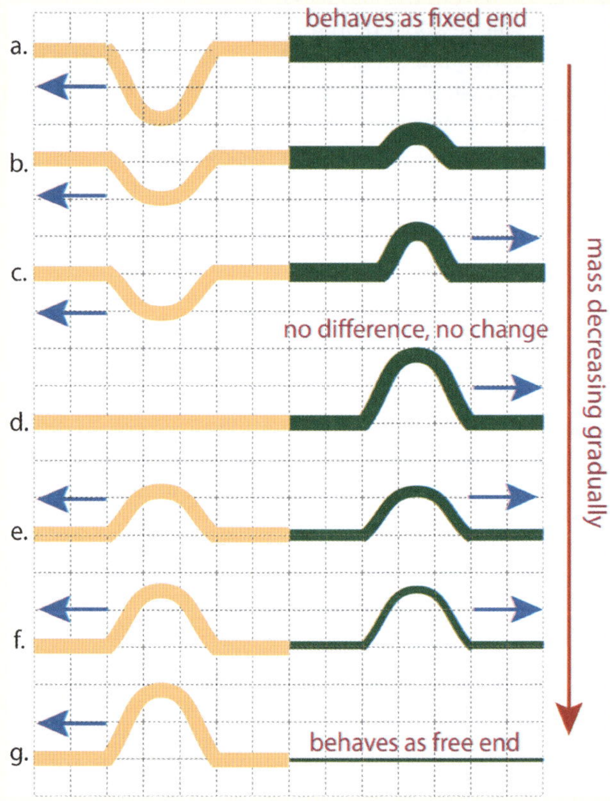

a. The extremely heavy string behaves like a fixed end. The reflected pulse is an inverted, same size pulse.

b. / c. As the mass of the string decreases, the transmission of the pulse begins. The transmitted pulse is slower and has a less width and amplitude than the incident wave. The reflected pulse has the same velocity and width. But the amplitude is smaller than the incident pulse.

d. The mass of the string is same for both strings. The pulse continues its motion without a change.

e. / f. The transmitted pulse is faster and has more width than the incident wave. The reflected pulse has the same velocity and width. But the amplitude is smaller than the incident pulse either for reflected and transmitted waves.

g. The extremely light string behaves like a free end. The reflected pulse has the same properties of the incident pulse.

This time we have two different waves on the same string. The waves are moving towards each other. What will happen when they meet? Will they stop their motion or continue?

Let's perform an experiment by sending two different waves from the opposite ends of the string. There are no barriers like a wall or a ring in between the ends of the string, so there is no reason for a change of the produced waves. The waves will continue to move in their own manner. The waves meet at a region of the string. Two waves are simultaneously present at the same region now. The displacements of the particles at that region is equal to the sum of the displacements produced by each wave. The vector addition of the displacements of a particle is called superposition. The superposition of two waves such as this case is called interference.

If two waves reach their maximum amplitude, their equilibrium position and the minimum amplitude at the same time, then they are in phase. The total displacement of the medium where the waves are overlapped is larger than the displacement that will be produced by an individual wave. Such an interference is constructive interference.

If as one of the waves reaches to its maximum amplitude, the other wave is reaching to its minimum amplitude or vice versa, then they are out of phase. The total displacement of the medium where the waves are overlapped is smaller than the displacement that will be produced by an individual wave. Such an interference is destructive interference. During the destructive interference, waves may completely cancel each other if they have the same shape and amplitude. This doesn't mean that the energy carried by the waves is vanished. There's no net displacement at any point of the medium at this instant but the string is actually moving rapidly. The energy of the waves is in the form of kinetic energy of the medium.

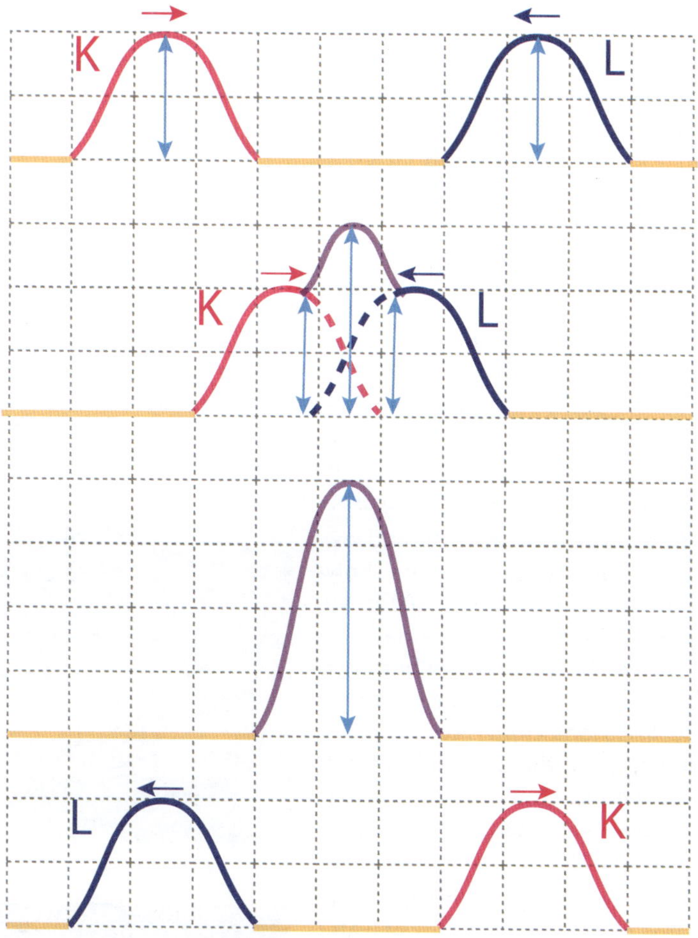

Constructive interference

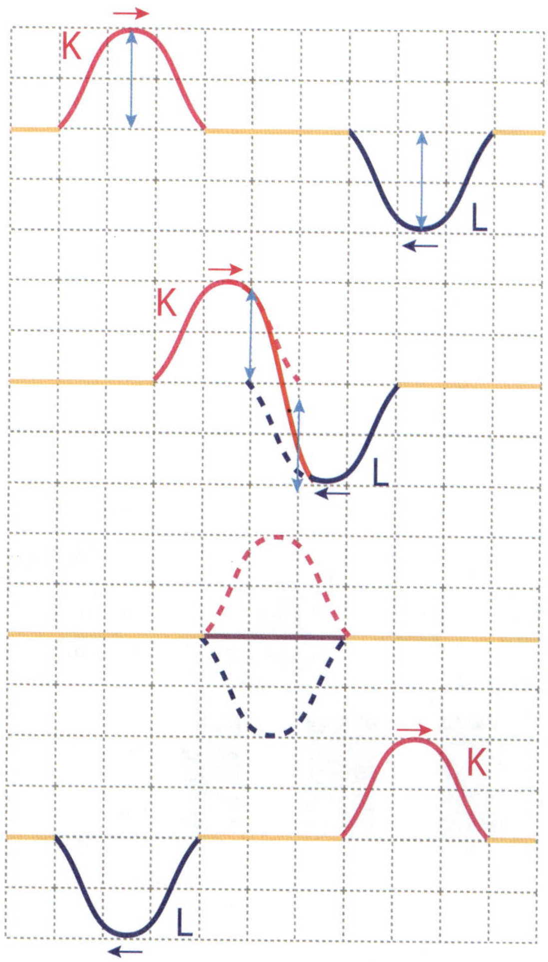

Destructive interference

Example

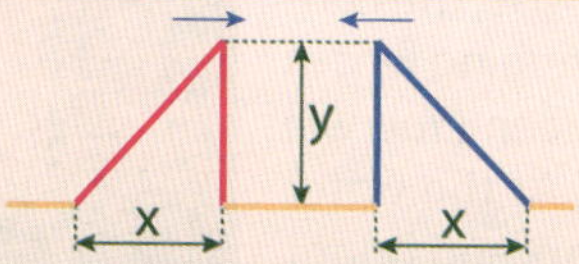

Two pulses formed on a stretched string are given in the figure.

How will the resultant pulse be when they overlap?

Solution

The waves will have a constructive interference. They have the same amplitude. So when they overlap, the amplitudes of the corresponding points add up.

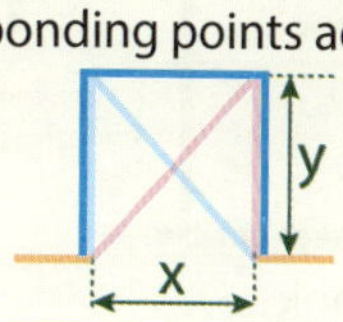

3.3. Water Waves

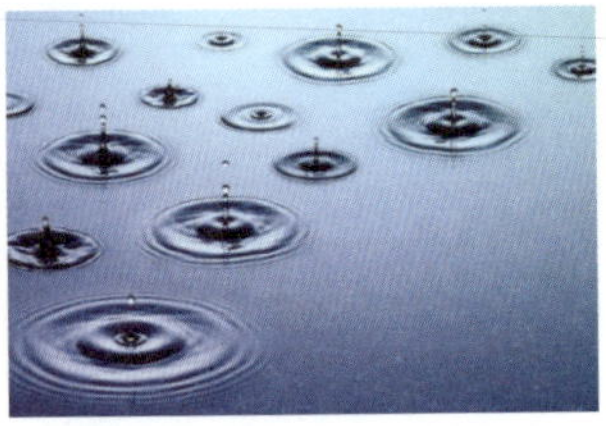

Rain drops falling onto the surface of calm water

So far we have been studying about waves formed on a string or slinky spring. They are *one dimensional* waves. Waves like ripples spreading in a pond are circular waves and they are *two dimensional* waves. These waves are referred as water waves. Sound, seismic and light waves move in *three dimension*.

In the path of a wave motion points that have the same displacements, like all the crests or all the troughs can be joined together to form a line. This line is called a wave front. The shape of a wave front is determined by the geometry of the source. A point source has wave fronts that are spheres whose centers are at the point source. A fluorescent tube would have wave fronts that are cylinders concentric with the tube itself. A very large sheet of material that is uniformly illuminated would generate wave fronts that are plane waves parallel to the sheet.

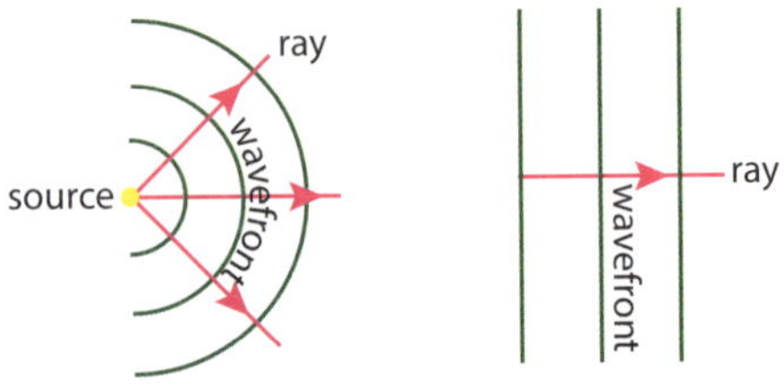

The direction of propagation of a wave is always perpendicular to the surface of the wave front at each point. The spherical wave fronts formed by a point source propagate radially outward. In the geometry of a sphere, the radius is always perpendicular to its circumference. A ray is a line that shows the propagation direction of the wave and it is perpendicular to the wave fronts. If you are very far away from the source, the radii of the spheres are so large that they look like plane waves to an observer. (Just as the Earth looks flat when viewed from a point near its surface.)

3.3.1. Linear and Circular Water Waves

When a rock hits the water, it bounces the surface down a bit and like a ball, the surface comes back up because of the surface tension of water. And just like a ball, it does this several times before gravity restores the water's equilibrium and the water is still again. When the rock hits the water and it is met with some resistance, the surface is displaced downwards and also outward as well. Then the disturbance is transferred from one water molecule to the next. A circular wave forms on the surface of water. This process continues to repeats as long as the center continues to *bounce*. Circular waves are two dimensional waves spreading on the surface of the water.

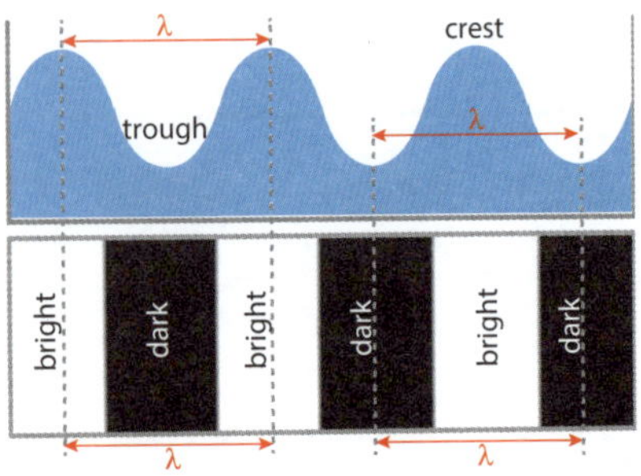

Highest points on a circular wave are crests and lowest points of the wave are troughs. Aside picture shows the circular ripples formed on the surface of water. Notice that the circular shadows are formed have dark and bright regions in the pattern. These dark and bright regions are formed on the ground because the crests and troughs on the wave act like a lens. Light is converged by the crests, thus bright regions of the pattern is formed. The dark regions are formed since the troughs are diverging the light. The distance between each successive bright (or dark) region is equal to one wavelength.

A ripple tank is used for the experiments about water waves. It has a large glass-bottomed tank of water. A light source shines upon the water from above and illuminates a screen directly below the tank. The convex surface of a crest of wave will behave like converging lens. Depending on the shape of the wave fronts, bright circles or lines will be observed on the screen. Likewise, the concave surfaces of troughs will behave like diverging lenses and dark regions will be formed on the screen (You will learn the details of lenses in unit 4, Optics). As the water waves move through the ripple tank, the dark and bright spots move as well. As the waves encounter obstacles in their path, their behavior can be observed by watching the motion of the dark and bright spots on the screen.

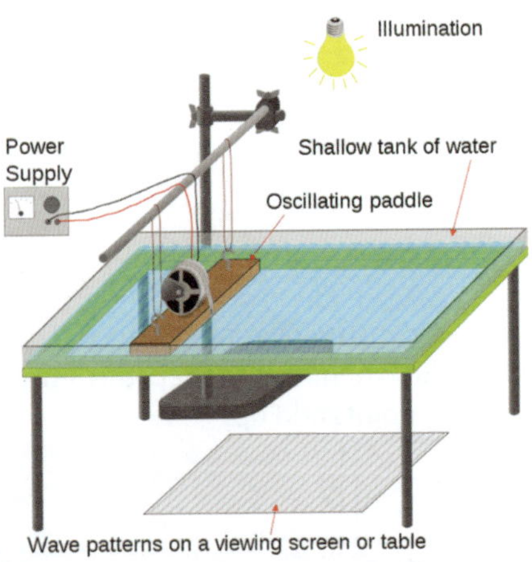

Ripple tank

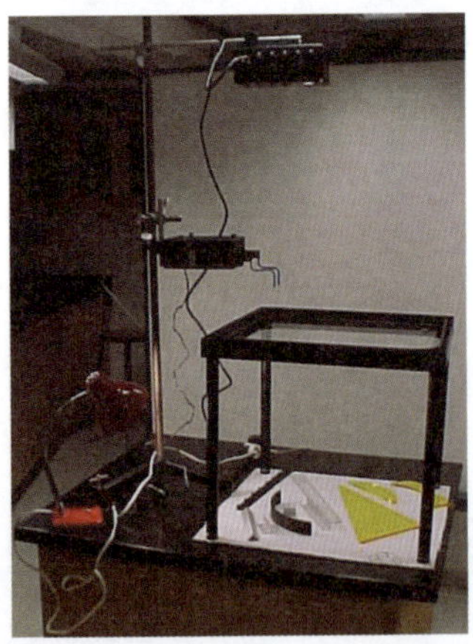

Plane waves can easily be formed on a ripple tank by using a ruler, or a linear object that can be attached to the ripple generator. Each bright line represents the crests and each dark line represents the troughs on the screen. The distance between each successive bright lines is equal to one wavelength. The same applies for the distance between two successive dark lines.

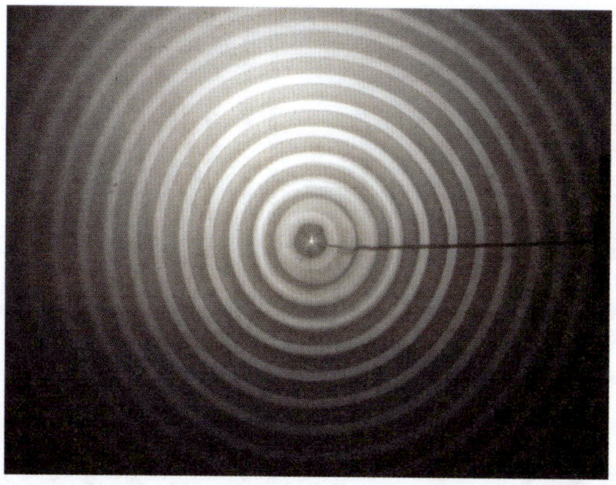

Circular waves can easily be formed on a ripple tank by using dropping water, or a point source that can be attached to the ripple generator. Each bright line represents the crests and each dark line represents the troughs on the screen. The distance between each successive bright lines is equal to one wavelength. The same applies for the distance between two successive dark lines.

3.3.2 Reflection of Water Waves from Plane and Parabolic Barriers

Just like the waves formed on strings, the water waves do reflect from an obstacle. We will consider the reflection of plane and circular water waves from plane and parabolic obstacles.

Reflection of water waves from a plane obstacle

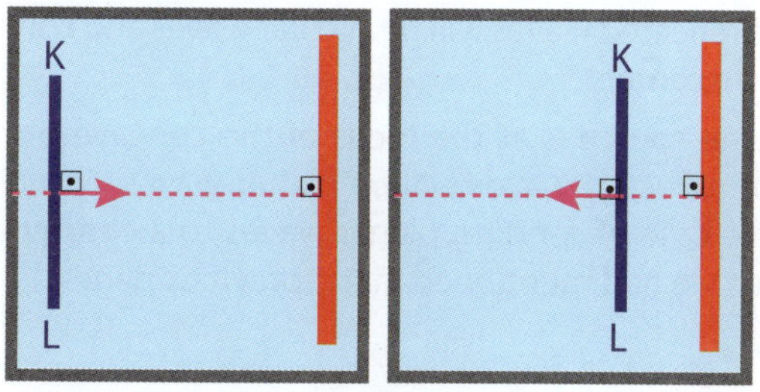

Reflection of plane water waves from a plane obstacle:

If a plane obstacle is placed parallel to a plane wave, the wave returns back on itself when it reaches to the obstacle. Reflected waves are still planar.

If the same obstacle is placed at an angle with the plane wave, then first incident end of the wave is reflected first. Then the following points on the wave get reflected from the obstacle. During the reflection; as some part of the wave is approaching to the obstacle, the other part of the wave gets away from it. Reflected waves are still planar.

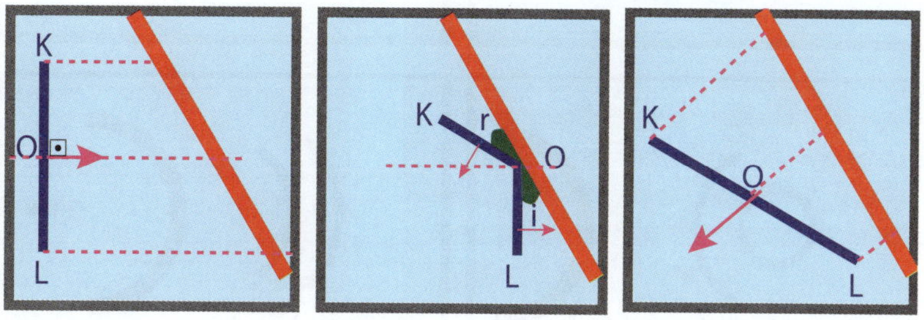

The angle that the incident wave makes with the reflecting surface, θ_i, is equal to the angle made by the reflected wave, θ_r ($\theta_i = \theta_r$). This is the law of reflection for waves.

Reflection of plane water waves from a concave obstacle:

The ends of the wave reaches to the obstacle first, so they get reflected first. Then the points towards the center will reach the obstacle and get reflected. The midpoint of the wave will be the final point to be reflected. The reflected waves are circular and they converge at a point in front of the obstacle and then diverge again from there as if there is a point source of wave at that point. This point is referred as the focal point (focus) of the concave obstacle.

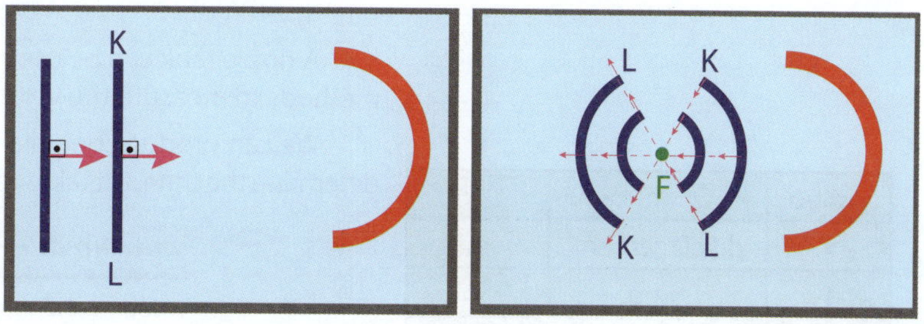

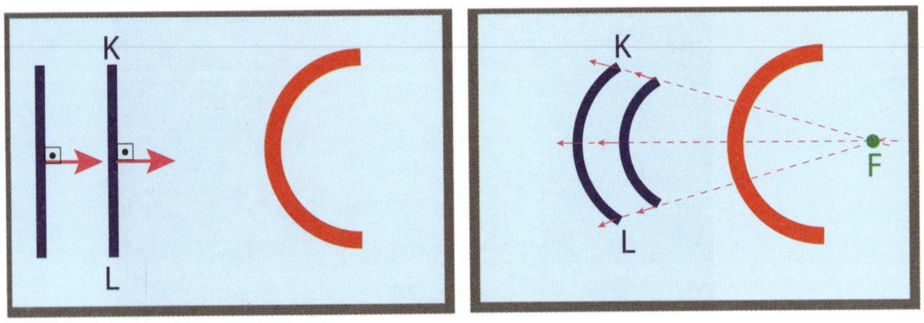

Reflection of plane water waves from a convex obstacle:

The midpoint of the wave reaches to the obstacle first, so it gets reflected first. Then the points towards the end will reach the obstacle and get reflected. The ends of the wave will be the final points to be reflected. The reflected waves are circular and they diverge as if there is a virtual point source of wave behind the obstacle. This point is referred as the focal point (focus) of the convex obstacle.

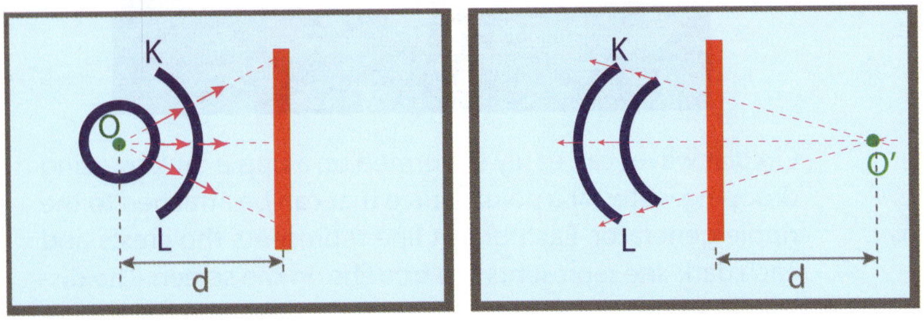

Reflection of circular water waves from a plane obstacle:

The midpoint of the wave reaches to the obstacle first, so it gets reflected first. Then the points towards the end will reach the obstacle and get reflected. The ends of the wave will be the final points to be reflected. The reflected waves are circular again, and they diverge as if there is a virtual point source of wave behind the obstacle. Both the source and the virtual source have the same distance to the obstacle.

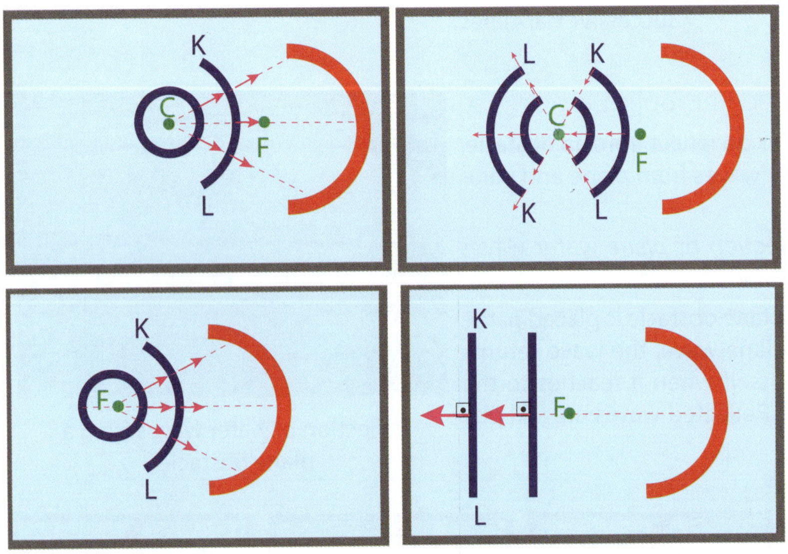

Reflection of circular water waves from a concave obstacle:

The behavior of the reflected waves depends on the location of the source with respect to the concave obstacle for circular water waves. We will limit our observations to two cases:

a. If the source is at the center of the concave obstacle, the reflected waves will be circular waves and converge at the center.

b. If the source is at the focus of the concave obstacle, the reflected waves will be planar waves. This is the reverse case of sending planar waves to a concave obstacle as we had mentioned in the previous page.

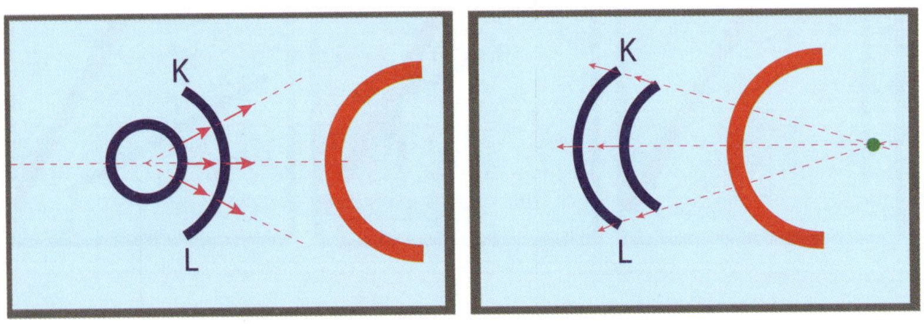

Reflection of circular water waves from a convex obstacle:

The reflected waves are circular and they diverge as if there is a virtual point source of wave at that point.

3.3.3. Velocity of Water Waves

A ripple tank can be used to measure the velocity of a water wave. There are several methods to measure the velocity of a water wave.

We can produce one plane wave in the ripple tank and using a stopwatch, we can determine the time, t, it takes to travel a distance of x. Then we can calculate the velocity by:

$$\frac{\text{measured} \leftarrow x}{\text{measured} \leftarrow t} = v \rightarrow \text{calculated}$$

Symbol	Quantity	Unit
x	displacement	m
t	time	s
v	velocity	m/s

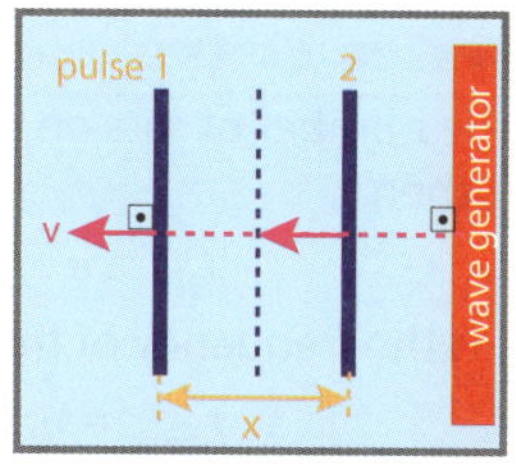

With the same method two successive pulses may be produced in the ripple tank within a time interval of t. The distance, x, between the two pulses remains the same. The velocity can be calculated by using the above equation again.

These two methods are simple in principle, but when it comes to practice, following the pulses and measuring the times and distances become very difficult.

Instead, we can generate pulses one after another in equal time intervals. The wave generator can be set to repeat its motion in an interval of T. Thus we have pulses produced with a period of T. Or we can say f number of pulses are produced in unit time, 1 second, in other words the frequency of the pulses is f. This repetitive movement of the wave generator results a periodic motion, thus a wave motion is generated.

Now let's concentrate on some point of the pattern that we have on the screen. The generated pulses move toward this point, and pass the point with the same frequency with which they left the source. We can count the number of waves passing through that point in some amount of time. For example if you are counting 20 pulses in 5 seconds, this means that in one second 4 pulses are passing from that point. So the frequency of the pulses, f, is 4 s^{-1}. The period was reciprocal of the frequency. Then for the period of the waves we generated:

$$T = \frac{1}{f} = \frac{1}{4} = 0.25\ s^{-1}$$

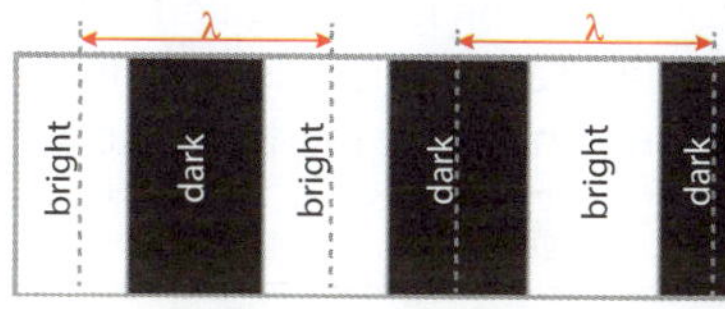

The time spent between two successive crests (or troughs) is the period of a wave, which is 0.25 s in our example. The distance between two successive bright lines or two successive dark lines of the pattern on the screen is equal to wavelength of this periodic motion. We can measure the distance between two bright lines with a ruler. Now we know the wavelength of the waves. Thus:

$$v = \frac{\lambda}{T} = \lambda . f$$

Symbol	Quantity	Unit
v	velocity	m/s
λ	wavelength	m
T	period	s
f	frequency	s^{-1} or Hz

The difficulty that may rise with this method would be measuring the distance between two bright lines since they are propagating continuously. Wouldn't it be easier to measure the wavelength, if they were at rest? The waves we are generating are travelling waves, so we cannot stop them. But we can use a stroboscope to view them at rest.

A hand stroboscope is a disc of hardboard or card with a simple pivot at its centre. The disc can be kept spinning by hand. There is a finger hole by the center to keep it spinning. There are narrow slits on its face, near the rim. These slits are evenly spaced. You can change the number of slits simply by sticking black tape over them. A stroboscope enables us to *freeze* the repetitive motions or to slow them down for closer study. Continuous ripples with higher frequencies can easily be observed at rest by using a stroboscope.

Hand stroboscope

Digital stroboscope

Sometimes, in a stage show, a special light is used, so that you observe the motion of the dancer frame by frame. That special light used there is actually an digital stroboscope. A digital stroboscope emits brief and rapid flashes of light. Usually a gas-discharge or solid-state lamp is used, because they are capable of emitting light nearly instantly when power is applied, and extinguishing just as fast when the power is removed. A digital stroboscope used in ripple tank experiments also have a digital frequency readout for accurate, quantitative data. Depending on the spinning frequency of a stroboscope, we may observe the travelling waves in a ripple tank at rest.

Strobelight photography of a dancer

If the generated waves are very fast, it is possible to have 2, 3, 4 or more number of waves passing within the rotation time for two successive slits. The waves can be observed at rest in this case also. To prevent this, the stroboscope is rotated very fast at the beginning of the measurements. Then its rotational speed is decreased gradually until the waves are observed at rest for the first time.

Physics Extra

If n is the number of slits on the stroboscope, $f_{stroboscope}$ is the spinning frequency of the stroboscope, f_{slit} is the frequency of the slits, then:

$$f_{slit} = n \cdot f_{stroboscope}$$

If f_{wave} is the frequency of the waves, then:

$$f_{wave} = n \cdot f_{stroboscope}$$

Thus if:

$f_{wave} = f_{slit}$ then the waves are observed as if they are at rest.

$f_{wave} > f_{slit}$ then the waves are observed as if they are moving forward.

$f_{wave} < f_{slit}$ then the waves are observed as if they are moving backward.

For the velocity of the waves:

$$v = \lambda \cdot f_{wave} = \lambda \cdot n \cdot f_{stroboscope}$$

This equation is not applied only for plane waves. We can use it for circular waves also. For the circular waves, the wavelength is measured along a radius. Even for plane or circular waves, as long as they are produced with the same conditions like, the same frequency and same medium, the waves will have the same wavelength and same velocity.

Example

A plane wave generator produces 30 waves in 5 seconds. A student trying to find the speed of these waves looks through a stroboscope which has 12 slits on it. The moment he first observes the waves at rest, he measures the distance between 4 crests as 9 cm.

a. Find the velocity of the waves.

b. Find the frequency of the stroboscope.

c. Find the number of rotations of the stroboscope in 6 seconds.

d. If waves were produced by a point source of same frequency, what would the area swept by the waves be in 2 seconds?

($\pi = 3$)

Solution

a. If 30 waves are generated in 5 seconds, then the frequency of the waves (waves generated in one second) is:

$$f = \frac{30}{6} = 5\ s^{-1}$$

The student observes 4 crests. The distance between 4 crests is equal to 3λ. So:

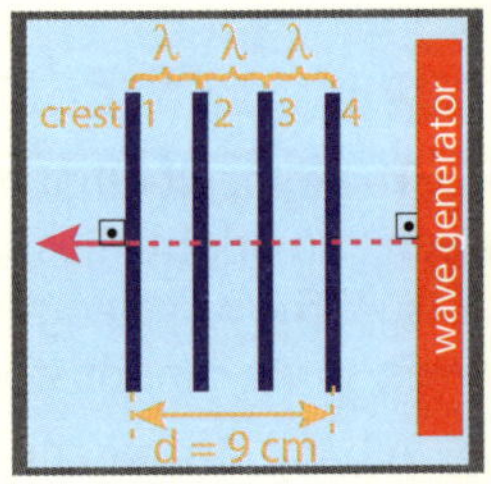

$3\lambda = 9 \Rightarrow \lambda = 3\ cm$

$v = \lambda \cdot f = 3 \cdot 6 = 18\ cm/s$

b. When the waves are observed at rest:

$f_{wave} = n \cdot f_{stroboscope}$

$6 = 12 \cdot f_{stroboscope}$

$f_{stroboscope} = 0.5\ s^{-1}$

c. If the stroboscope does 0.5 rotations in one second, then in 6 seconds:

$n = 0.5 \times 6 = 3$ rotations.

d. Since there is no change in the medium, the circular waves produced by the point source will have the same velocity and frequency. Then the radius of a circular wave in 2 s is:

$r = v \cdot t = 18 \cdot 2 = 36\ cm$

$A = \pi \cdot r^2 = 3 \cdot 36^2 = 3888\ cm^2 = 0.4\ m^2$

Example

A stroboscope with 4 slits on it does 240 rotations in one minute. The distance between two bright lines is measured as 3 cm.

Find the velocity of the wave.

Solution

$$\left.\begin{array}{l} 60\ s \rightarrow 240\ rotations \\ 1\ s \rightarrow f_{stroboscope} \end{array}\right\} f_{stroboscope} = 4\ s^{-1}$$

$f_{wave} = n \cdot f_{stroboscope} = 4.4 = 16\ s^{-1}$

$v = \lambda \cdot f_{wave} = 3.16 = 48\ cm/s$

We had studied that the waves produced in different thicknesses of strings were moving with different speeds. A change in the medium causes a change of speed for waves. This holds for water waves also. The velocity of the water waves changes with depth of water. In the deep regions of the seas or the oceans, waves moves rapidly. The waves move slowly when they are at shallow regions. As water waves approach the shore, it has contact with the seabed which slows it down. Waves break in shallow waters because the upper thick part of the wave moves faster than the lower thin part.

Waves breaking at shallow waters

3.3.4. Refraction of Water Waves

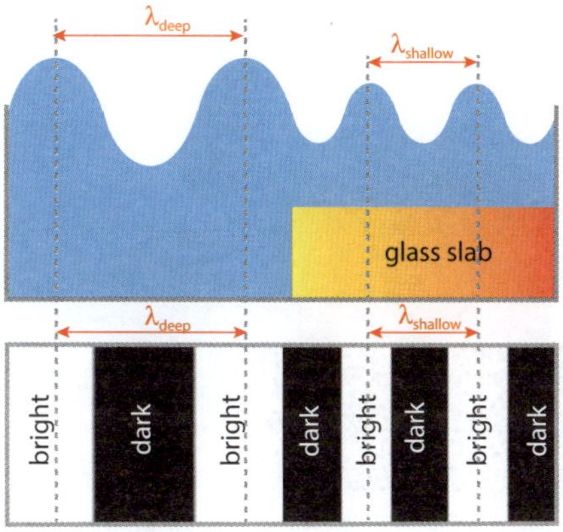

We can use the ripple tank again to investigate the velocity of waves passing from a deep region to a shallow one. If we place a thick glass slab to the base of the ripple tank, the depth will be changed. For an easier investigation we will work with plane waves.

If the pattern of the waves on the screen is observed with a stroboscope, with the right frequency, both the waves on deep and shallow regions will be observed as if they are at rest. The waves do not have different frequencies at different regions because the frequency is dependent on the source, not the medium.

$$f_{shallow} = f_{deep}$$

The distances between bright lines will be different for different depths. For the shallow region, the bright lines are closer to each other whereas in the deeper region they are more apart from each other.

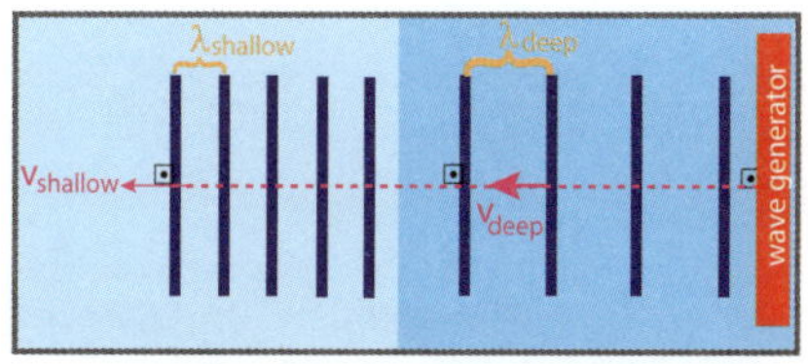

$$\lambda_{shallow} < \lambda_{deep}$$

Since $v = \lambda . f$ for the waves, the relationship between the velocities is:

$$v_{shallow} < v_{deep}$$

The behavior of waves in shallow water look alike to the waves formed on a heavy string and waves formed in deep water look alike to the waves formed on a light string.

If the glass slab we had placed is parallel to the wave generator, the only difference we will observe will be the difference between the wavelengths. What if the glass slab is not parallel to the wave generator, instead placed at an angle to it? How will the waves move?

Let's generate plane water waves in a deep medium and send them towards a shallow medium. The glass slab forming the shallow medium is placed at an angle to the incident plane waves. The points on the wave front that are first to reach the boundary will be passing to the shallow medium first. The waves in the shallow medium are slower, so they will have a smaller displacement as they propagate. The wave fronts will bend towards the boundary as they propagate. If the geometry of the figure is investigated, the incident wave front from the deep region makes a larger angle with the boundary than the wave fronts propagating in the shallow medium. The waves in the shallow medium are refracted waves. They refract because the velocity, thus the wavelength, of the waves decreases.

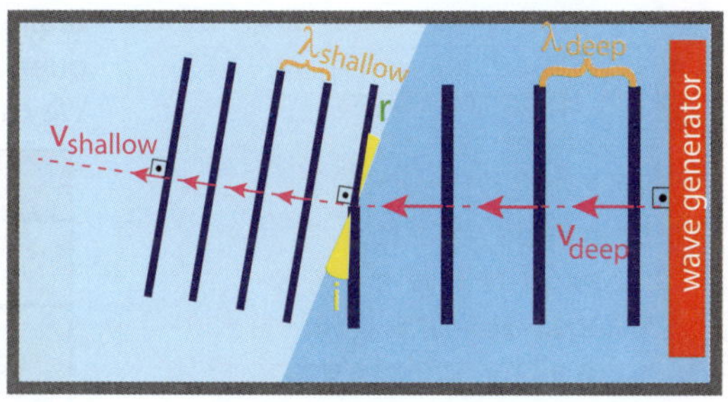

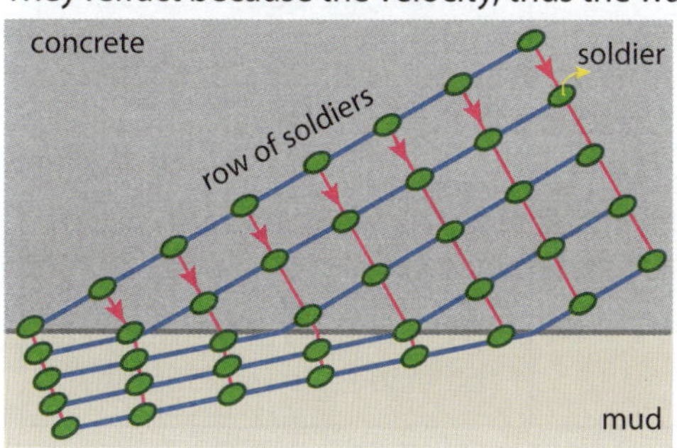

Let us think the wave front as a row of soldiers. The soldiers are marching from concrete ground into mud and hence they are slowed down at the boundary. The soldiers that reached to the mud as the first, are slowed down first and the row bends.

3.4. Sound Waves

3.4.1 Fundamental Concepts about Sound Waves

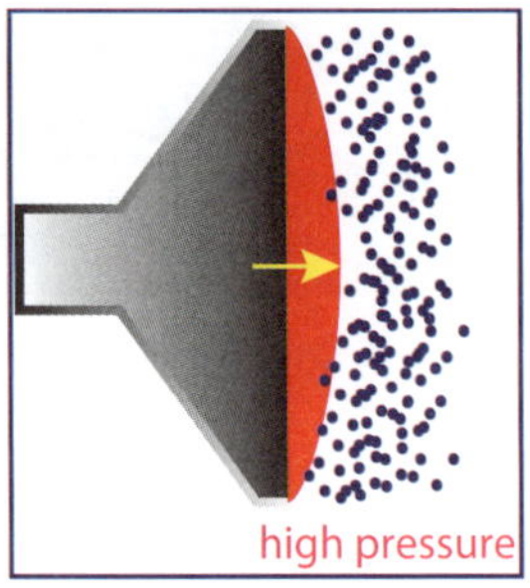

diaphgram moving forward

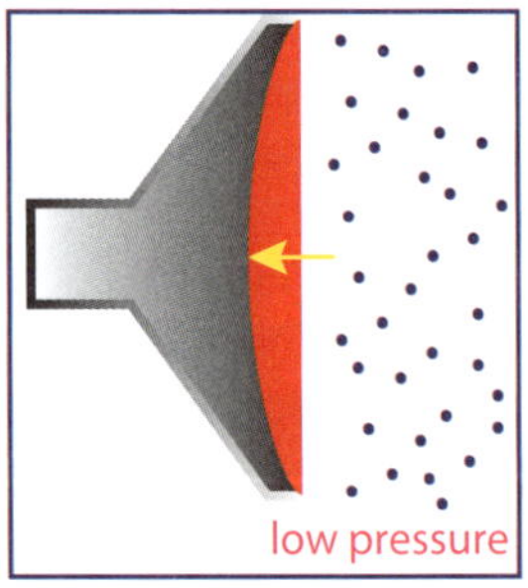

diaphgram moving backward

Sound is a mechanical wave. Like all mechanical waves, sound waves need a medium to propagate. They cannot be transmitted through a vacuum like the electromagnetic waves.

A source is needed for the formation of sound waves. A vibrating object like your vocal cords, the diaphragm of a loudspeaker, a drum membrane can be a source of sound. If we consider a speaker cone, as the diaphragm moves forward, it pushes the air molecules just in front of it. The pressure of air at that region increases immediately. Half cycle later, the diaphragm moves back. A partial vacuum is produced in front of it and the pressure decreases. These pressure fluctuations propagate as a wave. A sound wave can be thought of a pressure wave. Sound waves travel in the form of a longitudinal wave.

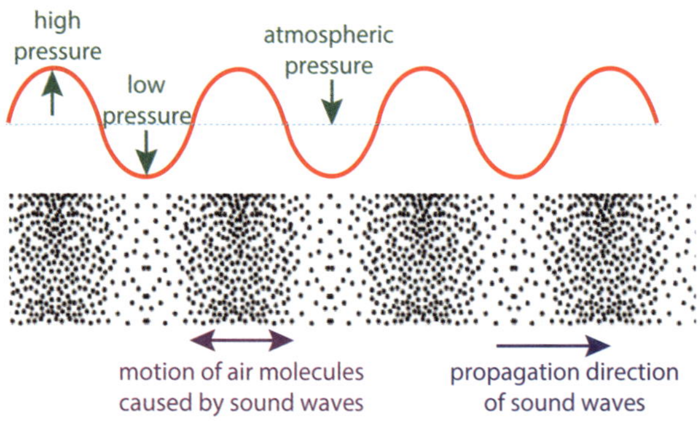

In the picture aside, when the guitar strings are vibrated, a sound wave is produced. Using an amplifier the intensity of the wave can be increased. The sound waves emerging from the loudspeaker can also cause a back and forth vibration of the metal that the boy is holding with the fluctuations of the air molecules.

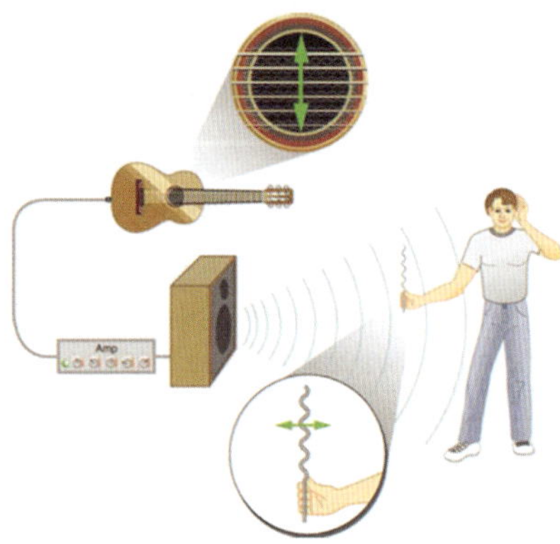

A sound wave is detected by an ear or a microphone. When the sound waves reach to your ear, the oscillating pressure of air causes your ear drum to vibrate. The vibration is transferred through the inner ear to the cochlea, then to the brain. Our ear is able to detect sounds with frequencies between 20 Hz and 20000 Hz (20 kHz). We can have three categories for sound waves.

INFRASONIC	AUDIBLE	ULTRASONIC
Below 20 Hz	20 Hz to 20 kHz	Above 20kHz
Can be produced during an earthquake / Elephants produce and use to communicate over great distances over like 6 km. Less energy is lost in transmission / Can be quite harmful, cause frictional heating of internal organs, irritation of the nervous system	Human ear can detect/	Dogs can detect/ Animals like bats, whales and dolphins use for navigation/ Bats emit loud chirps, detect the reflections by their large ears. They can sense the fine details in their surrounding/ Can travel very well in tissues. Interior images of the body can be made (ultrasonography)

"Information comes with the ear, the eye, the flesh, the taste, the smell, the feeling. You should accept, understand, realize, give a meaning, believe, implement, know."

Sound is characterized by loudness, pitch, frequency and timbre.

Producing sound requires only a small amount of energy. For example if 10000000 people were talking at the same moment, the energy that is released would be equal to the energy released by a flashlight. Like all waves, sound waves transfer energy in the medium. The sound energy transferred to a unit area in unit time is called the loudness or intensity of sound. The energy carried by a sound wave is proportional to its amplitude (squared). If the sound wave is large then the sound is referred as "loud". The energy carried by a sound wave can also be changed into other forms of energy. A telephone converts sound energy into electrical energy in a transmitter, and back into sound energy in the receiver. Sound energy is dissipated in the form of heat in air.

The range of intensities that the human ear can detect is so large. For such large ranges, physicists frequently prefer to measure it with a scale which is based on powers of 10. The scale for measuring the intensity (loudness of sound) is called the decibel (dB) scale. The threshold of hearing is set as 0 dB which is equal to 1×10^{-12} joules of energy delivered to 1 m^2 of area in 1 second (1×10^{-12} W/m^2). A sound that is 10 times more intense than 1×10^{-12} W/m^2 is equal to 10 dB, 10 x 10 = 100 times more intense than 1×10^{-12} W/m^2 is 20 dB and so on.... Sound intensities above 90 dB may cause harm to human ear.

The perception of a frequency is commonly referred to as the pitch of a sound. A *high pitch sound* corresponds to a *high frequency sound wave* and a *low pitch sound* corresponds to a *low frequency sound wave*. The sound produced by a bird is a high pitch one whereas the sound produced as a lion is roaring is a low pitch one. The pitch is the response of human ear to the frequency of the sound. The difference in voice quality of different individuals is because of their pitch.

A change in loudness means a change in the amplitude of the wave without a change in frequency whereas, a change in pitch means a change in frequency without a change in amplitude.

The quality of a musical sound is expressed by timbre. It is used for differentiating two sounds with the same pitch and loudness. For any sound to possess timbre, it must have one fundamental frequency (pitch of the note-lowest frequency waveform) and seven or more additional frequencies (harmonics).

As the temperature of air increases, the speed of sound also increases. Its speed in air at 0 °C is 331 m/s and at 20 °C it is 343 m/s. It also varies with the molecular mass in a gas at a given temperature. As the molecular mass increases, the speed of sound decreases. At room temperature speed of sound in helium is greater than that in air.

Gas molecules are far apart from each other so any fluctuations of pressure is not transmitted so quickly within the gas. The speed of sound waves is not dependent on the pressure of the medium for a gas. In liquids and especially in solids, the effects of local pressure fluctuations are transmitted much more quicker than in gases. The displacement of one atom can be felt by the neighboring atoms by the intermolecular forces.

Light and stiff substances such as diamond transfer sound faster (12000 m/s). Dense materials such as lead transmit sound less fast (5100 m/s). If we compare the speed of sound in states of matter;

$$v_{gas} < v_{liquid} < v_{solid}$$

If we strike an unmounted tuning fork with our hand, the sound will be heard hardly. But if we strike the same tuning fork against a table with the same force, a louder sound will be produced. The table has a larger surface so it will set more air molecules in motion when it is forced to vibrate like this. The table will be forced into vibration by a fork of any frequency. This is a case of forced vibration.

An object composed of an elastic material vibrates at its own special set of frequencies when it is disturbed. The set of these special frequencies together forms the special sound of the object. The least amount of energy is needed to produce forced vibrations of an object when it is vibrating with its natural frequency. Natural frequency depends on the size, shape and composition of the object.

AL-FARABI

Abû Nasr al-Fârâbî was probably born in 870 (CE) and died in 951 (CE). He had two main studies: Philosophy (logic in particular) and music.

His huge Kitâb al-musiqâ al-kabîr or Great Book of Music is the most important medieval musical treatise in Islamic lands and also includes sophisticated philosophical sections. His book is a study of the theory of Persian music of his day. But it has been introduced as a book on Arab music in the West. He contributed to the knowledge of musical notes and invented several musical instruments.

Al-Fârâbî was the first to explain that the sound waves were produced as a result of the vibrations of the air molecules. With the experiments he performed, he found that the vibrations were decreasing or increasing as a result of changes of the wavelength. His results lead him to discover the principles for the production of musical instruments.

It is told that he could play his instrument so well so that he can make people laugh or weep at will. Farabi's treatise Meanings of the Intellect dealt with music therapy, where he discussed the therapeutic effects of music on the soul.

If the frequency of the forced oscillations of an object matches the natural frequency of the object, the amplitude of the oscillation increases enormously. This is resonance. The word resonance comes from Latin and means to *resound; to sound out together with a loud sound.* In order for something to resonate, a force to pull it back to its starting position and enough energy to keep it vibrating are needed. The work done by the periodic force is used to overcome the friction in a resonating system.

Let's take two identical tuning forks, mount them on two separate hollow boxes and space them 1 meter apart from each other. The hollow boxes are for increasing the power of the sound wave generated by the vibrating tuning forks. If we hit the first fork and we can produce a sound wave. When a series of sound waves reaches to the second fork, each compression gives a tiny push on the ends of the fork. The frequency of these pushes corresponds to the natural frequency of the second fork. Thus these pushes will successively increase the amplitude of the vibration. This is because the pushes are occurring at the right time and in the same direction as the instantaneous motion of the fork repeatedly. If the forks are not adjusted to a matching frequency, the timing will not be correct and resonance won't occur.

A child's swing is a good example for a system with only one resonant frequency. If you give a tiny push on the swing, it comes back to you. If you continue to push it back, force it to vibrate, you can have a larger amplitude for the swing. Thus you have the swing vibrating at its resonant frequency. If you try to force the swing twice that the frequency, things may get harder. We can say that it is easy to get an object to vibrate at its resonant frequencies, hard at other frequencies. Swinging a child in a playground swing is easy since its natural frequency backs you up.

The vibrations produced by the motor of the electrical appliances may cause a vibration in a piece of furniture in another part of the same house if the natural frequency of the furniture matches with the frequency of the produced vibration. The furniture is in resonance with the motor. Resonance can also be observed in an automobile when the car is traveling at a certain speed. An object within the car like coins at a glass holder are in resonance with the vibrations of the engine at that speed. If the structural members of an airplane resonate at a frequency generated by the plane's engines, the resulting large amplitude deformations will produce cracks and cause failures.

Mechanical resonance can produce vibrations strong enough to destroy the object in which they occur. In 1831, troops marching across a footbridges near Manchester, England; unintentionally caused the bridge to collapse when they marched in rhythm with the bridges natural frequency. Since then it has become customary to order troops break step as they are crossing bridges. In 1940, at Puget Sound Narrows, Tacoma, Washington, a suspension bridge caused to vibrate at its natural frequency by the blowing wind and the bridge collapsed. The wind was blowing at 65 km/h when bridge literally vibrated itself to the point of collapse.

Tacoma bridge as it is collapsing

When you place a seashell up to your ear, you hear the familiar sound of the sea. Even in a very quiet room, there are sound waves with a range of frequencies. These sounds have very low intensities, so mostly they are inaudible. These background noise fills the seashell and causes vibrations within the shell. A seashell has its own set of natural frequencies at which it will vibrate. If one of the frequencies in the room forces air within the seashell to vibrate at its natural frequency, a resonance situation occurs. The result of resonance is always a larger vibration. Thus the sound is loud enough to hear now. The sound of the sea in a seashell is the amplification of one of the many background frequencies in the room in reality.

Electrical resonance is used to tune radios and television sets. Tuning consists of establishing a circuit with a resonant frequency equal to the assigned frequency of the desired station.

In music, resonance is used to increase the intensity (loudness) of a sound. The comparatively weak vibrations produced at the end of an organ pipe, for example, cause a column of air in the pipe to vibrate in resonance, thereby greatly increasing the loudness of the sound. This principle also applies to the human voice, in which the vibrations of the vocal cords are reinforced by resonant vibrations in the oral and nasal passages.

Brass instruments typically consist of a mouthpiece attached to a long tube filled with air. The metal tube serves as a container for a column of air. The vibrations of this column produces the sounds that we hear. The length of the vibrating air column inside the tube can be adjusted either by sliding the tube to increase and decrease its length or by opening and closing holes located along the tube in order to control where the air enters and exits the tube. Air is blown into a brass instruments through the mouthpiece. The vibrations of the lips against the mouthpiece produce a range of frequencies. One of the frequencies in the range of frequencies matches one of the natural frequencies of the air column inside of the brass instrument. This forces the air inside of the column into resonance vibrations. The result of resonance is a loud sound.

When a guitar player presses the string down on a fret to shorten its length or change the tension of the string, the natural frequency of a guitar string is easily changed. A guitarist controls the natural frequencies of the strings and plays a sequence and combination of sounds to make music.

Vibrating guitar strings

The fundamental frequency of a vibrating string is:

- inversely proportional to its length if the tension is kept constant; if the length of string is doubled, the frequency will be halved.
- directly proportional to (the square root of) tension; if the tension is quadrupled, the frequency will be doubled.
- inversely proportional to (the square root of) mass per unit length ratio; if the mass per unit length ratio is quadrupled, the frequency will be halved. Two strings made of same material and with the same length and tension, the thinner string will have a higher fundamental frequency.

If you shout into a well or a in valley, your sound comes back to you in a few moment. This is the reflection of sound waves and called echo. The sound waves you produced reflect off of the surface of the water at the bottom of the well or the wall on the far side of the valley and travel back to your ears. The length of time between the moment you shout and the moment that you hear the echo is determined by the distance between you and the surface that creates the echo. If you measure the time it takes for the echo to return and if you know the speed of sound, you can calculate the depth of the well fairly accurately.

The size of the reflecting surface should be large when compared to the wavelength of the sound wave. An ordinary sound wave has a wavelength of 1 meter usually. So a large building, the walls of a well are good reflecting surfaces for sound. The reflected sound wave should have enough intensity to be heard. If the reflecting surface is a large and rigid one, most of the sound energy is reflected back. But if the surface is a soft and irregular one, less amount of energy is reflected back. The unreflected portion of the energy is transmitted or absorbed by the surface.

Our ear is capable of detect any sound for about 0.1 seconds. This is known as the persistence of hearing. If the echo is heard within this time interval, the original sound and its echo cannot be distinguished. So for hearing an echo, the reflected sound should reach our ear only after a time interval of at least 0.1 second after the original sound dies off. Assuming that the average speed of sound in air is 340 m/s, the distance travelled by sound in 0.1 second is 34 m. During the reflection, the waves go forth towards the reflector and come back to the source. So the minimum distance between a source of sound and the reflector should be 34 m. If the obstacle is at a distance of 17 m at least, the reflected sound or the echo is heard after 0.1 second, distinctly.

If the sound waves reach to our ear in less than 0.1 s, reverberation occurs. The original sound wave is still held in the memory and there is not enough time delay between the perception of the sound wave and the original wave. This is why they tend to combine as one very prolonged sound wave. When you are speaking in an empty room or singing in the shower, you can experience the reverberation. The reflecting surfaces are much closer than 17 m and very reflective for the sound in such a case.

A wall isolated for reverberation

Echoes can be useful or a big problem. In a concert hall, echoes can ruin a performance if the walls and ceiling are not properly designed. If the walls are too hard, or too flat, they make good reflecting surfaces for the sound waves. If a trumpet plays on the stage, the sound can reflect off the back wall and return to the front of the seating still quite loud. Sometimes this sound can be heard by the audience as an echo. To overcome this problem, absorbs can be put on the rear wall to stop the sound reflecting.

Sound isolation and absorption work together to reduce the sound. The idea of sound isolation is resisting the sound waves through reflection or sending them back into the air. By this way the noise is reduced. When sound absorbed, the energy carried by the sound wave is transferred to heat energy, and the noise is absorbed. In your home, soft furnishings like carpets, absorb the sound and reduce the interior echoes. Soundproofing insulation materials can be added between the walls or directly on to drywall. A cork board, a fiberboard insulation, a rock wool insulation flooring and specialized acoustical wall coverings like acoustic tiles are among the sound isolation materials.

Nowadays many people are suffering from the results of noise pollution. Unwanted and harmful sound created by industrialization, poor urban planning, social events, transportation, construction activities, household chores causes noise pollution. Noise pollution may result hearing problems such as damage of ear drums and loss of hearing, psychological health issues, sleeping disorders, cardiovascular disease and stress related heart problems, communication troubles, and even problems on wildlife since they are more dependent on sound.

Sometimes even though there is no source of sound in the environment you hear a buzzing, ringing, whistling or hissing sound. This is noise originating in the ear, not from the environment and it is called tinnitus. Tinnitus is a symptom, not a specific disease. When you are trying to sleep, tinnitus may be very disturbing. The experience of tinnitus changes from person to person. Some people are highly disturbed by it whereas it is bearable for others.

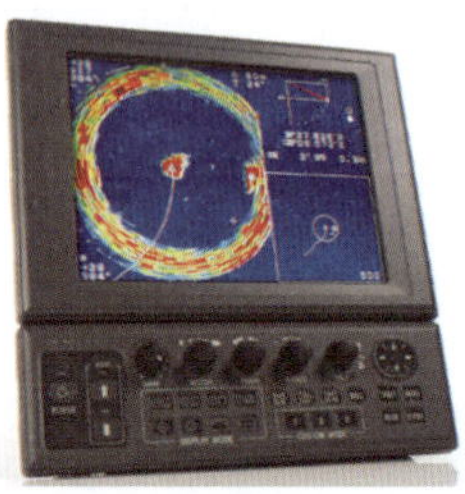

Sonar

3.4.2 Uses of Sound Waves In our Daily Life

Animals such as bats and dolphins use echoes to find their way to or around their prey. This is called echolocation. With echolocation they can detect the size and positions of the objects around them. A bat sends out tiny, high pitched chirps. They reflect back from the objects in the bat's flight path. The bat uses echoes to adjust its course to avoid the obstacles. Many bats have very large ears to catch as much of the reflected sound as possible.

Seabed is scanned with sonar

We can use the science behind echoes, in our devices. A sonar (SOund NAvigation Ranging) device sends out high frequency sound waves from a ship to find out how close the vessel is to the sea bed.

Electric signals can be changed into ultrasonic sound waves by electrical devices (like a loudspeaker). An ultrasound scanner, has many uses in medicine and industry.

For pre-natal scanning, with a receiver, the time it takes to detect the ultrasound waves is compared and this information is then used to build up a picture of the baby in the womb and it is displayed on a computer screen. Likewise the flaws and cracks in metal are detected.

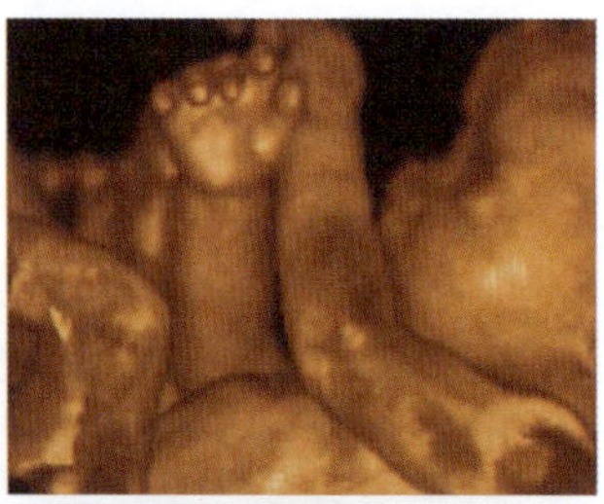

Image of a baby formed by an ultrasound device

For cleaning delicate instruments without having to take the instrument apart ultrasound waves can be used. The instrument is placed in a liquid. Then the ultrasound waves clean the surface of the equipment by making the liquid particles to vibrate at a high frequency.

3.4. Seismic Waves and Their Properties

3.4.1 Formation of Seismic Waves

Earthquakes occur when one mass of rock suddenly moves against another, releasing a huge amount of energy in the form of seismic waves. Seismic waves may also be generated by the explosions of volcanos, explosions and landslides.

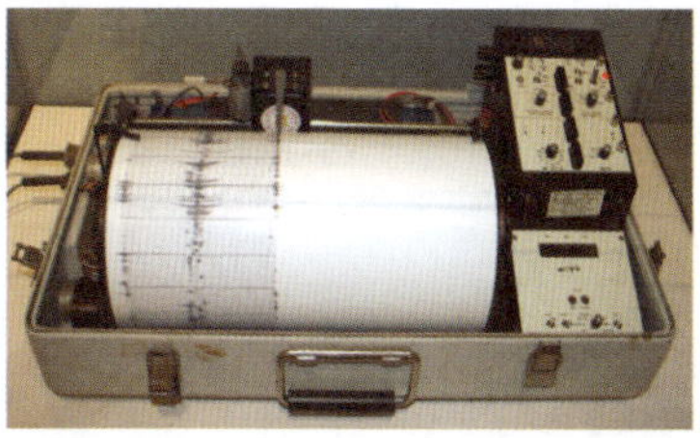

The earthquakes are detected and measured by a seismograph. The total energy from an earthquake includes energy required to create new cracks in rock, energy dissipated as heat through friction, and energy elastically radiated through the earth. The radiated energy from a seismic wave shakes buildings and it is recorded by a seismogram. Here at this point it is useful to note that a seismogram is a record on a paper or computer and it is recorded by an instrument called the seismograph.

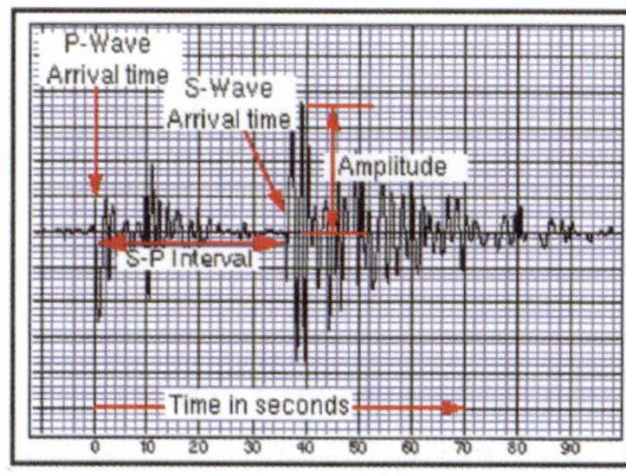

A seismogram, recording of earthquake waves

The Richter scale is a standard scale used to compare earthquakes. It is a *logarithmic scale*. The numbers on the scale measure factors of 10. So, for example, an earthquake that measures 4.0 on the Richter scale is 10 times smaller than one that measures 5.0. On the Richter scale, anything below 2.0 is undetectable to a normal person and is called a microquake. Microquakes occur constantly. Moderate earthquakes measure less than 6.0 or so on the Richter scale. Earthquakes measuring more than 6.0 can cause significant damage. The biggest quake in the world since 1900 scored a 9.5 on the Richter scale. It occurred in Chile on May 22, 1960.

Bent railways at Adapazarı, after the Marmara earthquake, 1997

Faults formed after earthquakes

Collapsed buildings

Earthquakes are accompanied with tsunami's if they occur in deep water. The word tsunami comes from Japanese and translated as *harbor wave* in English. Tsunamis are sometimes referred to as *tidal waves* by the public, but they are not a result of the imbalanced extraterrestrial gravitational influences of the moon, the Sun, and planets. Tsunamis normally occur in Pacific Ocean, and are relatively rare in areas surrounding the Indian subcontinent.

Tsunamis can be generated when the sea floor suddenly deforms and vertically displaces the overlying water. Earthquakes, mass movements above or below water, volcanic eruptions and other underwater explosions, landslides, large meteorite impacts and testing with nuclear weapons at sea all have the potential to generate a tsunami.

Tectonic earthquakes are associated with the earth's crustal deformation. When they occur beneath the sea, the water above the deformed area is displaced from its equilibrium position. Waves are formed as the displaced water mass, acting under the influence of gravity, tries to regain equilibrium. When large areas of the sea floor elevate or subside, a tsunami can be created.

As a tsunami leaves the deep ocean and travels toward the shallow coast, it transforms. A tsunami moves at a speed relative to the water depth, therefore the tsunami slows as the water depth decreases. The tsunami's intensity is dependent on its wave speed and wave height and it remains nearly constant. As a result, the tsunami's speed decreases as it travels into shallower water, and its height increases. Because of this shoaling effect, a tsunami, non obvious at sea, may heighten to several meters or more near the coast. When it reaches the coast, it may appear as a rapidly rising or falling tide or a series of breaking waves.

The moment tsunami was generated it consists of one wave. But in a short period of time it splits to 3-5 waves and begins to spread outwards. The first and the last of these waves are weak when compared to the second wave. Abnormal change in the water level by the coast is a warning for the first gentle wave. This change points out the more powerful coming waves.

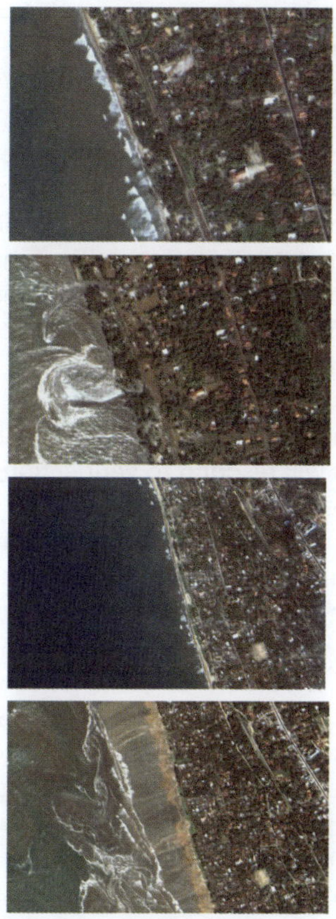

Before and after pictures of Tsunami

SAFETY PRECAUTIONS IN CASE OF AN EARTHQUAKE

Although there are warning systems for the tsunamis, scientists are still searching for a method to know when and where the earthquake will happen. For now, best we can do is to get ready for an earthquake and take the safety precautions.

Before an Earthquake

Research the area you live and find out if you live near an active fault line. If you're in a high risk area, reinforce your house. Secure any furniture such as bookshelves and cabinets to the walls to minimize risk of falling over during a quake. Secure cabinet doors to help keep dishes and other contents from falling out.

Create a disaster plan to protect yourself and your family. Locate a place in each room of the house that you can go to in case of an earthquake. It should be a spot where nothing is likely to fall on you. Designate a safe place to meet outside of the house after the shaking stops.

Know how to turn off your gas and water mains.

Put together an emergency kit. Your kit should include non-perishable food, water, first aid supplies, flashlights, extra batteries, blankets and any personal items you may need (medications, toiletries, clothing). If you have pets, make sure they also have adequate supplies. Plan for a week's worth of supplies for each person.

During an Earthquake

Remain calm as the quake occurs – others will respond to your actions. A cool head can prevent panic.

Stay away from windows and furniture that could potentially fall over.

Take cover in a safe place in your house. Drop, cover and hold on under a desk, a table or along an interior wall. Protect your head, neck and face. Stay under cover until the shaking stops and debris settles. If you can't take safe cover, protect your head and neck with your arms.

Do not try and go outside until after the shaking stops. If you are already

Solved Questions

Question

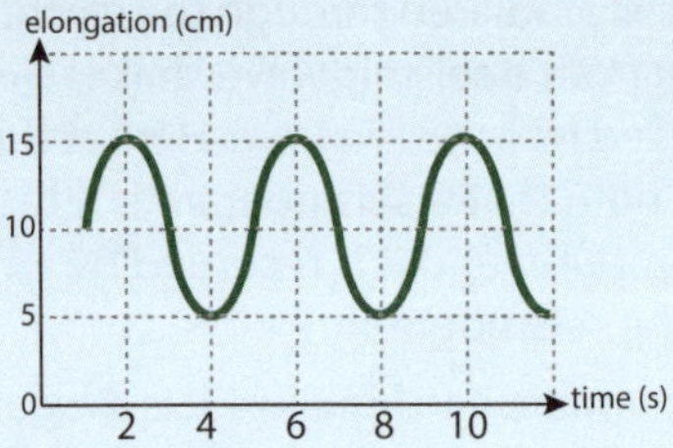

Displacement versus time graph of periodic oscillations is given in the figure. The velocity of this wave is 8 cm/s.

a. What is the amplitude of the wave?

b. Find the frequency of the wave.

c. Find the wavelength of the wave.

Solution

a. The difference between the highest and the lowest points of the wave is 10 cm. This distance is twice that of the amplitude. So the amplitude is 5 cm.

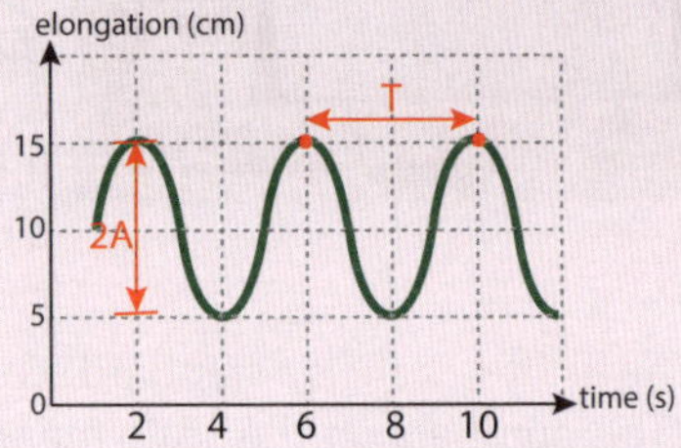

b. In one period of time one complete wave is formed. The time spent between two successive identical points, like two crests, is the period of the motion.

$T = 6 - 2 = 4\ s$

Frequency is the number of waves produced in unit time and is equal to the reciprocal of the period. Then:

$$f = \frac{1}{T} = \frac{1}{4} = 0.25\ s^{-1}$$

c. $\lambda = v.\ T = 8\ .\ 4 = 32\ cm$

Question

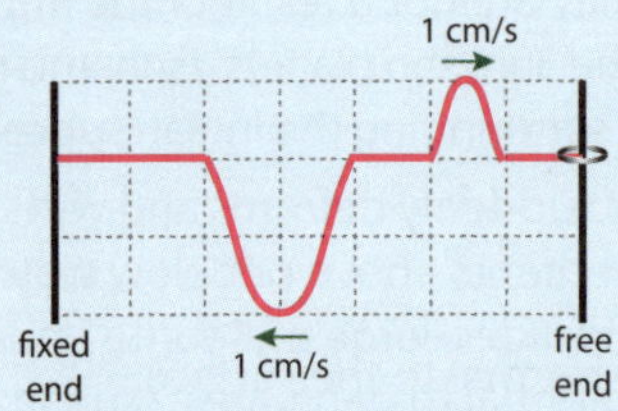

One end of a string is fixed to a rod and its other end is tied to a ring which can freely move on a rod. Two pulses are created on the string. They are moving with 1 cm/s in the given directions. The figure show their position at t = 0.

Draw the final positions of the pulses at t = 5 s.

(each square is 1 cm long)

Solution

Pulses reaching the fixed end will be inverted. The reflected pulses from the free end will remain unchanged. Point reaching to the end will be the leading point when reflected.

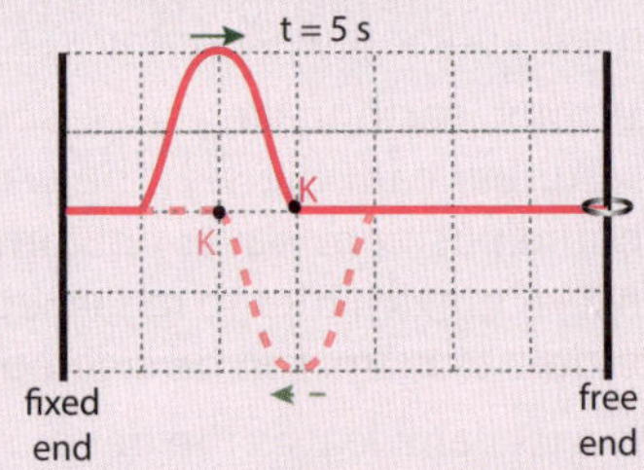

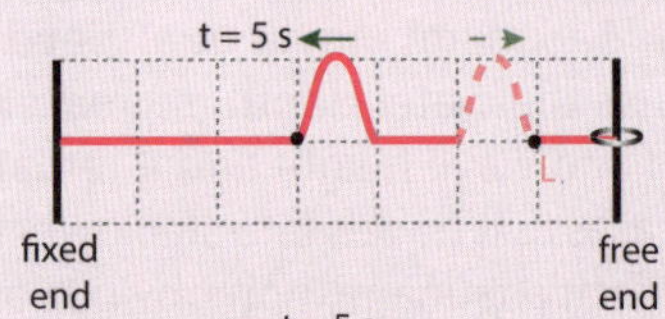

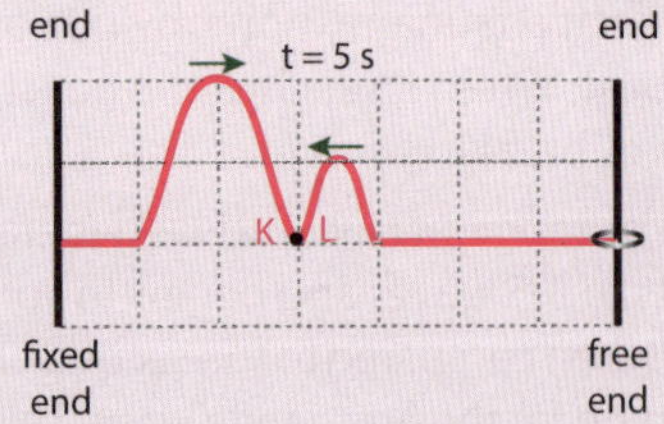

Question

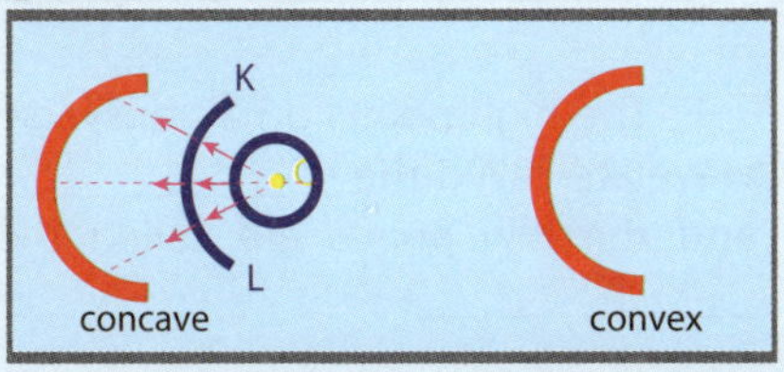

A point wave source is placed at the center of a concave obstacle. The waves produced are first being reflected from the concave obstacle, then from the convex obstacle.

Draw the reflections of the waves from the concave and convex obstacles.

Solution

Waves sent from the focus are reflected as circular waves from the concave obstacle.

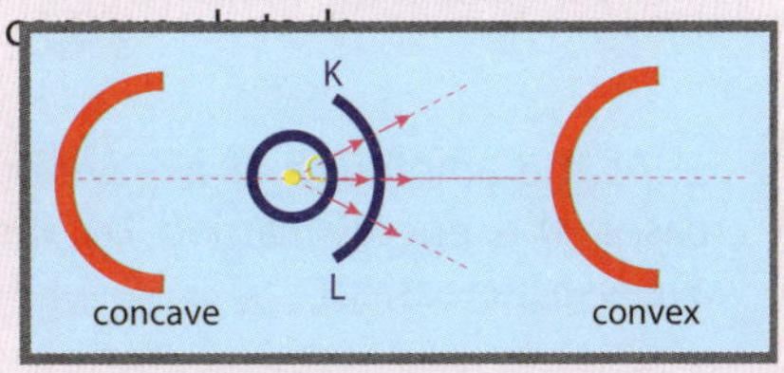

Plane waves will be reflected as if they are diverging from a point source behind the obstacle.

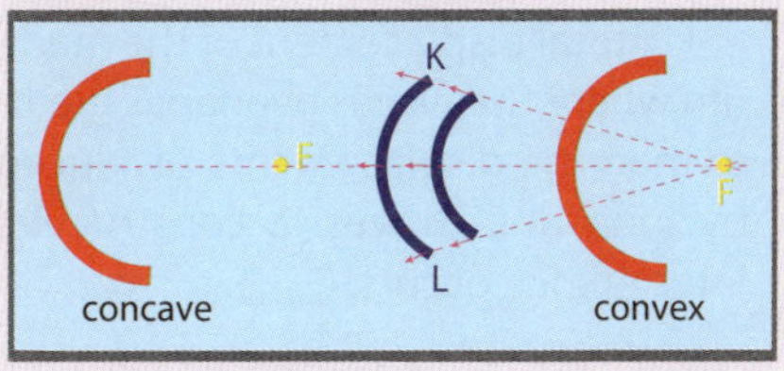

Question

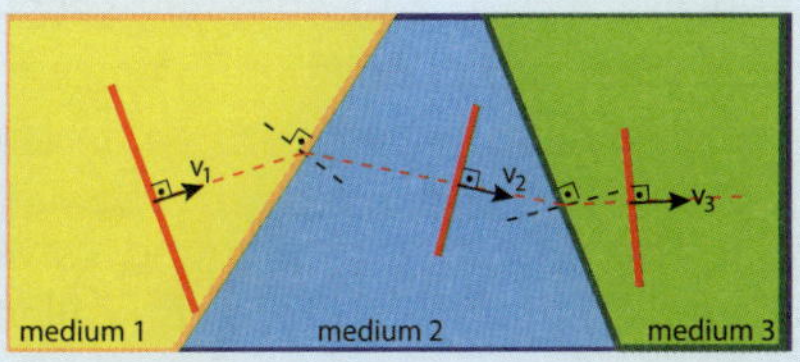

The propagation direction of plane waves in three different depths is given is given in the figure.

a. What is the relationship between the velocity of the waves in these media?

b. What is the relationship between the depths of water?

Solution

a. The incident wave front from the deep region makes a larger angle with the boundary than the wave fronts propagating in the shallow medium. If the geometry of the figure is investigated, the angle made with the boundary is also equal to the angle between the propagation direction and normal to the boundary. It is easier to keep track of the angle made with the propagation direction and the normal to the surface.

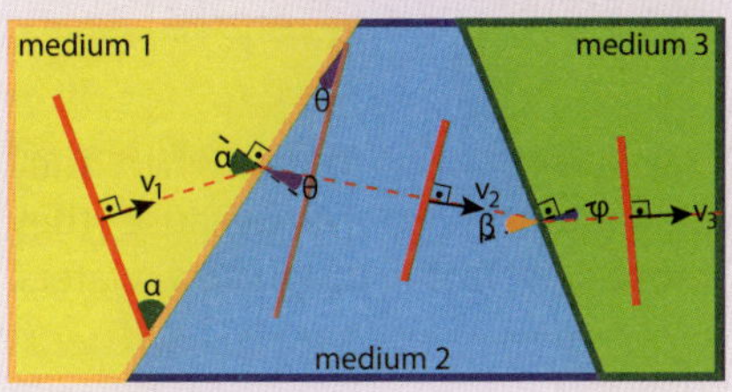

$\alpha > \theta$ so $v_1 > v_2$

$\beta > \phi$ so $v_2 > v_3$

As a result $v_1 > v_2 > v_3$

b. Velocity decreases as water gets shallower; so:

$h_1 > h_2 > h_3$

indoors, you are safer taking cover inside than attempting to leave your house during an earthquake you could be hit by falling debris as you're trying to get out.

If you are outdoors, move to an open area away from falling hazards such as trees, power lines, and buildings. Drop to the ground and cover your head and neck.

If you are in a car, slow down and drive to a clear place. Stay in the car until the shaking stops.

After an Earthquake

Earthquakes are often followed by aftershocks additional quakes that follow the main event. These can last for days or even weeks after a major earthquake.

Remain calm and reassuring. Check yourself and other for injuries. Do not move injured people unless they are in danger.

Be ready to act without electricity or lights. Know how to move around your work area and how to exit in the dark. Know how to access and use your emergency supplies. Be aware of objects that have shifted during the quake.

Check your gas lines and make sure there are no leaks. If you smell gas leaking, turn off the gas if possible and call the gas company. Do not use an open flame in your house until you are sure it is safe.

Turn off the power if you see damaged wiring in your house.

Keep your shoes on. You may have broken glass or spilled chemicals on the floor of your house as a result of the earthquake. Don't walk around barefoot until you're sure the floor is clean and safe.

If you must leave a building, use extreme caution. Continually assess your surroundings and be on the lookout for falling debris and other hazards. Take your keys, personal items and emergency supplies with you if safe to do so. Do not re-enter damaged buildings until an all-clear is given.

Use telephones only to report a life-threatening emergency. Cell and hard-line phone systems will be jammed. Text messages take less band width and may go through when voice calls can't be made.

Conceptual Questions

1. An astronaut wants to listen to music as he is working outside the space station. In order to do that he fixes his mini speaker to the arm of his suit. How high should he set the volume to hear it?

2. The sonar device on a fishing boat is used to locate fish. Would you expect sonar to be a longitudinal or a transverse wave?

3. The ocean waves splash onto the shore continuously. Explain why the beach is not completely submerged and why the middle of the ocean has not yet been depleted of its water supply.

4. Describe how the fans in a stadium must move in order to produce a longitudinal stadium wave.

True - False

1. Transverse and longitudinal waves can propagate in fluids. (___)

2. As the amplitude of a wave increases, the energy carried by the wave also increases. (___)

3. A pulse will be inverted as it is being reflected from a free end. (___)

4. If total displacement of the medium where the waves overlap is greater than the displacements produced by individual waves, a constructive interference occurs. (___)

5. Sound moves faster in solids. (___)

6. Electromagnetic waves can propagate through a vacuum. (___)

7. The distance between a rarefaction and compression in a longitudinal wave is its wavelength. (___)

8. Troops break march when they are passing through a bridge in order to prevent resonance. (___)

9. Water waves break when they reach to the shore, because they propagate faster in shallow waters. (___)

10. Tsunamis are result of tides occurring as a result of gravitational forces between earth and moon. (___)

Fill in the Blanks

1. The high points on a transverse wave are called the __________ and the low points are called the __________.

2. The number of waves produced in one second is called __________.

3. The distance between two successive crests is called the __________.

4. In passing from a heavy string to a light string the velocity of a pulse __________.

5. Water waves propagate between the __________ of water and air. This is why they are called __________ waves.

6. The reflection of a sound wave from a smooth surface is called __________.

7. A device for measuring the intensity of an earthquake is called __________. The intensity is measured with respect to __________ scale.

8. Whales and dolphins use __________ for navigation.

9. Microwaves, light, radio waves are examples of __________ waves.

10. Water waves propagate faster in __________ waters.

Conceptual Questions

1. An astronaut wants to listen to music as he is working outside the space station. In order to do that he fixes his mini speaker to the arm of his suit. How high should he set the volume to hear it?

Zero to maximum, doesn't matter. He won't hear anything. Sound is a mechanical wave and needs a medium to propagate. It cannot be transmitted in a vacuum.

2. The sonar device on a fishing boat is used to locate fish. Would you expect sonar to be a longitudinal or a transverse wave?

Transverse waves cannot propagate in fluids. Moreover a sonar uses sound waves to scan the seabed. The wave used by the sonar should be a longitudinal one.

3. The ocean waves splash onto the shore continuously. Explain why the beach is not completely submerged and why the middle of the ocean has not yet been depleted of its water supply.

Ocean waves do not transport water. Ocean waves can only bring energy to the shore; the particles of the medium (water) oscillate about their fixed position. Thus, water does not pile up on the beach.

4. Describe how the fans in a stadium must move in order to produce a longitudinal stadium wave.

The fans will need to sway side to side. As the wave travels around the stadium they would be moving parallel to its direction of motion. If they rise up and sit down, then they would be creating a transverse wave.

True - False

1. Transverse and longitudinal waves can propagate in fluids. (_F_)

2. As the amplitude of a wave increases, the energy carried by the wave also increases. (_T_)

3. A pulse will be inverted as it is being reflected from a free end. (_F_)

4. If total displacement of the medium where the waves overlap is greater than the displacements produced by individual waves, a constructive interference occurs. (_T_)

5. Sound moves faster in solids. (_T_)

6. Electromagnetic waves can propagate through a vacuum. (_T_)

7. The distance between a rarefaction and compression in a longitudinal wave is its wavelength. (_F_)

8. Troops break march when they are passing through a bridge in order to prevent resonance. (_T_)

9. Water waves break when they reach to the shore, because they propagate faster in shallow waters. (_F_)

10. Tsunamis are result of tides occurring as a result of gravitational forces between earth and moon. (_F_)

Fill in the Blanks

1. The high points on a transverse wave are called the crests and the low points are called the troughs.

2. The number of waves produced in one second is called frequency.

3. The distance between two successive crests is called the wavelength.

4. In passing from a heavy string to a light string the velocity of a pulse increases.

5. Water waves propagate between the boundary of water and air. This is why they are called surface waves.

6. The reflection of a sound wave from a smooth surface is called echo.

7. A device for measuring the intensity of an earthquake is called seismogram. The intensity is measured with respect to Richter's scale.

8. Whales and dolphins use ultrasound for navigation.

9. Microwaves, light, radio waves are examples of electromagnetic waves.

10. Water waves propagate faster in deep waters.

Problems

1.

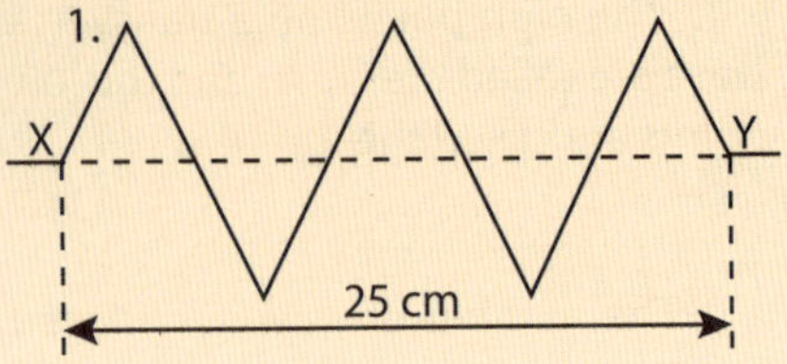

A source is generating the periodic waves given in the figure. The distance between XY is 25 cm and the time spent in moving from X to Y 15 s.

a. Find the period of this wave.

b. Find the frequency of this wave.

c. Find the wavelength of this wave.

d. Find the velocity of this wave.

2. A wave source makes 5 vibrations in one second. The distance between two successive troughs is measured as 12 cm.

Find the velocity of this wave.

3.

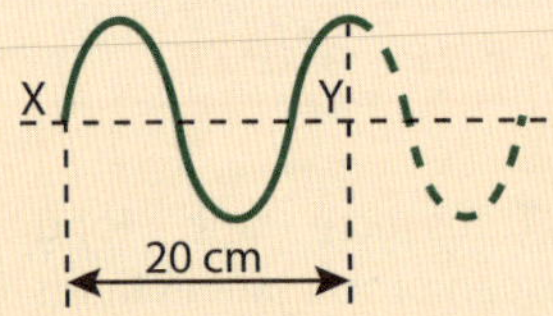

A wave source generates waves with a period of 8 s.

Find the velocity of this wave.

4.

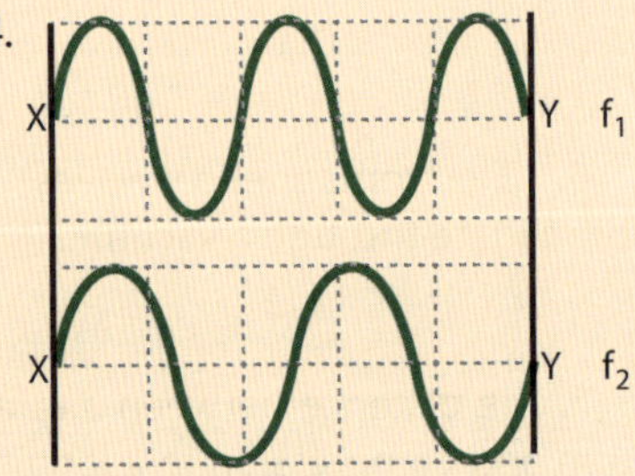

Two identical strings are stretched between points X and Y. Two different wave sources have generated the waves with frequencies f_1 and f_2.

What is the ratio of f_1/f_2?

5.

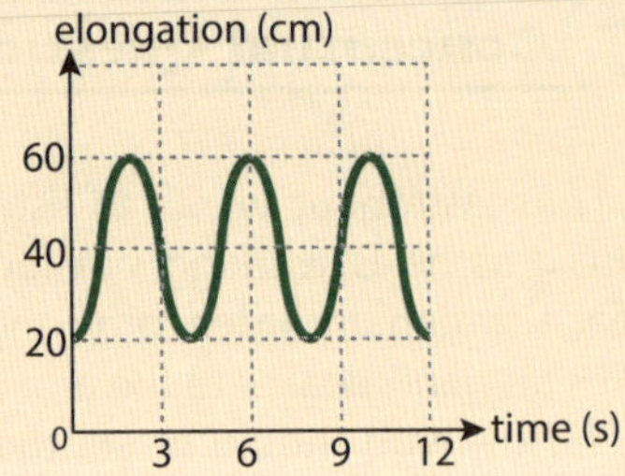

Displacement versus time graph of a periodic oscillation is given in the figure. The velocity of this wave is 6 cm/s.

a. What is the amplitude of the wave?

b. Find the frequency of the wave.

c. Find the period of the wave.

d. Find the wavelength of the wave.

6.

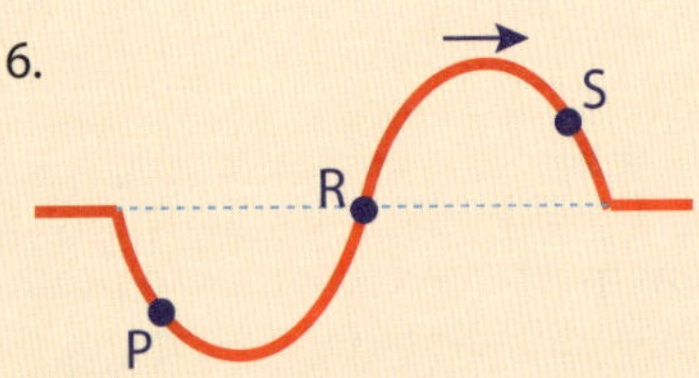

Show the direction of motion of the points P, R, S on the propagating wave given in the figure.

1. a. 6 s; b. 1/6 s^{-1}; c. 10 cm d. 5/3 cm/s /2. 60 cm/s /3. 2cm/s /4. 5/4 /5. a. 20 cm; b. 0.25 s^{-1}; c. 4 s; d. 24cm/s /6. ↑↓↑

7.

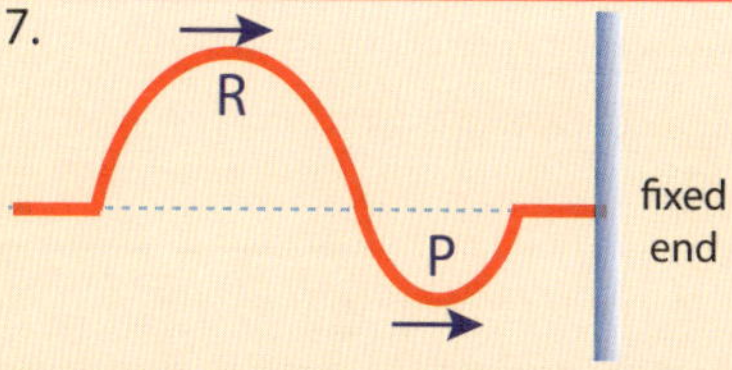

Draw reflections of the pulses P and R from the fixed end.

8.

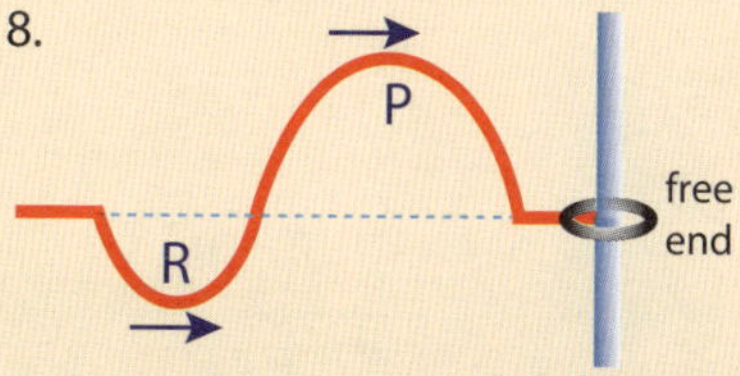

Draw reflections of the pulses P and R from the free end.

9.

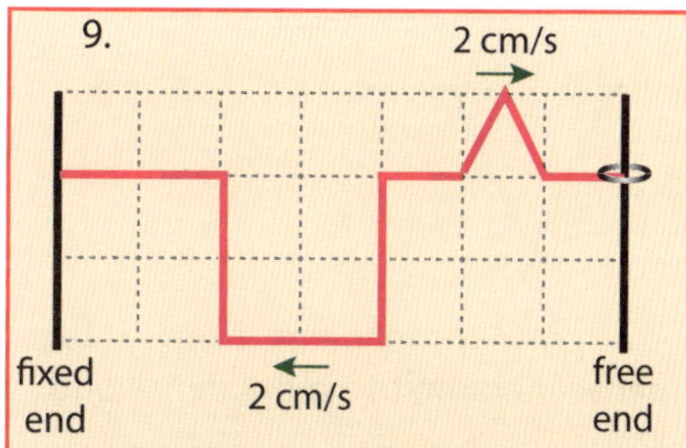

The pulses are moving with a speed of 2 cm/s in the given directions.

How will they look like at the end of 3rd second?

(each square is 1 cm long)

10.

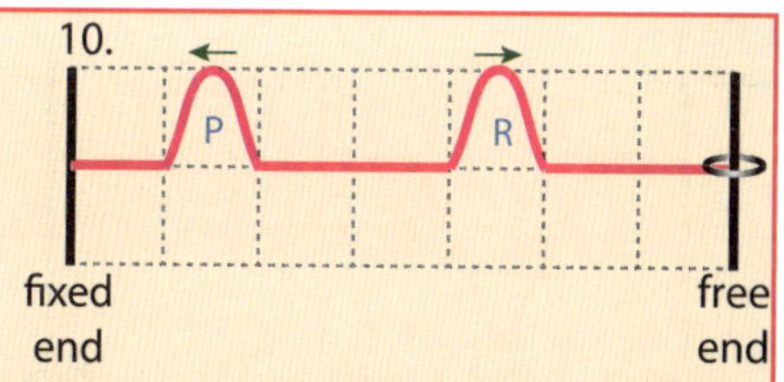

The pulses are moving with a speed of 1 cm/s in the given directions.

a. In how many seconds will the pulses damp each other?

b. How will they look like at the end of 12 s?

(each square is 1 cm long)

7. 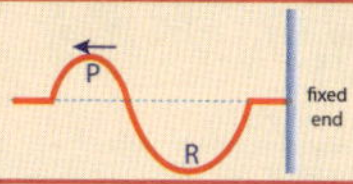

/8. 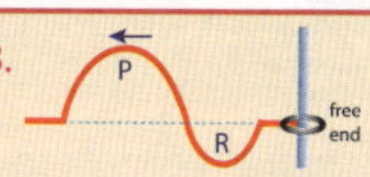

/9. 
/10. a. 5.5 s; b. 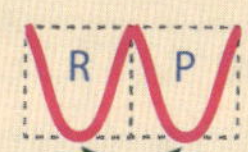

11.

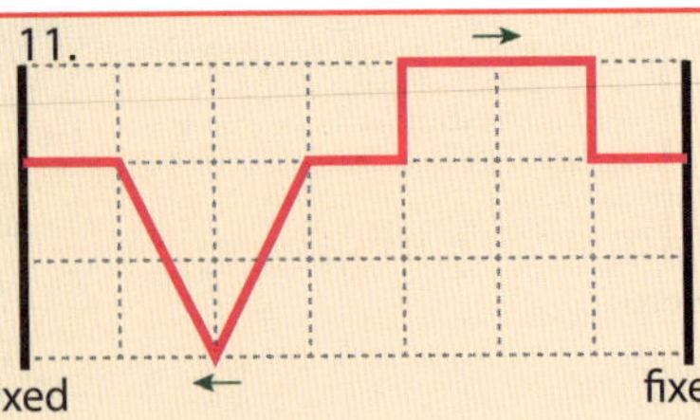

The pulses are moving with a speed of 1 cm/s in the given directions.

a. In how many seconds will the pulses overlap?

b. How will they look like when they overlap?

(each square is 1 cm long)

12.

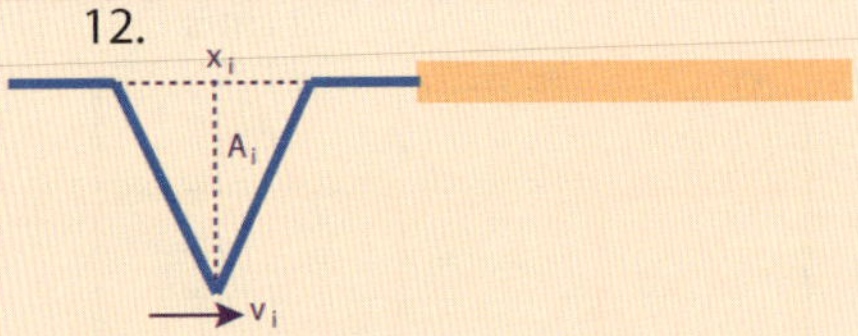

A pulse propagates from a less dense medium to a more dense one.

a. Investigate the shape of reflected and transmitted pulses.

b. Compare the width, amplitude, velocity and frequency of the reflected pulse with the incident pulse. Give your reasonings.

c. Compare the width, amplitude, velocity and frequency of the transmitted pulse with the incident pulse. Give your reasonings.

13.

A pulse propagates from a less dense medium to a more dense one.

a. Investigate the shape of reflected and transmitted pulses.

b. Compare the width, amplitude, velocity and frequency of the reflected pulse with the incident pulse. Give your reasonings.

c. Compare the width, amplitude, velocity and frequency of the transmitted pulse with the incident pulse. Give your reasonings.

11. a. 5.5 s; b. 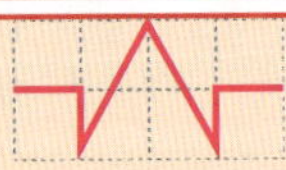 /12. a. 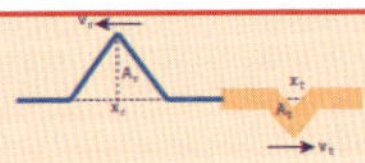 b. $A_i > A_r$; $v_i = v_r$; $f_i = f_r$; $x_i = x_r$; c. $A_i > A_t$; $v_i > v_t$; $f_i = f_t$; $x_i > x_t$

/13. a. 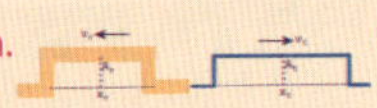 b. $A_i > A_r$; $v_i = v_r$; $f_i = f_r$; $x_i = x_r$ c. $A_i > A_t$; $v_i < v_t$; $f_i = f_t$; $x_i < x_t$

14.

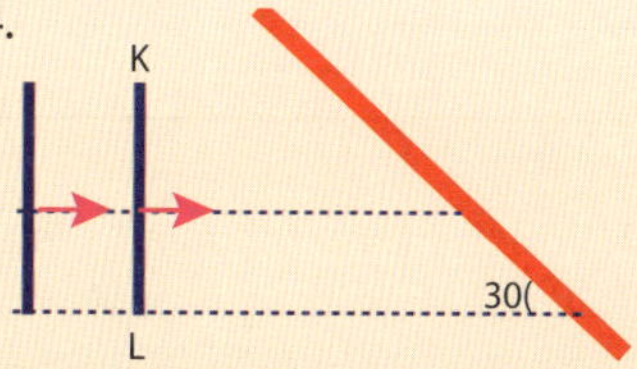

A plane wave propagates towards a linear obstacle as given in the figure.

Draw the reflected wave.

15.

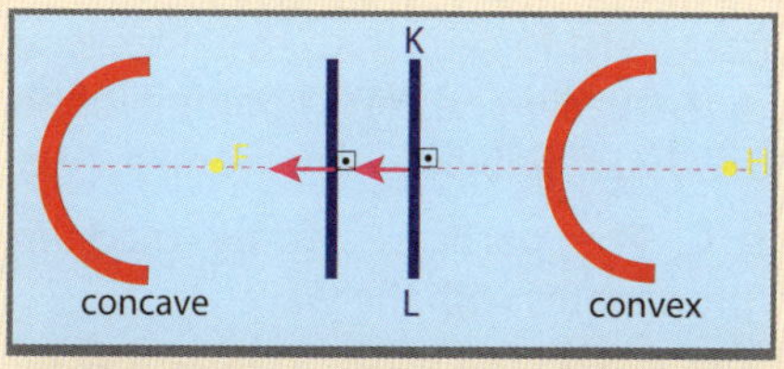

Plane waves produced in between a concave and convex obstacles. The waves are first reflected by the concave obstacle, then from the convex obstacle.

a. Draw the reflected waves from the concave obstacle.

b. Draw the reflected waves from the convex obstacle.

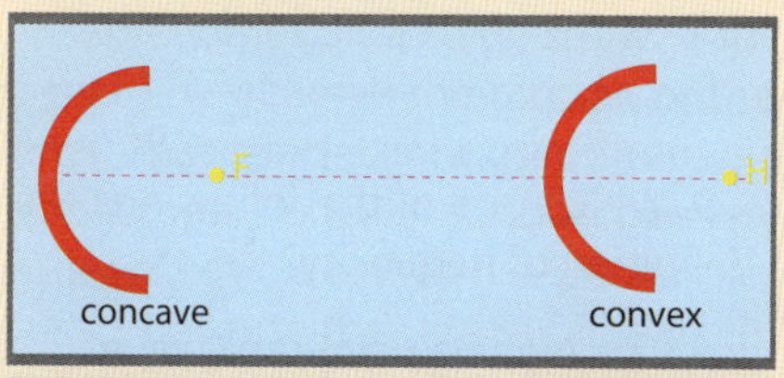

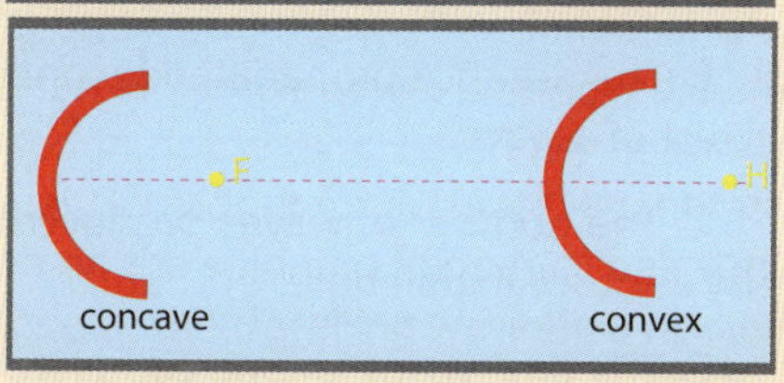

16.

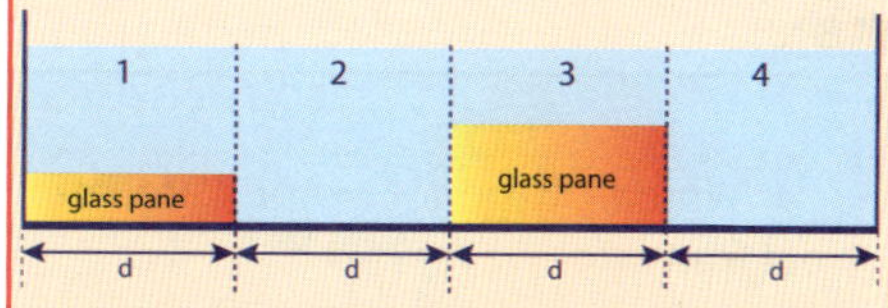

The cross-section of a water tank is given in the figure.

Draw the wave pattern that will be observed on a screen.

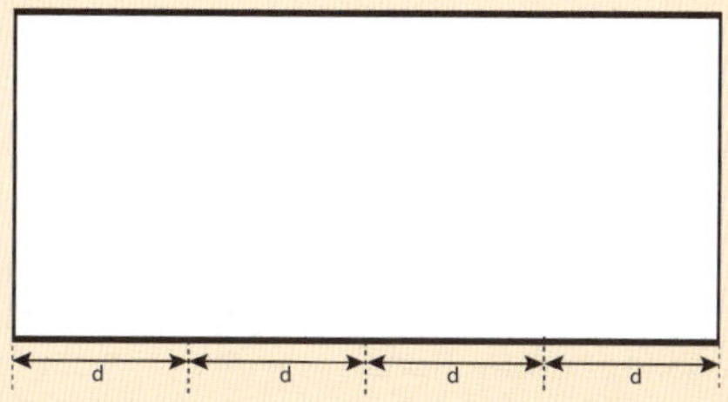

17.

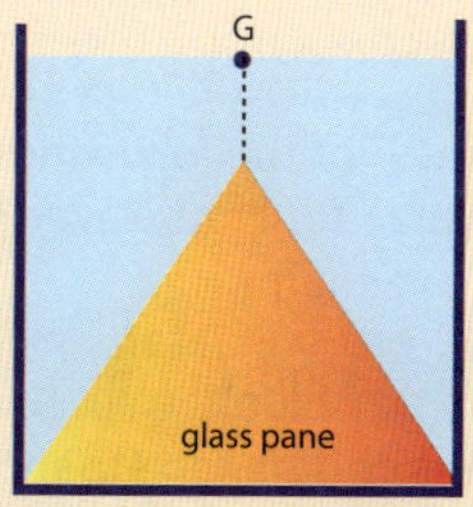

The cross-section of a water tank is given in the figure. G is a point wave generator.

Draw the wave pattern that will be observed on a screen.

18. A plane wave source generates waves with a frequency of 36 s^{-1}. A stroboscope with 6 slits on it is used to observe the waves.

What should be the maximum frequency of the stroboscope so that the waves are observed at rest?

19. A stroboscope with 8 slits on it is used to observe the water waves in a ripple tank. The observer sees that the waves are at rest when the stroboscope rotates with a maximum frequency of 4 s^{-1}.

a. What is the frequency of the waves? b. The distance between two bright lines is measured as 5 cm on the screen.

What is the wavelength of the waves?

14

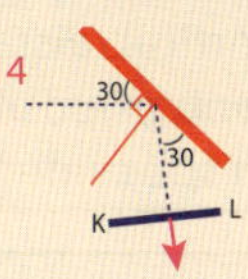

/15. a. Circular waves emerging from F; b. Circular waves as if emerging from a point behind convex obstacle
/16. $\lambda_2 = \lambda_4 > \lambda_1 > \lambda_3$ / 17. Concentric circular waves with increasing distances with each other
/18. 6 s^{-1} /19. a. 32 s^{-1}; b. 160 cm/s

Multiple Choice Questions

1. The wavelength, amplitude and the periods of the sound waves propagating in media P, Q and R are given in the table.

Medium	λ (m)	A	T (s)
P	15	r	0.03
Q	30	2r	0.06
R	45	3r	0.09

What can be said about the properties of these media?

A. P and R may be the same, but Q is different.

B. P and Q may be the same, but R is different.

C. Q and R may be the same, but P is different.

D. All are different.

E. All are the same.

2. Which of the followings is used for measuring the intensity of earthquake waves?

A. Thermograph B. Heliograph

C. Kymograph D. Seismograph

E. Oleograph

3. Which of the followings is false for electromagnetic waves?

A. They are transverse waves

B. They propagate with the speed of light in vacuum.

C. They are produced as a result of pressure fluctuations in air.

D. Radio waves, visible light, micro waves are a few examples of electromagnetic waves.

E. They are produced by the accelerated motion of the electric charges.

4. Which of the followings are correct for water waves?

A. They propagate at the surface of water as longitudinal waves.

B. As they propagate from a deeper region to a shallower one, they speed up.

C. If they come over an obstacle in their path, they stop moving.

D. The water molecules do elliptical motion at the surface.

E. The high points on the wave are called rarefactions.

5. Which of the followings is a constant value for light waves propagating in a vacuum?

A. Wavelength B. Speed

C. Period D. Frequency

E. Energy

6. Which of the followings is false for mechanical waves?

A. Sound and string waves are mechanical waves.

B. The particles move back and forth with the wave.

C. They can propagate in solids, liquids and gases.

D. They can propagate in vacuum.

E. They have relatively slow speeds when compared with electromagnetic waves.

7. Which of the followings is an example of electromagnetic waves?

A. Water waves B. String waves

C. Sound waves D. Radio waves

E. Rayleigh waves

8. Which of the followings is true for the waves?

A. The particles of the medium moves perpendicular to the direction of motion for transverse waves.

B. All kinds of waves need a medium to propagate.

C. The particles of the medium propagates with the wave.

D. The crests and troughs of the water waves move in opposite directions.

E. Waves carry matter and energy with them.

9. Which one of the followings is a definition of frequency?

A. The highest point on a wave.

B. The distance propagated by one crest of a wave.

C. The number of waves produced by a source in unit time.

D. The time spent for one complete wave to be formed.

E. The vertical distance between the equilibrium position and the maximum displacement of a wave.

1.E 2.D 3.C 4.D 5.B 6.D 7.D 8.A 9.C

10. Which of the following statement(s) are true for the speed of a wave on a spring?

I. As the tension increases, the velocity increases.

II. As the linear density of a spring increases, the velocity increases.

III. As the amplitude of the wave produced increases, the velocity increases.

A. Only I B. Only II C. Only III

D. I and II E. II and III

11. 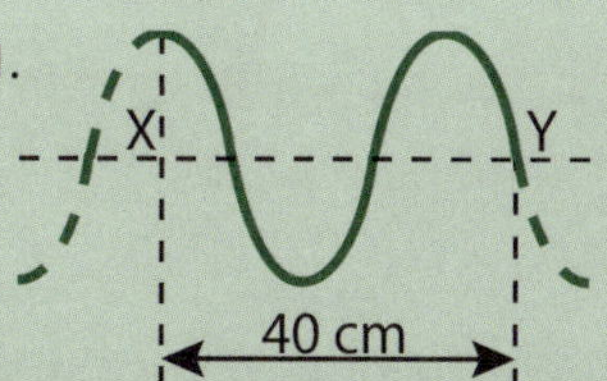

A source of wave generates 6 waves in 3 seconds. The distance between X and Y is 40 cm for the produced waves.

Find the velocity of the waves.

A. 4 B. 8 C. 16 D. 32 E. 64

12. 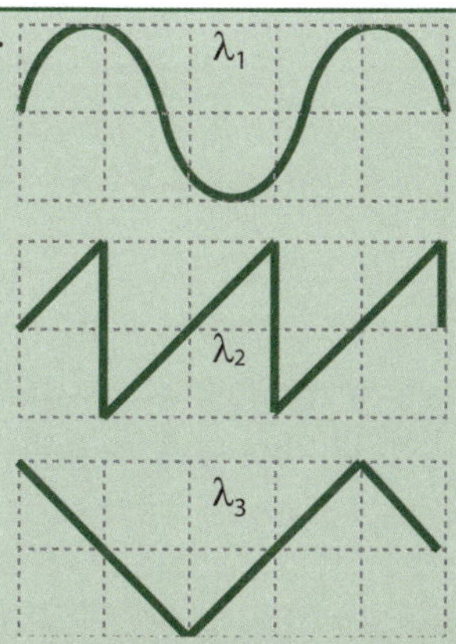

What is the relationship between the wavelengths of the waves given in the figure?

A. $\lambda_1 < \lambda_2 < \lambda_3$ B. $\lambda_3 < \lambda_1 < \lambda_2$

C. $\lambda_1 < \lambda_3 < \lambda_2$ D. $\lambda_3 < \lambda_2 < \lambda_1$

E. $\lambda_2 < \lambda_1 < \lambda_3$

13. 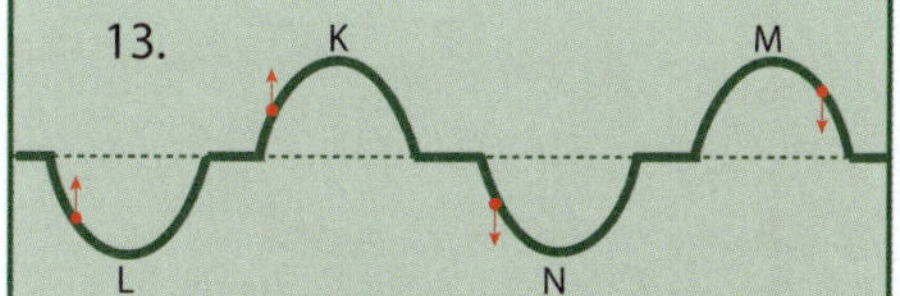

Four pulses are produced on a stretched string. The direction of motion of some points on the pulses are given in the figure.

Which pulses propagate in the same direction?

A. K, M and N B. K, L and N

C. M, L and N D. M and L

E. L and N

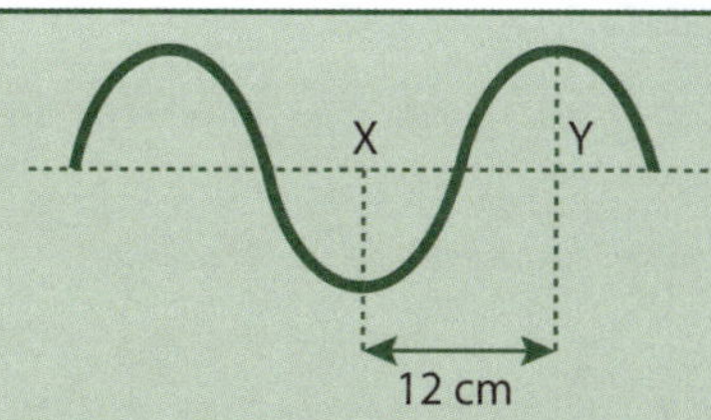

Answer questions 14 and 15 with respect to the given figure.

14. The distance between points XY is 12 cm.

Find the wavelength of the waves in cm.

A. 30 B. 24 C. 18 D. 12 E. 6

15. The wave generator produces 3 crests in every 8 seconds.

Find the velocity of the waves.

A. 3 B. 4 C. 6 D. 8 E. 12

10. D 11. E 12. E 13. A 14. B 15. C

16.

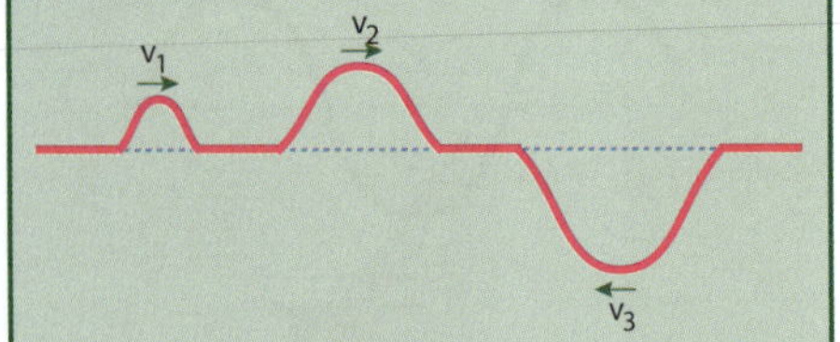

Three pulses are formed on a stretched spring.

What is the relationship between the velocities of the pulses produced?

A. $v_1 < v_2 < v_3$ B. $v_1 = v_2 = v_3$

C. $v_3 < v_2 < v_1$ D. $v_2 < v_1 < v_3$

E. $v_1 = v_2 < v_3$

17.

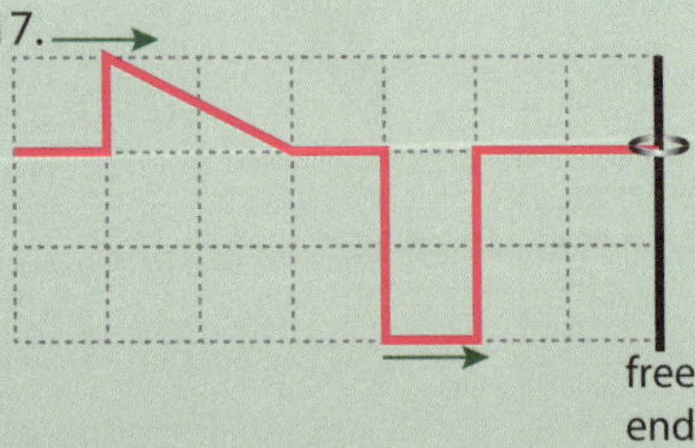

Which of the followings show the reflection of the pulses from the free end?

A.
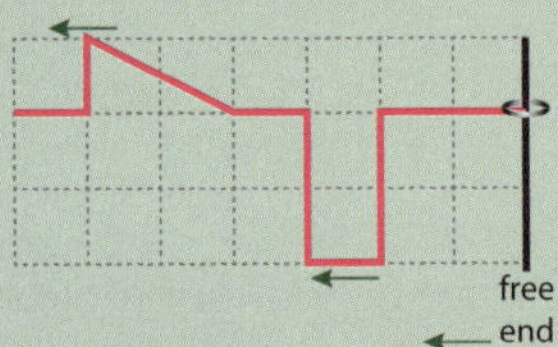

B.
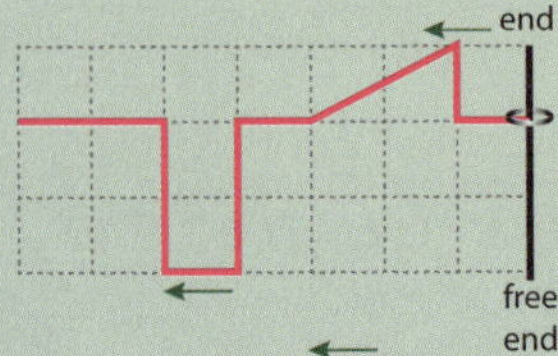

C.
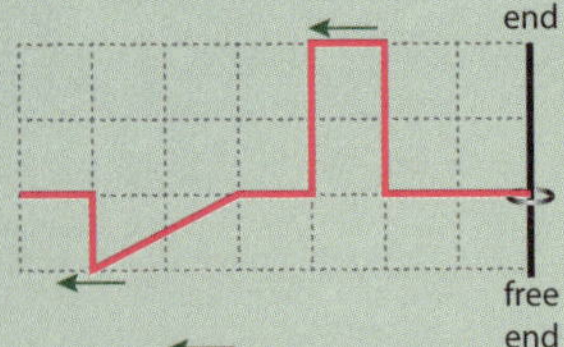

D.
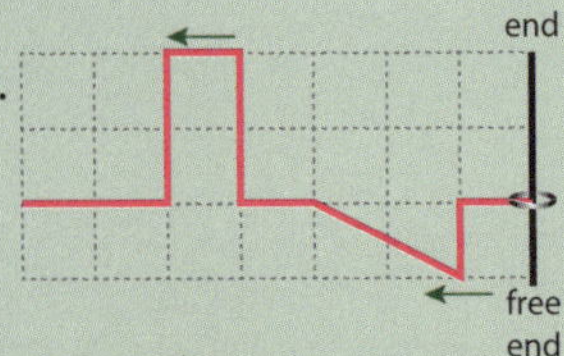

E.
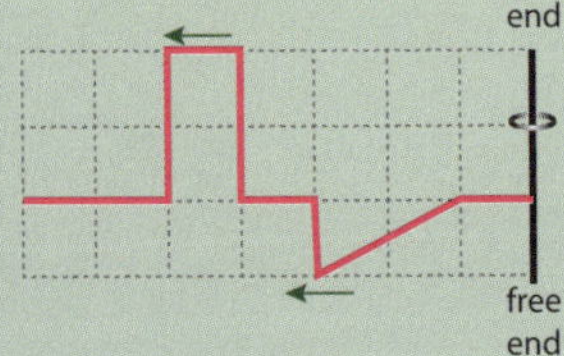

18.

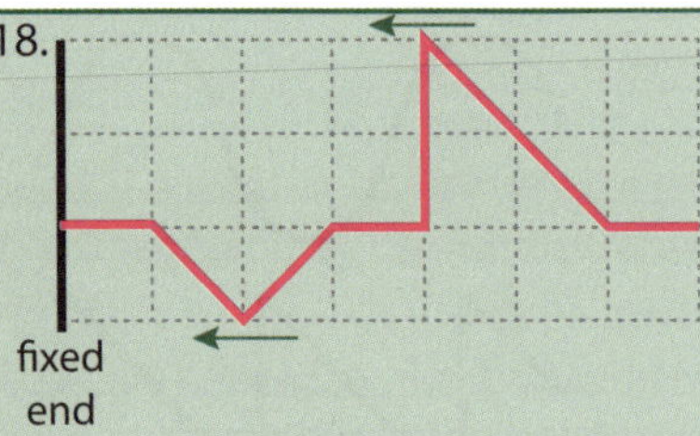

Which of the followings show the reflection of the pulse from the fixed end?

A.
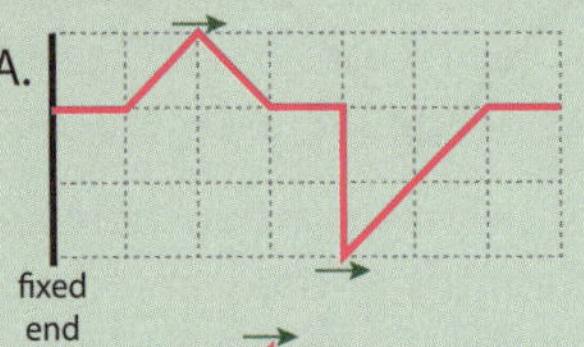

B.
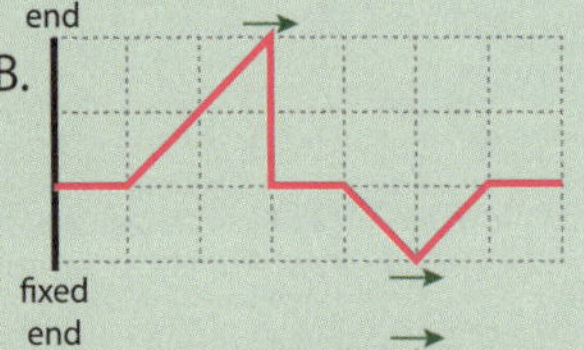

C.
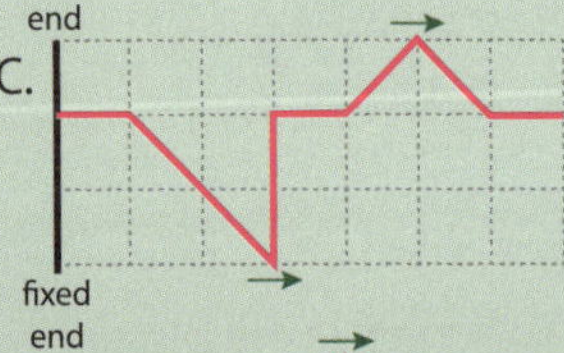

D.
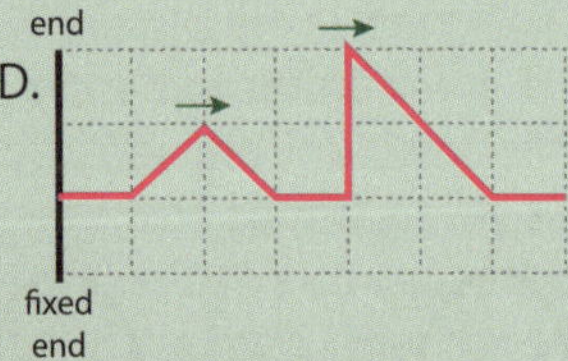

E.
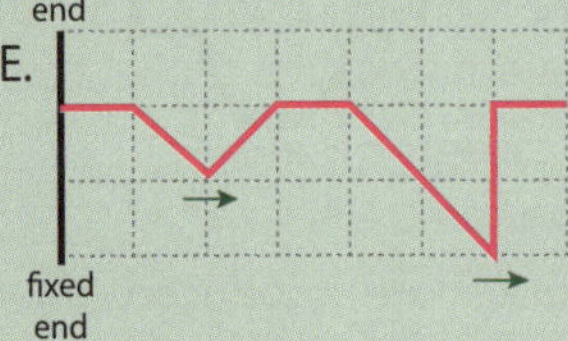

19. A child shouts to his friend on the other side of a valley. He hears his echo 0.4 s later.

Assuming that the average speed of sound in air is 340 m/s, how many meters is the distance between the child and the wall?

A. 34 B. 68 C. 102 D. 136 E. 170

20.

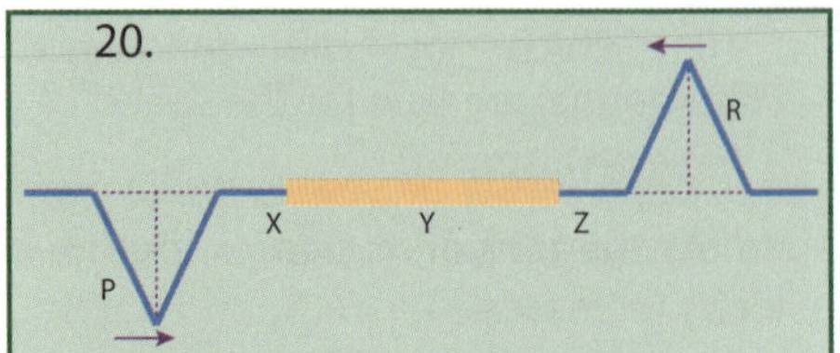

X and Z are identical strings. They are attached to the ends of a heavier string, Y. Pulse P is formed on X and pulse R is formed on Y. P and R are moving towards each other.

Which of the followings show the resultant pulse on string Y?

A.

B.

C.

D.
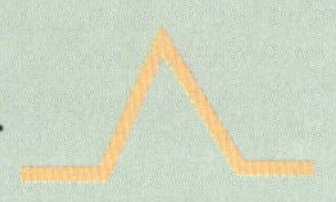

E.
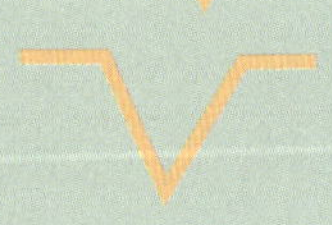

21.

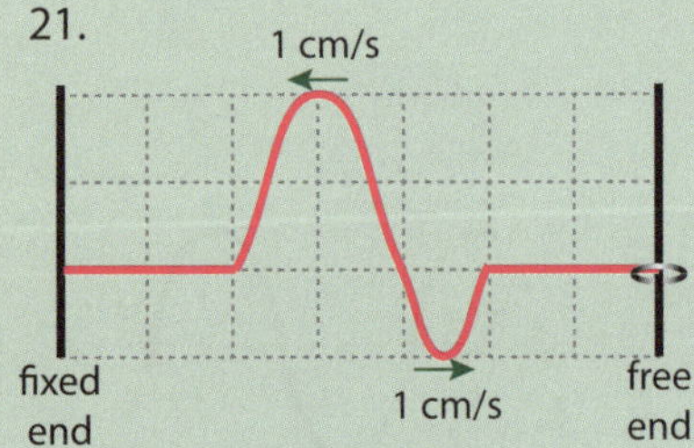

The pulses are moving with a velocity of 1 cm/s in the given directions.

Which of the followings show the resultant pulse formed when the pulses first overlap?

(each square is 1 cm long)

A.
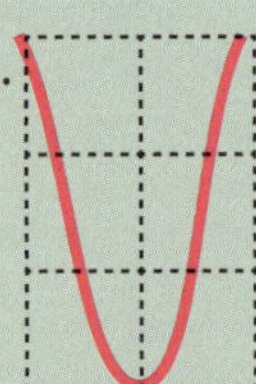
B.
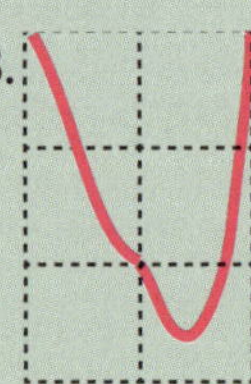

C.
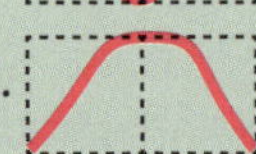
D.
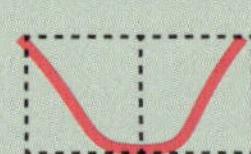

E.
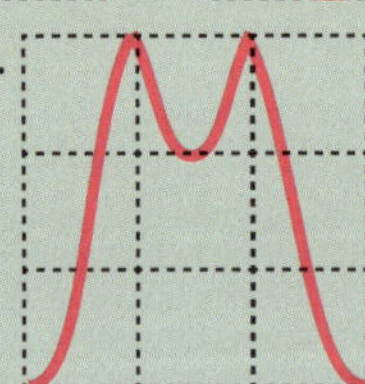

22.

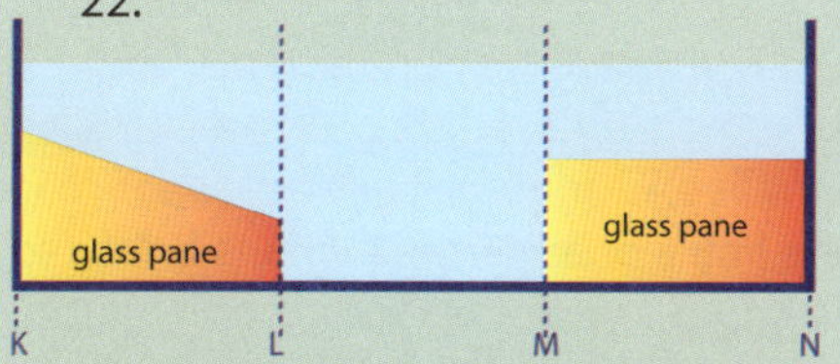

Cross section of a ripple tank is given in the figure.

I. The wavelength decreases between KL.

II. The wavelength is largest between LM.

III. The waves have the highest frequency between MN.

Which of the statement(s) is true?

A. Only I B. Only II C. Only III

D. I and II E. II and III

23.

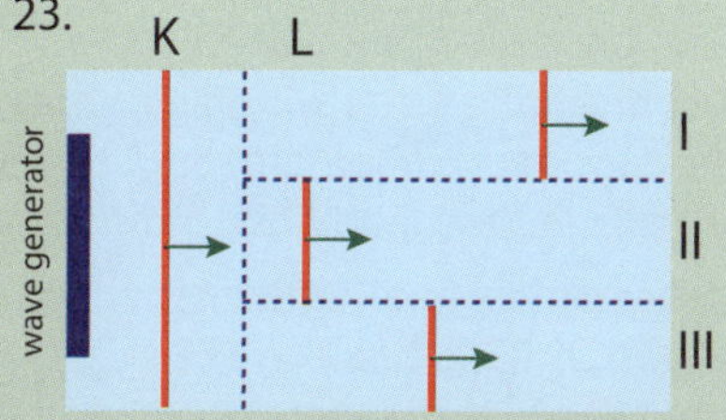

A plane wave is produced in medium K at t=0. Then the wave propagated to medium L. The final position of the waves at t= 2 s is given in the figure.

Which can be said about the depth of regions I, II and III in medium L.

	I	II	II
A.	Deep	Shallow	Very deep
B.	Deep	Very deep	Shallow
C.	Shallow	Deep	Very deep
D.	Shallow	Very deep	Deep
E.	Very deep	Shallow	Deep

24.

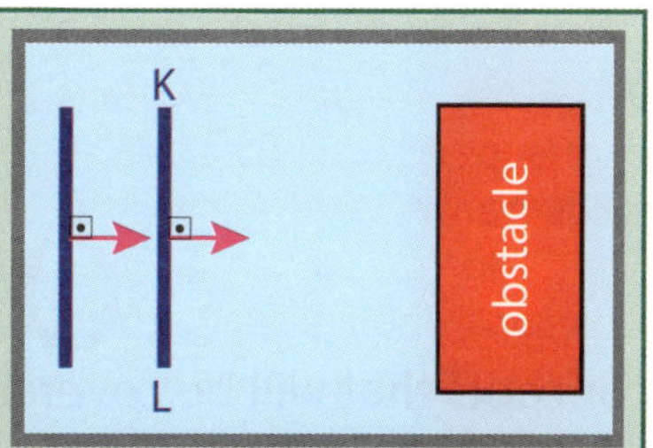

Figure 1

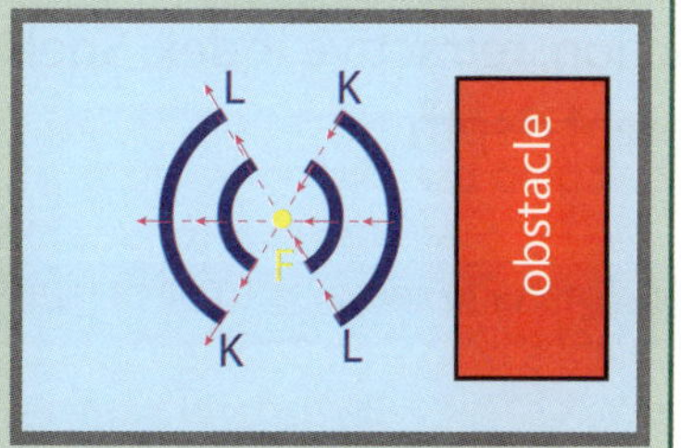

Figure 2

Plane waves are sent to an obstacle. in figure 1. Their reflections are shown in figure 2.

Which can be the shape of the obstacle?

A. Planar B. Triangular

C. Concave D. Trapezoid

E. Convex

25.

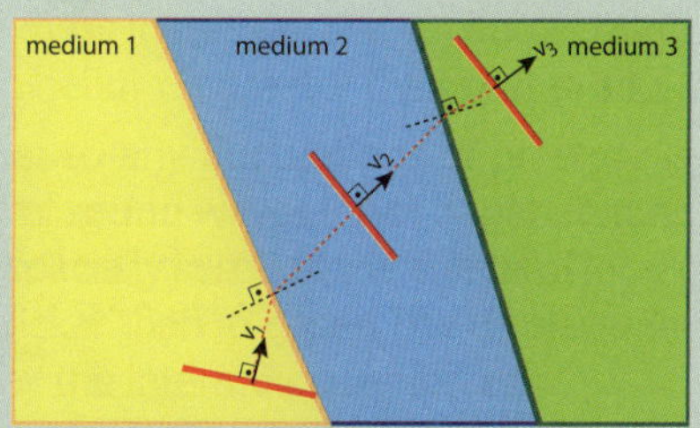

The propagation direction of a plane wave in three different depths of water is given in the figure.

Which is the relationship between the depths of the media?

A. $v_3 < v_2 < v_1$ B. $v_1 = v_2 = v_3$

C. $v_1 < v_2 < v_3$ D. $v_2 < v_1 < v_3$

E. $v_1 = v_2 < v_3$

26.

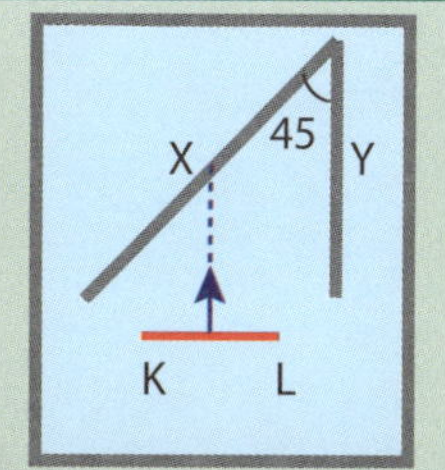

Planar obstacles X and Y are placed at 45° to each other as given in the figure.

How will the plane wave KL be reflected from obstacle Y?

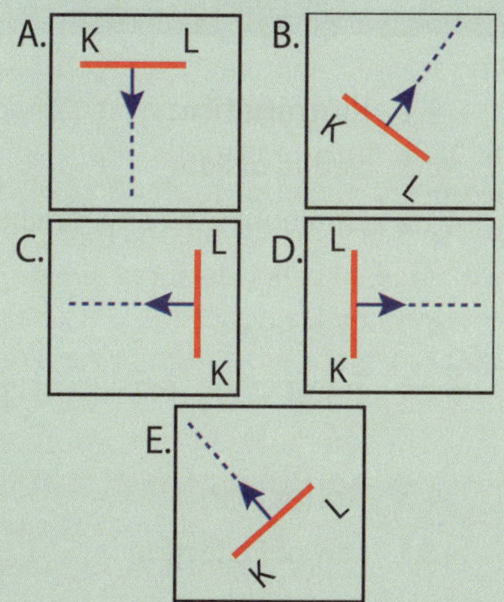

27. A stroboscope with 10 slits on it is used to observe the water waves in a ripple tank. The observer sees that the waves are at rest when the stroboscope rotates with a frequency of 3 s^{-1}. The distance between two bright lines is measured as 7 cm on the screen.

What is the wavelength of the waves in terms of m/s?

A. 0.9 B. 1.2 C. 1.5 D. 1.8 E. 2.1

UNIT 4 OPTICS

The concepts that will be covered in this unit are:

Luminous intensity, luminous flux, Illumination, umbra, penumbra, reflection, refraction, refractive index, Snell's law, total internal reflection, critical angle, apparent depth.

CONTENTS		PAGE

Diamonds are crystals of pure carbon that form under crushing pressures and intense heat. They mostly form in the Earth's mantle, the layer beneath the crust or surface layer, at a depth of about 150 km. Diamond is the hardest material known natural material on Earth. Diamond conducts heat 5 times faster than copper, which is why it feels cold to the touch. Diamond can be an insulator or a conductor, so it can either allow electricity to pass through or block it. Diamond can survive intact in environments that will destroy other materials. It can survive severe physical, chemical and radioactive forces. If you place diamonds inside the human body they will not trigger an immune response. Pure diamond is the most transparent material known. It reflects visible light, ultra-violet light, and infrared light, and all the frequencies in between.

But how and why a diamond shines with the colors of the rainbow?
The answer comes with the concepts that you will learn throughout this unit: Reflection, refraction and finally total internal reflection.

OPTICS

4.1. Illumination

4.1.1. Nature of Light

Light has always been a great phenomenon to the mankind. Why is the sky blue? How can the light from the sun reach to our eyes? How do we see? Why is the deeper regions of seas appear in a darker blue whereas the shallow regions appear turquoise? Optics, the study of behavior of light in different situations tries to find answers to such questions. Today we know the answer to these questions, but the path lead to these answers was a long and confusing one. The Greek philosophers Phytagoras, Democritus, Plato, Aristotle and others developed several theories of light. But the greatest accomplishments about the nature of light were received in the seventeenth century.

As we had mentioned in the previous chapter, light is an electromagnetic wave. Although wave perspective is successful in explaining many behavior of light such as reflection from shiny surfaces or passage through transparent media, considering light as a wave is not enough to explain some other phenomenons such as the photoelectric effect.

Newton believed that light was consisting of corpuscles which were very small, light and fast particles. The corpuscles were travelling in straight lines. Around the same times some other scientists like Robert Hooke, Christian Huygens were strongly opposing to Newton and were introducing light as a wave. The acceptance of wave theory took some time in the world of physics for some several reasons. Water and sound waves were known at the time but they were in need of a medium to propagate. But the light was propagating through the vacuum between the sun and the Earth. Moreover water waves were observed to *bend* around the obstacles (diffraction); but light waves were not bending around the corners. Now we know that light does bend around the corners of obstacles. The wavelength of light waves is very small so it is very hard to observe this phenomenon. Although Francesco Grimaldi provided experimental evidence, the wave theory was rejected by most scientists for more than one century. This was mostly because of the great reputation of Newton as a scientist.

Water waves passing through a small slit bending behind the obstacles

A bright ring of light surrounding the moon created by diffraction of light around cloud droplets

In 1801, Young performed his famous double slit experiment and provided concrete data about the wave nature of light. In the coming days Jean Foucault showed that speed of light in liquids was smaller than the speed of light in air. According to corpuscle theory, the speed of light was supposed to be higher in glasses and liquids than in air. In 1873, James Clerk Maxwell introduced that light is an high frequency electromagnetic wave. His theory predicted the speed of light in vacuum to be equal to 3×10^8 m/s. Eight years after his death, in 1887, Heinrich Rudolph Hertz, provided experimental confirmation to Maxwell's theory by producing and detecting electromagnetic waves. With the electromagnetic waves, he was able to perform reflection, refraction and all other properties of waves.

But, still there were difficulties in explaining some phenomenon related with light by using the wave theory. Hertz discovered the photoelectric effect of light, the ejection of electrons from a metal surface exposed to light. His experiments showed that the kinetic energy of an ejected electron was independent of the intensity of light. Einstein proposed an explanation to this phenomenon in 1905 by using the concept of quantization developed by Max Planck in 1900. According to quantization approximation, light occurred as energy packages called the photons. Einstein awarded with the Nobel prize in 1921, for his discovery on the law of photoelectric effect.

Just like chameleons changes its colors to adapt themselves to the environment, light changes its behavior or character in different situations. Under some circumstances light acts like particles travelling in straight lines. If the circumstances changes, light shows wave-like behavior. Another change in circumstances causes light to display characteristics of both wave-like or particle-like characteristics.

Wave model of light is most widely applicable model of light. Electro-optical devices, LASERs, most of the technologies of 21st century are best explained by using the wave model. The study of light as a wave is called wave optics.

Ray model of light assumes that light travels in straight lines called the light rays. According to Newton, light rays are trajectories of particle like corpuscles of light. The behavior of light in prisms, lenses, mirrors, optical devices such as microscopes are best explained by ray model of light. Ray model of light is the basis of geometrical optics.

Modern technology of our day is reliant on quantum physics. In the quantum world, light behaves like neither a wave nor a particle. Light consists of photons that have both wave-like and particle-like properties. This is the *photon model* of light.

4.1.2. Luminous Intensity, Luminous Flux and Illumination

If you are not seeing the details clearly, you bring the object closer to a light source. As you get away from the light source, the *brightness* of the surface gets less. Although this is something natural for you, what is the physics behind this behavior?

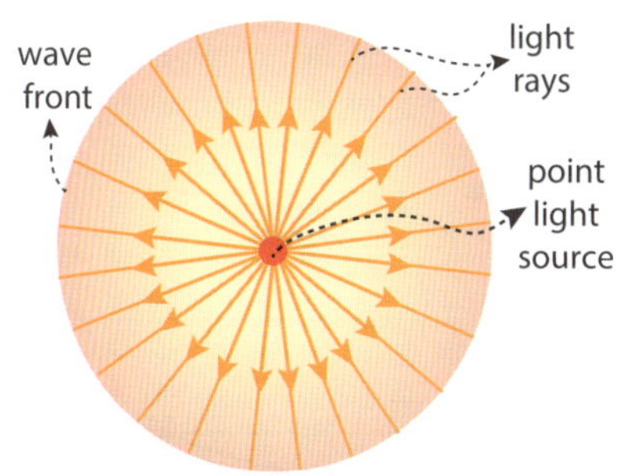

Wave fronts emerging from a point light source are spherical. The direction of propagation of the wavefront can also be represented by light rays.

The energy carried by a wave is directly proportional to the amplitude of the wave. But scientists usually prefer to use luminous intensity to define the energy carried by a light wave. Luminous intensity can be referred as the brightness of a light source. It is measured as the rate at which light energy is delivered to a unit of surface, or energy per unit time per unit area. An easier modelling of luminous intensity can also be made by using the number of photons or light rays leaving a light source.

The SI unit of luminous intensity is candela (cd), which is a basic unit of SI.

The rate of total visible light that a light source emits is called the luminous flux of a light source. Its unit is lumen (lm).

Symbol	Quantity	Unit
I	luminous intensity	candela (cd)
Φ	luminous flux	lumen (lm)

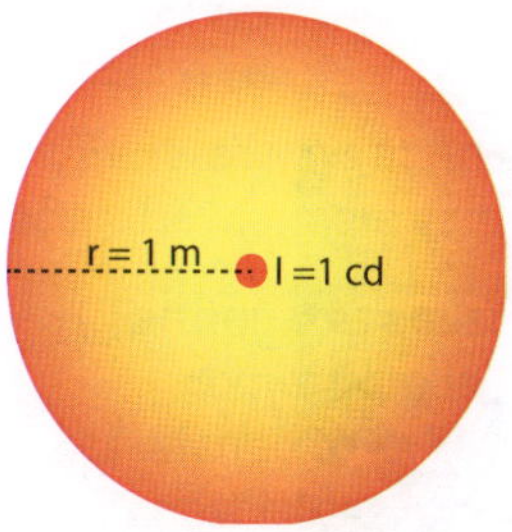

A point light source with a luminous intensity of 1 cd is at the center of a sphere with a radius of 1 meter.

The surface area of the sphere is equal to:

$4\pi r^2 = 4\pi . 1^2 = 4\pi$

The luminous flux of the light source at the center of this sphere is:

$4\pi . 1 = 4\pi$

If the luminous intensity of the light source is I, then the luminous flux is:

$\Phi = 4\pi . I$

Just like the quills of a hedgehog, the rays emerge radially outwards from a point light source. Now, let's concentrate on the figure aside: α is the solid angle we are considering for the portion of the light rays emerging from the light source, S. As we get away from the source, the distances between the light rays increases as a result of spherical symmetry. No matter how far we are away from the source, as long as the angle is kept the same, the number of light rays in the region will remain the same. However if we concentrate on the number of light rays passing through a unit area, the number of light rays will decrease as we get away from the source. For the unit area represented by A in the figure, as the unit area propagates outwards from the source S, the number of light rays passing through A decreases. If there is no loss of energy to the medium, then the energy carried by the waves decreases as the wave front gets away from the source. Thus the brightness decreases.

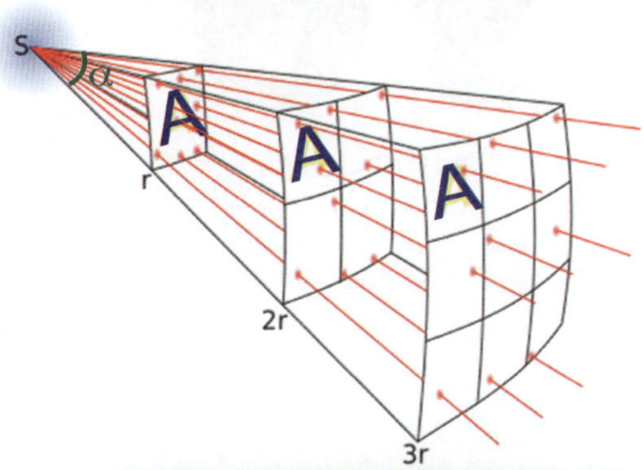

The number of light rays gives us an idea about the illumination of the surface. As the number of light rays passing through the area increases, the illumination on the surface also increases. This is why as you get closer to a light source the surface gets *brighter* in some sense. The luminous flux that reaches to a unit area of a surface is called illumination. Its unit is lm/m^2 which is equal to lux(lx).

$$E = \frac{\Phi}{A} = \frac{4\pi I}{4\pi r^2} = \frac{I}{r^2}$$

Symbol	Quantity	Unit
I	luminous intensity	candela (cd)
r	distance to the light source	meter (m)
E	illumination	lux (lx)

Here, we reach to another inverse square law in physics. The illumination on a surface is directly proportional to the luminous intensity of the light source and inversely proportional to the square of the distance between the light source and the considered surface area.

The illumination upon a surface also varies with the orientation of the surface with respect to the light source. Imagine a parallel beam of light rays incident upon a rectangular screen. How will you have the maximum illumination on the screen? When it is at right angles to the beam of light or when it is parallel to it? What if it makes an angle to the beam of light?

Let's imagine the screen as an empty frame where light rays can easily pass through. When the screen is oriented perpendicular to the beam of light, all rays of light can easily pass through it. But when the frame is parallel to the beam, none of the light rays will pass through the frame. Rather they will sweep the surface of the screen. For all other orientations of the screen which is from parallel to perpendicular to the light waves, the number of light rays will be less than the first case (perpendicular to the beam) and more than the second case (parallel to the beam)

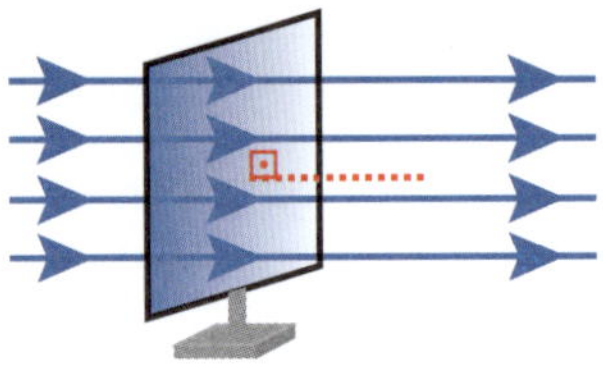
Screen perpendicular to the light beam

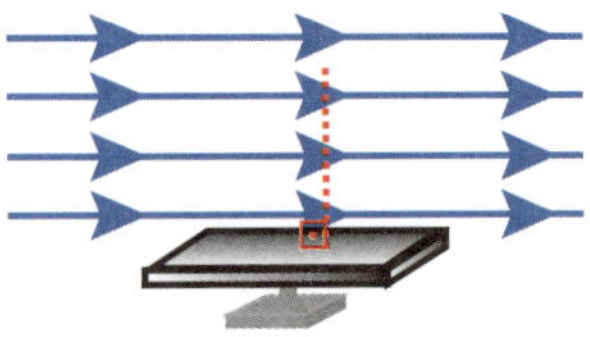
Screen parallel to the light beam

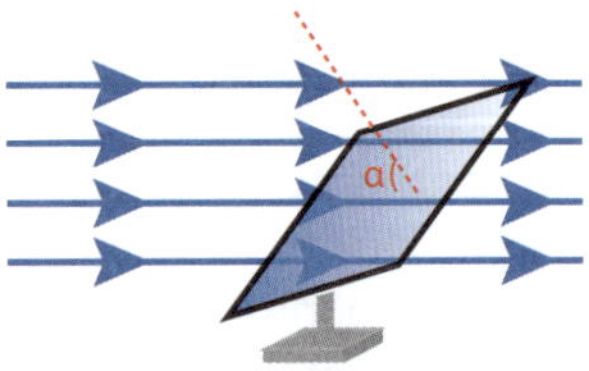

Screen at an angle to the light beam

If the light rays and the surface are perpendicular to each other, than the illumination is maximum, whereas it is minimum if they are parallel to each other.

$0 < \alpha < 90$ then $\Phi_{max} > \Phi > 0$

The luminous intensity of an unknown light source can be evaluated by comparing it with a light source of known intensity.

A white sheet of paper with a greasy spot at the center is placed between a known and unknown light source. The position of the unknown light source is adjusted so that the paper and the grease seem equally illuminated when viewed from both sides. Under this condition the illuminations of both sides are equal. Then:

$$E_1 = E_2$$

$$\frac{I_1}{r_1^2} = \frac{I_2}{r_2^2}$$

Such a set up is known as Bunsen's Photometer.

Digital photometer (or lightmeter)

Today digital photometers are widely used by photographers to measure the intensity of light.

Any object that gives out light is called a light source or a luminous object (lumen means light in Latin). Light sources can either be *natural* such as sun and the stars; or *artificial* such as lamps. A luminous object radiates light in all directions. Objects other than the luminous objects are non luminous objects. They become visible only when light from a source falls upon them and the light falling upon them reaches to our eyes. Trees, your book, doors are examples of non luminous objects.

For some cases, a non luminous object may became luminous. The filament of a light bulb is non luminous as long as no electric current passes through it. But once the electric current begins to flow, it begins to give out light beginning from a dim red to bright yellow and finally to a white glow. When solids are heated above 800 °C (and as they are melting), they begin to give out light and become a source of light. They are known as incandescent light sources. A light bulb is an incandescent light source. It becomes very hot as it glows, and the color and brightness of the light changes with the increasing current.

Neon lights or fluorescent lamps do give light when current passes through them. But they remain quite cool as the current passes through them. If the current is increased for a fluorescent light source, the brightness increases but the color of the light doesn't change. The color of the light is dependent on the kind of gas in the tube.

A great deal of light reaches our eyes from non luminous objects.

4.2. Shadows

4.2.1 Transparent, Translucent and Opaque Objects

Substances do not always behave same to the light falling upon them. Materials such as water and glass lets all the visible light to pass through them in straight lines. A substance through which light can easily pass is called transparent. Depending on its thickness or depth; water and glass are transparent to visible light.

Even for the same kind of substance, the response to the falling light changes according to the frequency of light falling upon it. When a light wave is incident upon a material, the electrons in the material are set into vibration. The natural frequency of electrons of glass corresponds to the frequency of light in ultraviolet (UV) region of the electromagnetic spectrum. When UV rays fall upon a piece of glass, resonance occurs within the glass, and molecules vibrate with a large amplitude. Glass is not transparent to UV rays.

For some cases, not all of the light falling upon a piece of glass is transmitted through it. Some of the light is scattered so that the objects beyond cannot be seen clearly. Frosted glass or a thicker piece of glass do not let all the light pass through it. Such substances are called translucent substances. Clear images cannot be observed through translucent materials.

Opaque materials do not let any of the light to pass through them. All light rays are blocked by the material. They absorb the light falling upon them. The energy of light waves is given to the molecules of the material. The kinetic energy of the molecules thus the internal energy of the substances increases. This is why, when light falls upon an object it gets warmer. Most things around us are opaque objects. Walls of your house, door of your classroom are just a few examples of opaque objects.

Earth's atmosphere is transparent to visible light and some part of the infrared rays (IR). It is quite opaque to high frequency UV rays. Sunburns are a result of small amount of UV rays passing through the atmosphere. Clouds are semi-transparent to UV rays. This is why you may get a sun burnt in cloudy weather.

When light is blocked by the opaque objects, back side of the objects cannot be illuminated. The black region formed behind an illuminated opaque object is called a shadow. Shadows usually look alike to the objects that are forming them.

The garbage formations in the aside photos don't resemble anything meaningful. But once light is casted upon them, from a certain point of projection, Michael Jackson as he is dancing and the marvelous sailing ship magically appears.

Shadows are a result of straight line propagation of light. The observation of sharp-edged shadows led Newton to propose the corpuscles of light. We will use the ray model of light to explain the formation of shadows.

Observation of a sharp or a fuzzy shadow depends on the size and also distance to the light source. Most objects illuminated by sunlight are closer to the ground than we are to their shadows. Aside pictures of shadow are viewed from half, one time and two times the object distance. As you get away from the shadow, the fuzziness of the shadow decreases, and it becomes sharper.

Glass is transparent to visible light

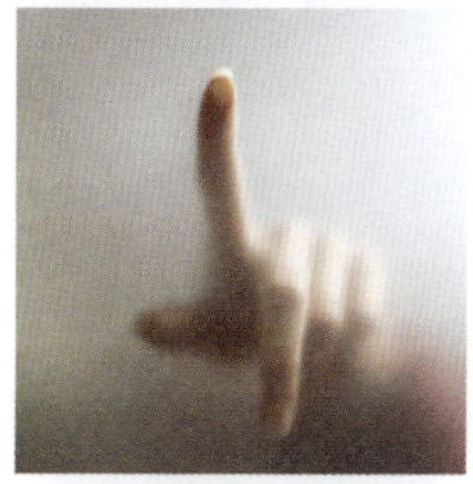

Frostedglass

Translucent white concrete; allows light to pass through and maintains the strength of normal concrete

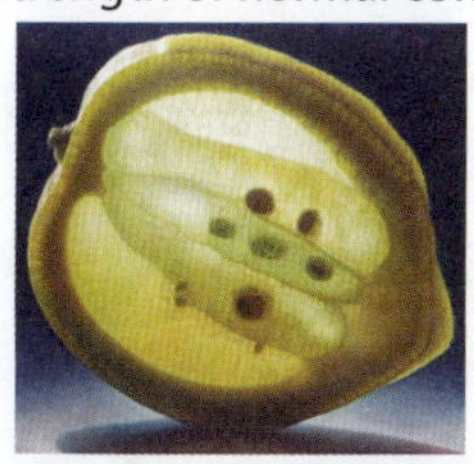

Even a thin slice of lemon can be translucent

As the number of plastic glasses increase, they become opaque rather than being translucent.

"Those who are running after shadows cannot leave the shadow play."

"Shadows are there since there is light."

Although it is hypothetical to talk about a point source of light, a very distant and small light bulb can be assumed as a point source of light. If an opaque object is placed between a point source of light and a screen, a sharp shadow can be observed on the screen. Although the light source radiates light rays in all directions, we will use only the *borderline* rays to construct the shadow on the screen.

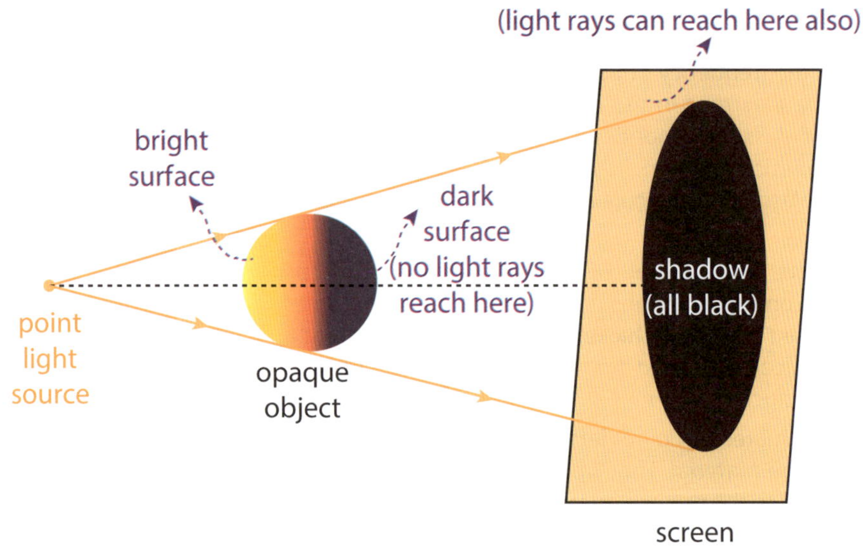

A large source of light can be considered as combination of millions of point light sources. If we draw the shadow that will be formed by a large source of light:

"Don't let your shadow go ahead of you."

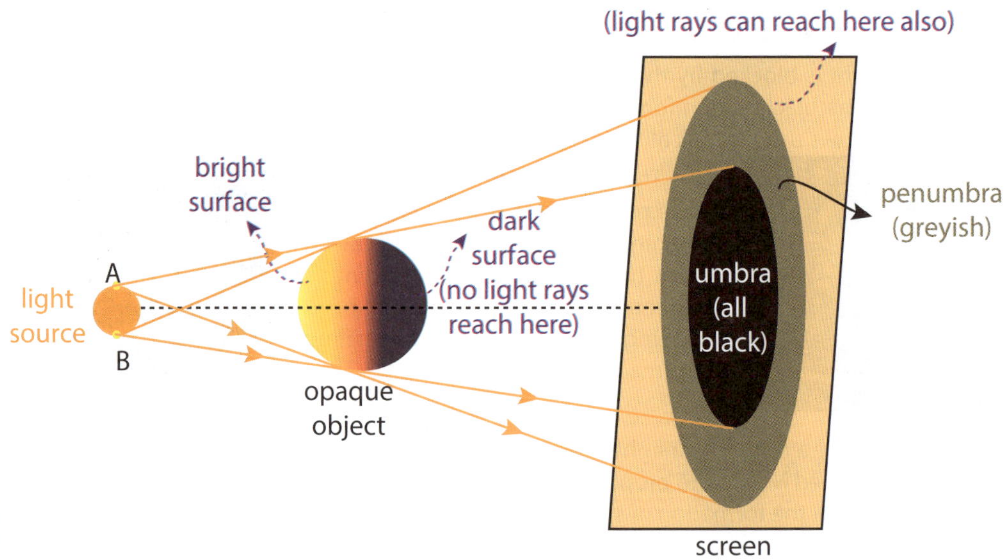

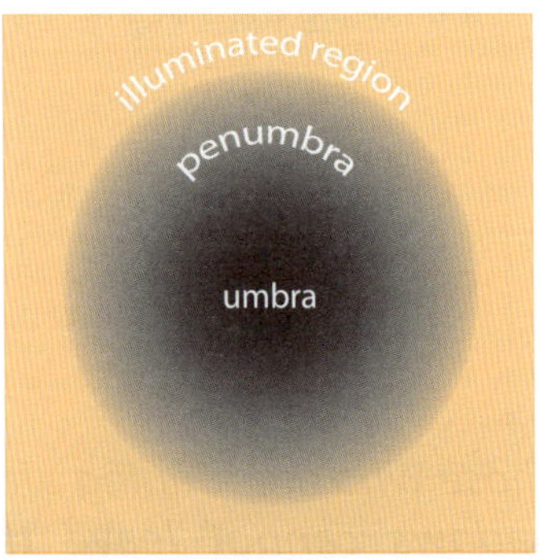

Front view of the screen

A and B are the top and bottom points of the light source. It is easier to draw the shadow of these two point light sources. Then we can form the shadow formed on the screen by joining them. The shadow formed on the screen won't be sharp and completely black as it is with the point source of light. The center of the shadow is black and called the umbra. The shadow becomes greyish on the screen. This is because some light rays can reach to upper part of the shadow from point A, and some can reach to below part of the shadow from point B. This greyish shadow is called the penumbra. Penumbra becomes less and less noticeable outwards from the center. The rest of the screen is illuminated with the light source and appears bright.

Physics Extra

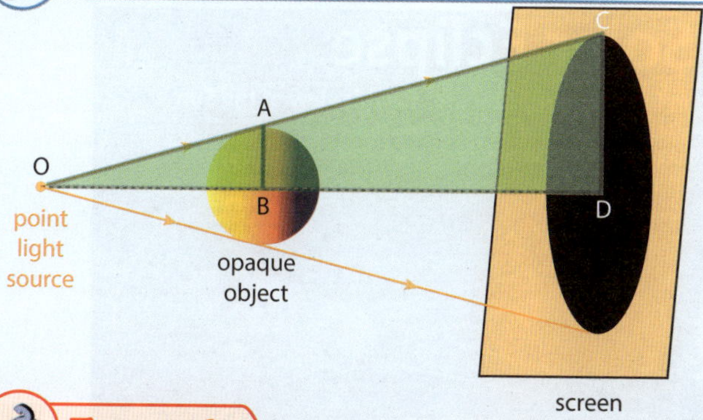

The size of the shadow can be found by using the similarities of triangles. The triangles AOB and COD are similar to each other in the figure. Thus:

$$\Delta AOB \approx \Delta COD$$

$$\frac{OA}{OC} = \frac{OB}{OD} = \frac{AB}{CD}$$

The size of the umbra and penumbra can be calculated by using the similarities of triangles again.

Example

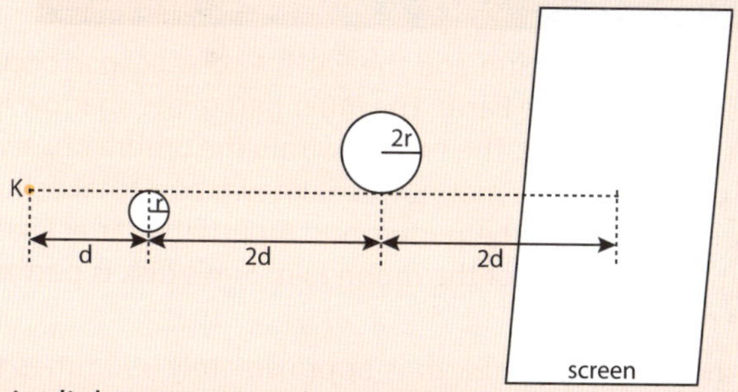

Point light source K, spherical objects with radii r and 2r are placed in front of a white screen at distances given in the figure.

a. Draw the shadow formed on the screen.

b. Find the ratio of areas of the shadows formed on the screen, A_1/A_2.

Solution

a.

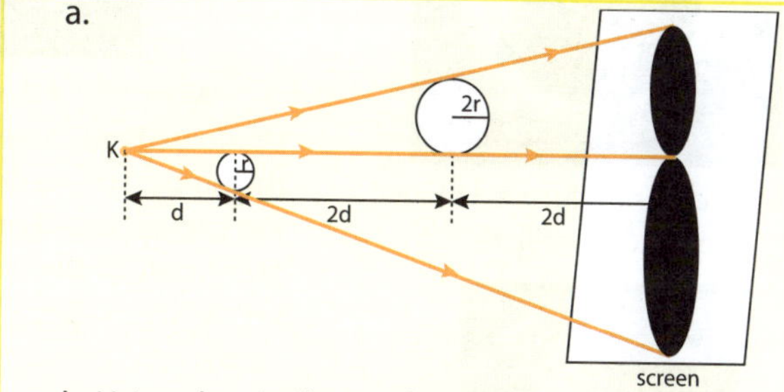

b. Using the similarity of triangles:

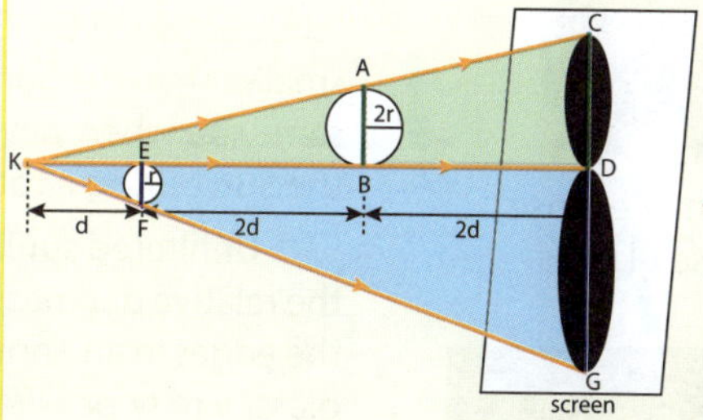

$\Delta AKB \approx \Delta CKD$

$$\frac{3d}{5d} = \frac{4r}{2r_1} \Rightarrow r_1 = \frac{10r}{3}$$

$\Delta EKF \approx \Delta DKG$

$$\frac{d}{5d} = \frac{2r}{2r_2} \Rightarrow r_2 = 5r$$

$$\frac{A_1}{A_2} = \frac{\pi\left(\frac{10}{3}r\right)^2}{\pi 5r^2} = \frac{4}{9}$$

Example

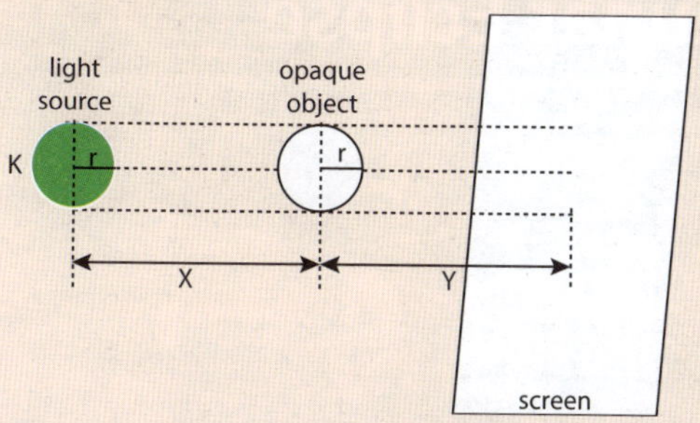

An opaque object is placed in front of a spherical light source as given in the figure. Umbra and penumbra are formed on the screen.

a. Draw the shadow formed on the screen.

b. If penumbra is 15 times larger than the umbra find the ratio of X/Y.

Solution

a.

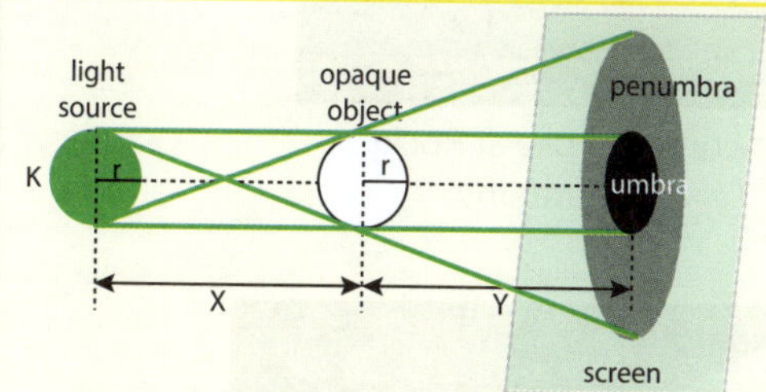

b. If penumbra is 15 times larger than the umbra; then

$P + U = 15U + U = 16U$

The area of a circle is πr^2. The radii are proportional to the square root of the areas. $\sqrt{\frac{U}{16U}} = \frac{r_U}{r_P} \Rightarrow \frac{r_U}{r_P} = \frac{1}{4}$

Using the similarity of triangles:

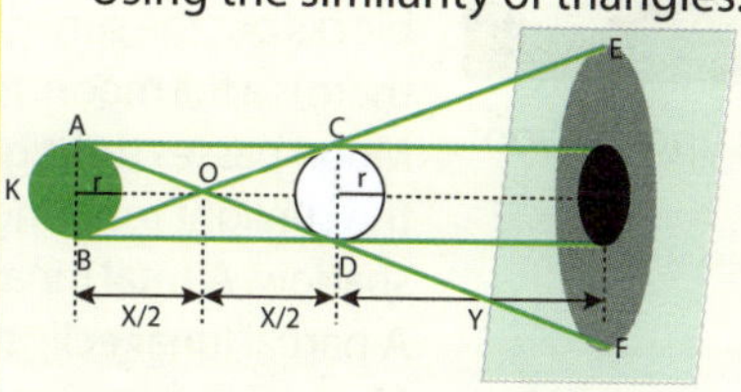

$\Delta AOB \approx \Delta COD \approx \Delta COD$

$$\frac{\frac{X}{2}}{\frac{X}{2}+Y} = \frac{2}{8} \Rightarrow \frac{4X}{2} = \frac{X}{2} + Y$$

$$\frac{3X}{2} = Y \Rightarrow \frac{X}{Y} = \frac{2}{3}$$

Shadows formed during a partial solar eclipse

Diamond ring formed in a total solar eclipse

Observe a solar eclipse with eclipse glasses

Lunar eclipse at Rocky Mountains

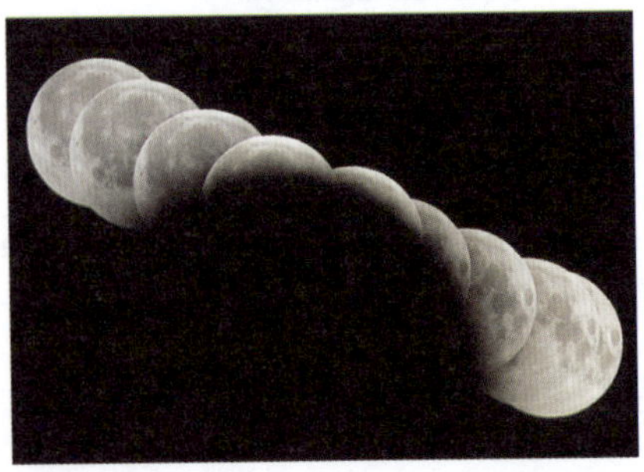

Shadow of Earth falling upon Moon

The formation of shadows explains an exciting phenomenon of nature: The eclipses. An eclipse occurs when three celestial bodies become aligned.

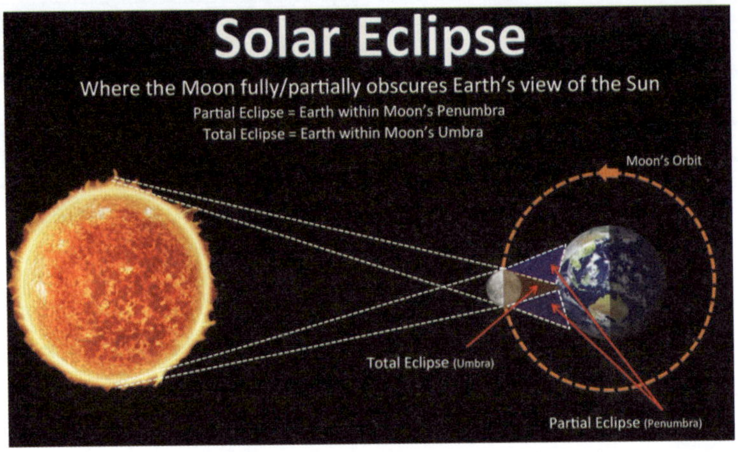

When the moon moves between the Sun and the Earth a solar eclipse occurs. The shadow of moon casts across the face of the Earth. The inner part of the shadow is called the umbra where no sunlight can penetrate. The people within the umbra observes that the sun appears completely covered by the moon. This is known as a total eclipse. The outer part of the moon's shadow is called the penumbra. A partial solar eclipse is observed by the people where the penumbra of the moon falls. Sun's disk is partially covered by the penumbra of the moon.

The distance between the Sun and the Earth is approximately 390 times the distance between the Earth and the Moon. The moon's diameter is approximately 400 times smaller than the Sun's diameter. This ratio causes the Sun and the Moon appear to be the same size when viewed from the earth, thus the moon can totally block out the light from the Sun during a total solar eclipse.

Unfiltered sunlight sears your eye's retina irreparably. Your eye's iris will be fooled by the relative darkness of the moon's silhouette. Dangerous direct sunlight appears around the edges in an annular solar eclipse and can hurt you badly without precautions. Eclipse glasses may provide protection for your eyes, but for a short period of time. Smoked glass, brown or green glass, polarized sunglasses are not safe either. Just because the sun looks dark through the substance does not mean that risky invisible IR or hazardous UV is not reaching your delicate retina.

When the Earth is between the Sun and the Moon, a lunar eclipse occurs. The Earth blocks out the sunlight and casts its shadow over the Moon. A lunar eclipse only occurs when there is a full moon. A full moon only occurs when the moon is directly opposite the Sun. If the Moon passes through the Earth's shadow a lunar eclipse takes place. The Earth blocks out all the sunlight reaching the Moon when the moon is within the umbra, the inner region of the shadow. A total lunar eclipse is observed when the entire Moon passes through the umbra. A partial lunar eclipse is observed when part of the moon passes the umbra shadow. As the Moon moves across the penumbra, the outer shadow where the Earth partially blocks the sunlight, a penumbral lunar eclipse is observed. A lunar eclipse does not occur at every full moon. This is because the Moon's orbit is tipped 5 degrees to the Earth's orbit around the sun. The moon passes either above or below the Earth's shadow most of the time.

4.3. Reflection

The light rays used in the ray model of light, travels forever unless they interact with matter. Light can either be scattered or absorbed within a material. At an interface between two materials, light can either pass through the interface to a second medium or bounce back to the incident medium.

The light rays *bounce back* from an object in other words *reflect* and reach to our eye. They enter to our eye through the pupil and the eye focuses a bundle of light rays. So, to see the objects:

1. Light must fall upon the object;

2. The light from the object should reach to our eye.

Objects reflect light rays incident upon them in every direction. Such a reflection is called diffuse reflection. Objects are seen as a result of diffuse reflection. Reflection from a shiny surface like polished metals or mirrors is called specular reflection (speculum = mirror in Latin). Images are formed as a result of specular reflection. When you are driving at night, if the weather is dry; light coming from the headlights of oncoming cars is scattered off the road in different directions. You will have a better vision as a result of diffuse reflection of light on the road. However, on a rainy night; the irregularities of the road is filled by water, the incident light undergoes specular reflection on the smooth surface of the road.

Diffuse reflection on water's surface

Specular reflection on water's surface

4.3.1 Law of Reflection

Light behaves like water waves as it is reflecting from a surface. The angle of incidence and reflection are again equal to each other in the reflection of light waves. Instead of using wave fronts we will use the propagation direction of light waves, in other words we will use the light rays to show the reflection from a surface. With the waves we were measuring the angles with respect to the surface and the wave front. A major difference rises when we use the light rays. In optics calculations we always measure the angle from the normal to the surface to the light ray. Normal to the surface is the direction at right angles to that surface.

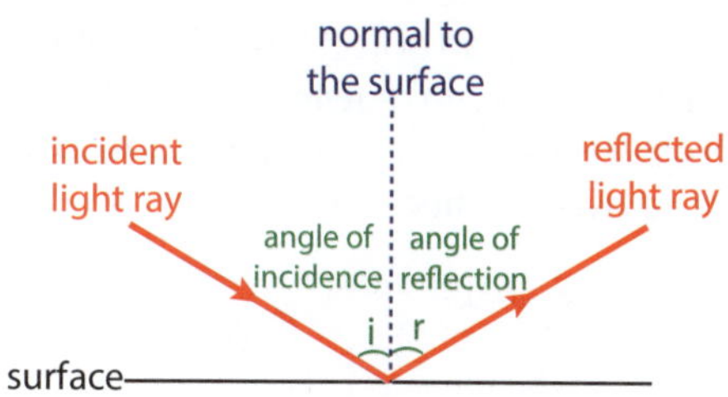

The laws of reflection state that:

1. The incident light ray, the normal to the surface and the reflected light ray are in the same plane.

2. Angle of incidence is equal to the angle of reflection.

$$i = r$$

When light reflects from a rough surface, the law of reflection still holds, but the angle of reflection varies. If a parallel beam of light rays is sent upon a rough surface, the reflected light rays are not parallel to each other. But if the same beam of light rays were sent upon a mirror like surface, they would be parallel to each other again after the reflection.

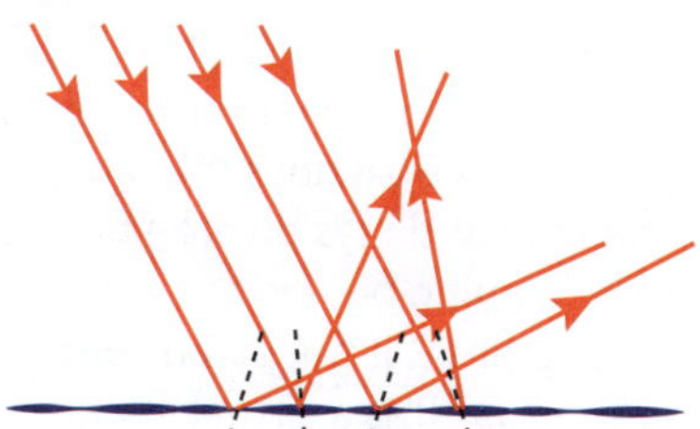

rough surface (diffuse reflection)

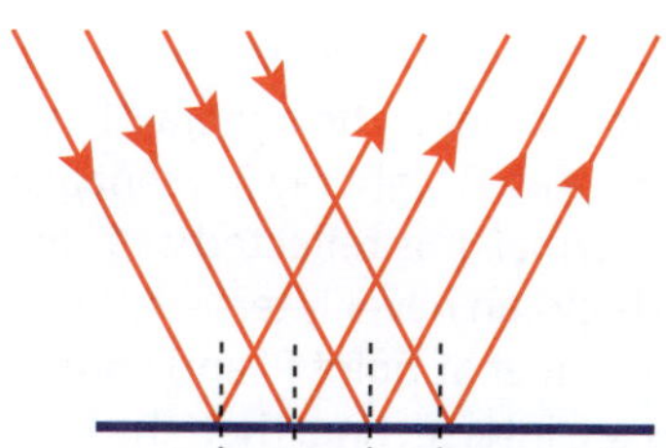

smooth surface (specular reflection)

Example

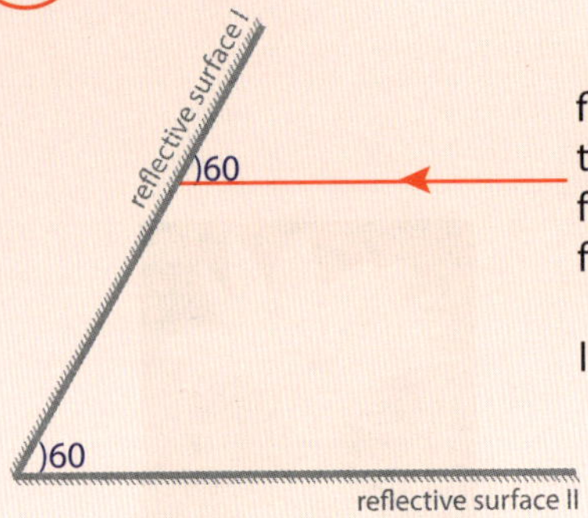

Light ray K is first reflected from the reflective surface I and then from the reflective surface II.

Draw the path followed by the light ray K.

Solution

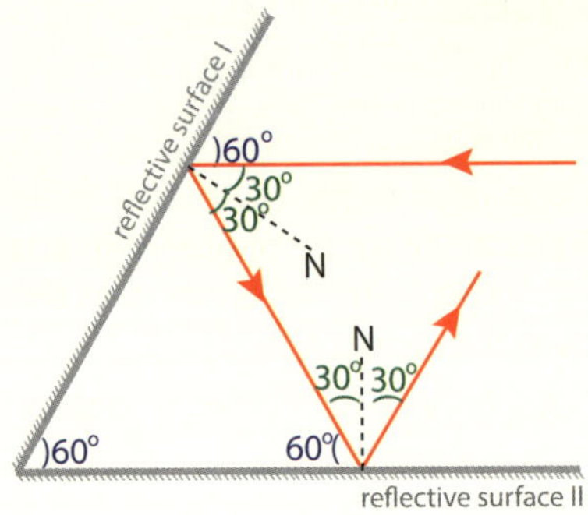

Example

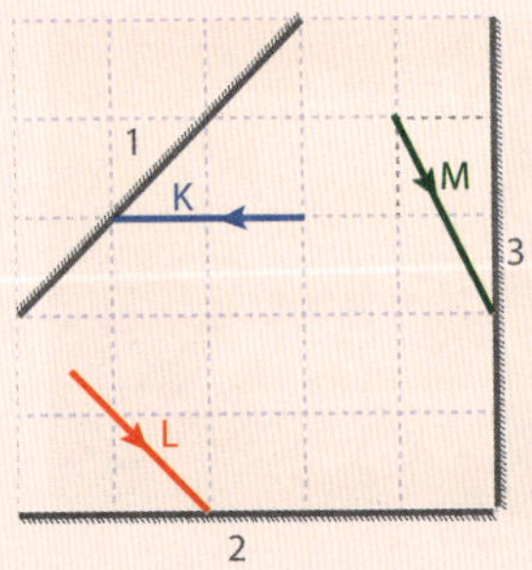

Light rays are K, L and M are sent to the reflective surfaces 1, 2 and 3 as given in the figure.

Which rays will reflect back on themselves?

Solution

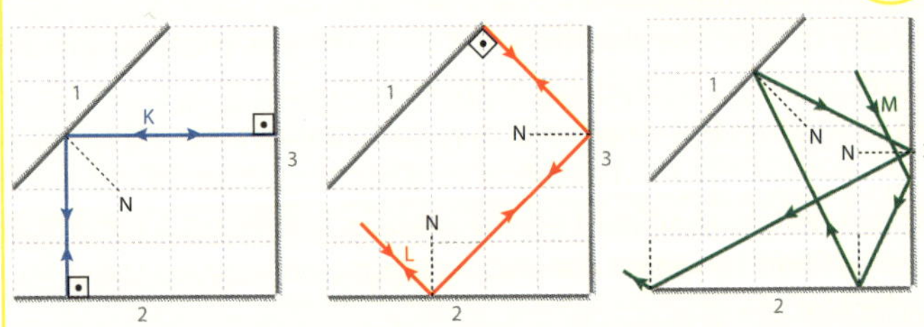

The paths followed by the light rays is shown in the figure. K and L are reflected back on themselves.

4.4. Plane Mirrors

4.4.1 Images Formed by Plane Mirrors

A smooth plane reflecting surface is a plane mirror. A good plane mirror is made by putting a highly reflective metal coating on one surface of a very flat piece of glass. Plane mirrors are widely used at our homes, such as a bathroom mirror or a dressing mirror.

Let's use the laws of reflection to construct the image of an object by a plane mirror. A ray diagram will be helpful to us. Just as we did while we were drawing the shadow of an object, we will use only a few light rays to construct the image by a plane mirror. The light rays incident on the plane mirror will obey the laws of reflection.

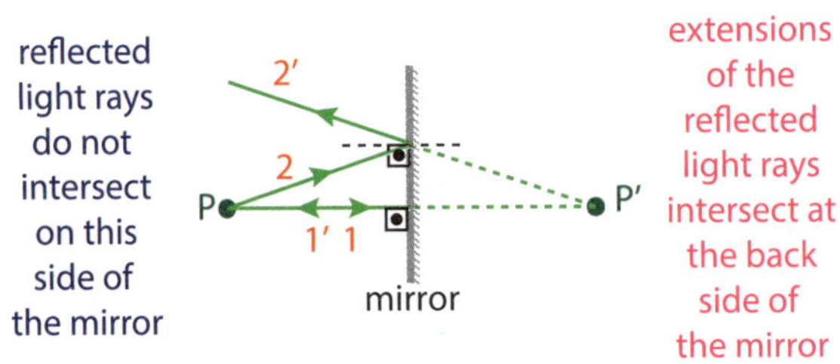

To have a point, at least two lines must intersect. In the figure above, two light rays are sent from point P. The light ray 1, is in the normal direction to the mirror, so it reflects back on itself as 1'. Light ray 2 is sent at an angle to the normal. So it does the same angle with the normal. It is reflected as 2' in the figure. The reflected rays diverge from each other, so they can never intersect. But they appear to be coming behind the mirror; their extensions meet at point P' and construct the image of point P. If the extensions of the light rays are intersecting, then the image is said to be virtual. P' is a virtual image.

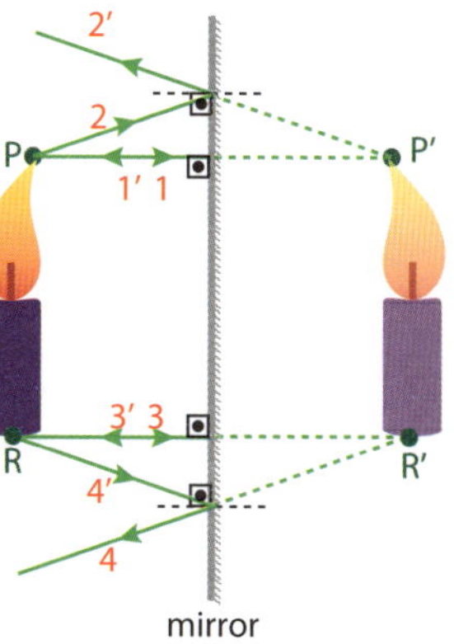

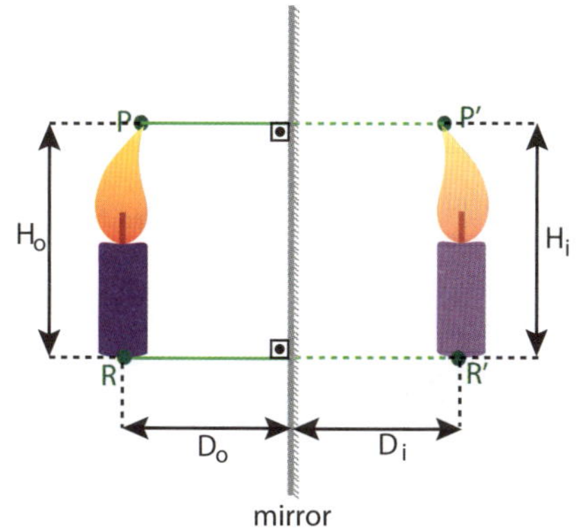

The image of an object can be formed by using the same method. But this time P, a point on top and another point R, at the bottom of the object are chosen. Their images are formed with respect to the mirror. Then these image points P'R' are joined to each other. As you can see from the drawings, the image is alike to the object. If you have noticed, the flame bends to right for the object whereas it bends to left for the image. This is called a mirror symmetry. Your left hand appears as like your right hand with a plane mirror.

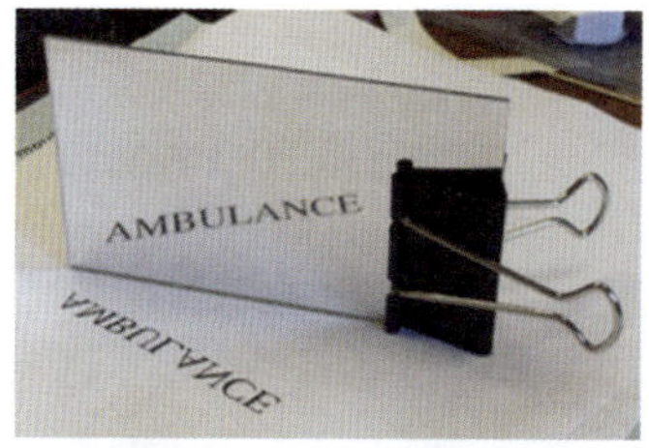

Images formed with a plane mirror have symmetry with respect to the mirror.

The properties of the image formed	
1.	The image has the same height as the object ($H_o = H_i$).
2.	The distance of the image to the mirror is equal to the distance of the object to the mirror ($D_o = D_i$).
3.	The image is upright. However it displays mirror symmetry with the object.
4.	The image is virtual. The extensions of the reflected light rays intersect.

Example

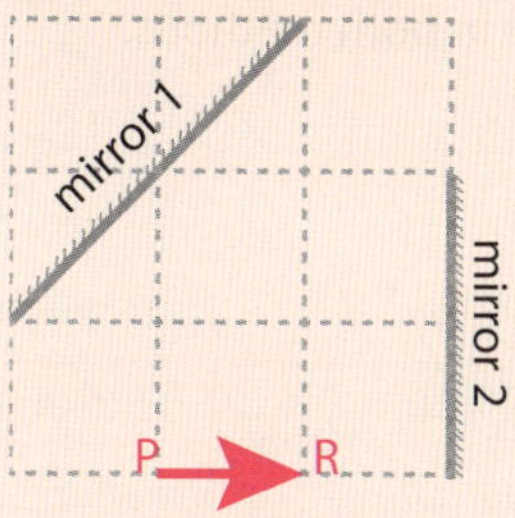

Plane mirrors 1 and 2, and an illuminated object PR are placed as given in the figure.

a. Draw the image of object PR if the light rays are first reflected from mirror 1 and then from mirror 2.

b. Draw the image of object PR if the light rays are first reflected from mirror 2 and then from mirror 1.

Solution

P'R' is the first formed image of PR.
P''R'' is the second formed image of PR.
a. A mirror can be lengthened for having the image of an object. We will need to lengthen mirror 2 in order to find the image of P'R'.

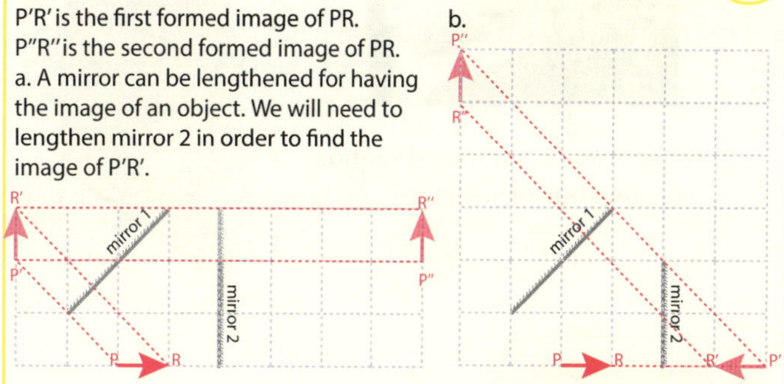

Example

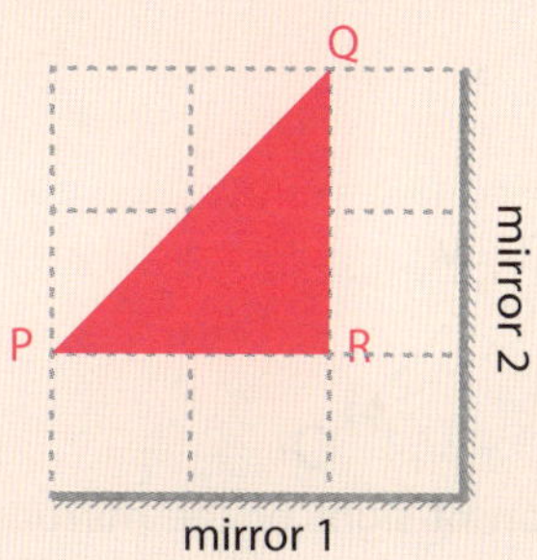

Plane mirrors 1 and 2, and an illuminated triangular object PQR are placed as given in the figure.

a. Draw the image of object PQR if the light rays are first reflected from mirror 1 and then from mirror 2.

b. Draw the image of object PQR if the light rays are first reflected from mirror 2 and then from mirror 1.

Solution

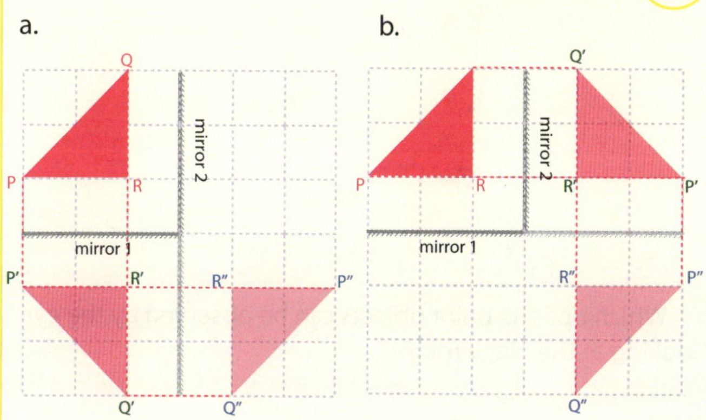

For a person to see herself or himself completely in a plane mirror, the mirror's length should have half the height of that person and the bottom line of the mirror should be a height which is half the distance between the person's eyes and feet. If the person is not seeing herself or himself completely in the mirror; getting close to the mirror or far from it will not change the size of the image observed.

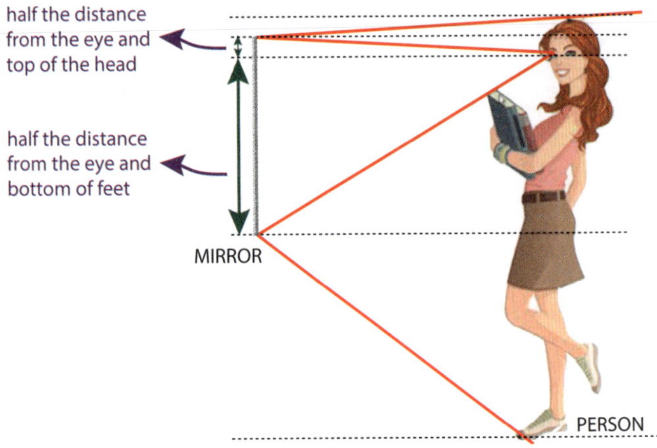

What you see in a mirror depends on your position....

The size of the mirror is important for the region you can see from a plane mirror. In addition to the size of the mirror, this region is also dependent on the position of the observer. As you change your position, things that you observe will also change. Don't forget that, for seeing an object, light rays from the object should reach to your eye. The region between the rays reflected from the very ends of a plane mirror is called the region of sight.

An observer at point O is looking to the plane mirror KL. To find the region of sight of the plane mirror KL, we can either send light rays from the observer to the ending points (KL) of the plane mirror or we can find the image of the observer with respect to plane mirror KL and send light rays from the image of the observer to the ending points of the mirror KL. The region between the reflected light rays is the region of sight.

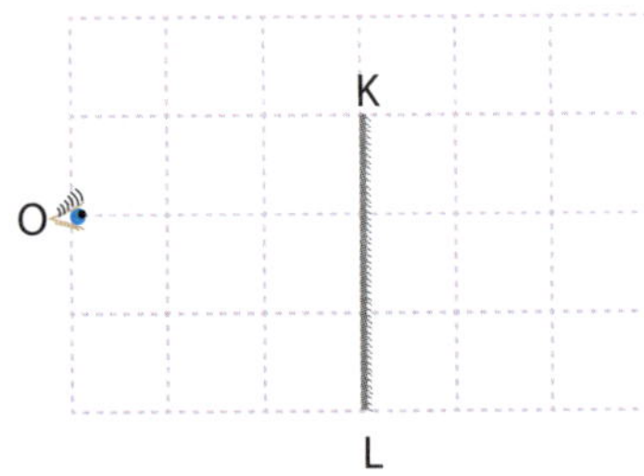

Method 1

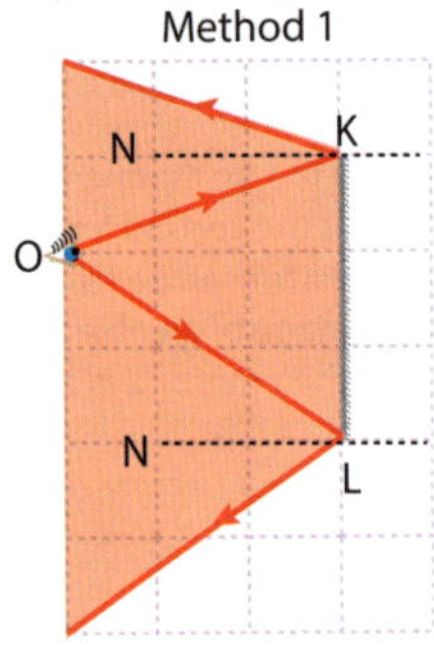

Method 2

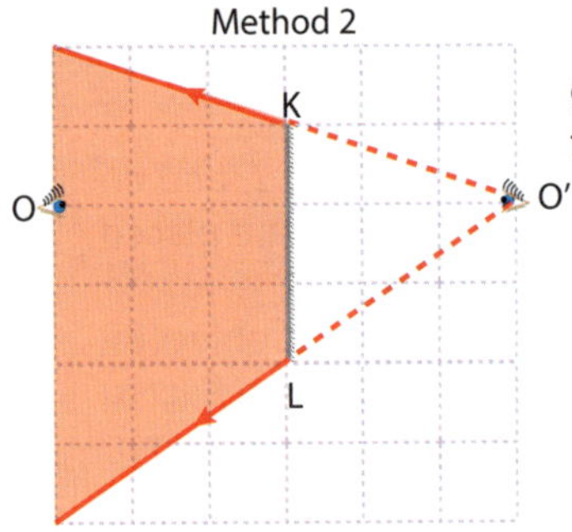

The shaded areas are the region of sight of mirror KL. They are the same in both methods.

Example

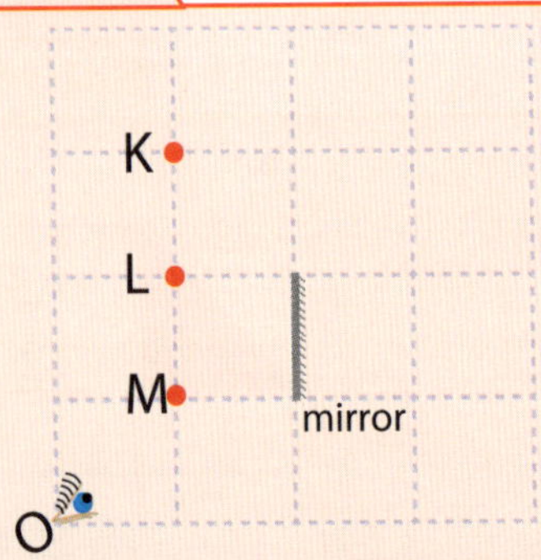

Which opaque point objects can be observed by the eye looking at the plane mirror?

Solution

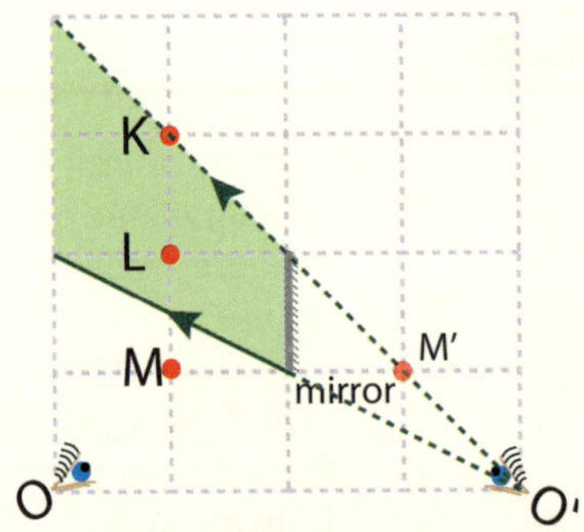

K and L are found in the region of sight of the mirror. However, the image of M blocks the light rays towards K. Thus, it cannot be observed. Only L will be seen by the observer.

Example

An opaque ball and a plane mirror are placed as given in the figure. An observer is looking at the mirror from point O.

Which points of the wall AF cannot be seen from the mirror?

Solution

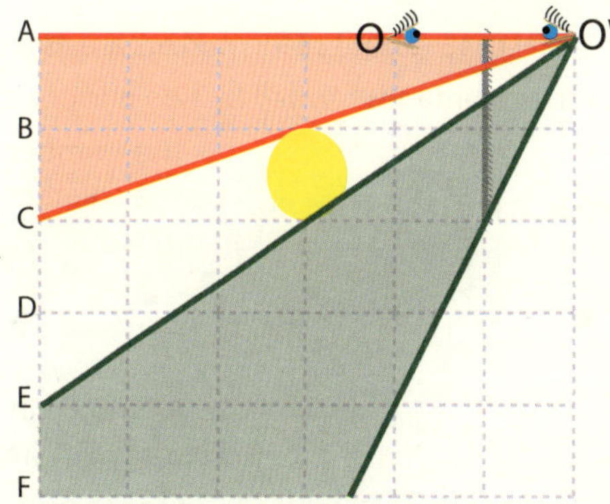

The green and red shaded areas can be seen by the observer. Th points between AC and EF are seen by the observer. CE cannot be seen by the observer since the opaque ball blocks the light rays.

Example

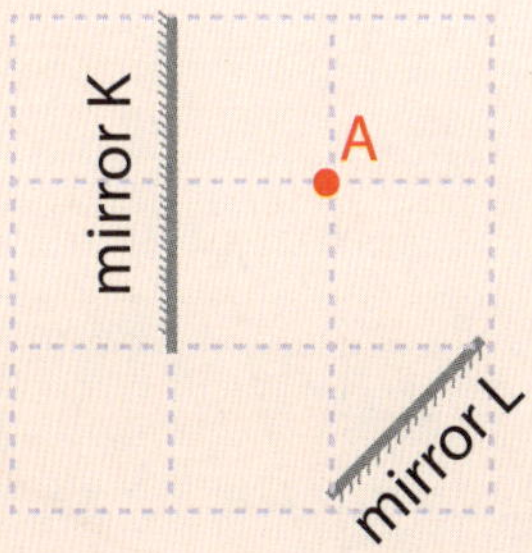

How many images of point A will be formed between plane mirrors K and L?

Solution

The images will be formed as long as there is a shiny side of a mirror within the region of sight. The region of sight of each image formed is drawn with respect to the mirror forming the image.

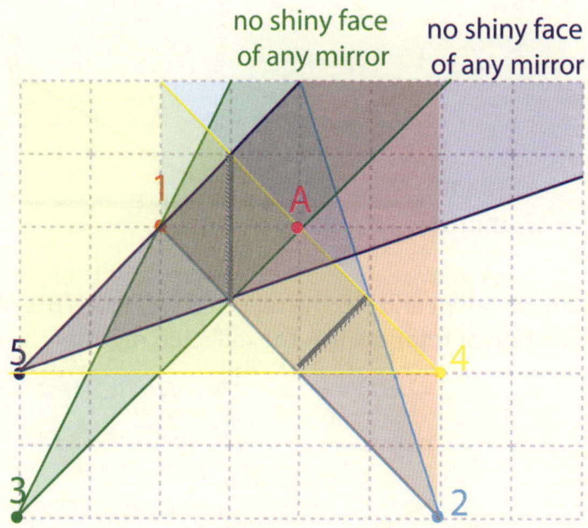

1 is the image of A formed by mirror K first.

2 is the image of 1 formed by mirror L.

3 is the image of 2 formed by mirror K. No more shiny faces in the region of sight of image 3. Thus no more images are formed.

4 is the image of A formed by mirror L first.

5 is the image of 4 formed by mirror K. No more shiny faces in the region of sight of image 5. Thus no more images are formed.

Overall, 5 images of A is formed by mirrors K and L.

Physics Extra

Some special cases about plane mirrors are as follows.

- Number of images formed by two mirrors at an angle:

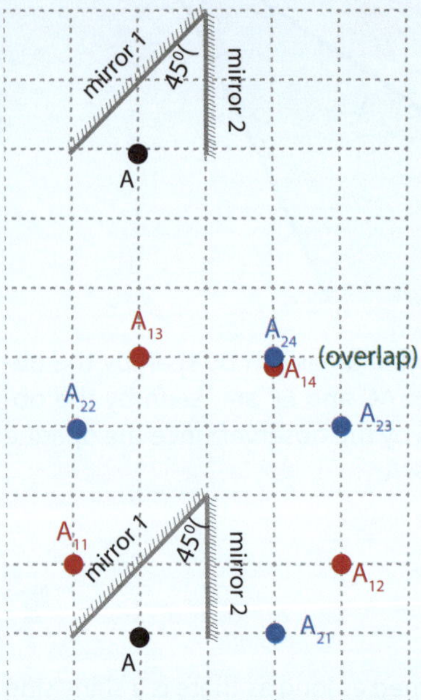

The number of images formed depends on the angle in between the mirrors and given by the equation:

$$n = \frac{360}{\alpha} - 1$$

As shown in the aside drawing one of the images formed by individual mirrors overlap. That's why 1 image is subtracted in the equation.

For the aside drawing:

$$n = \frac{360}{\alpha} - 1 = \frac{360}{45} - 1 = 7 \ images$$

- Relationship between the reflected rays when a plane mirror is rotated:

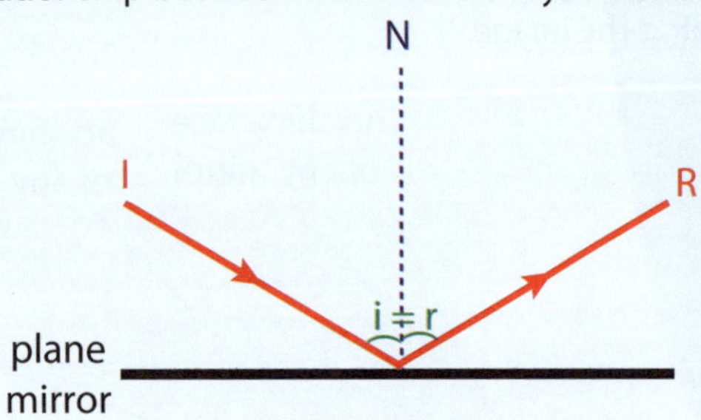

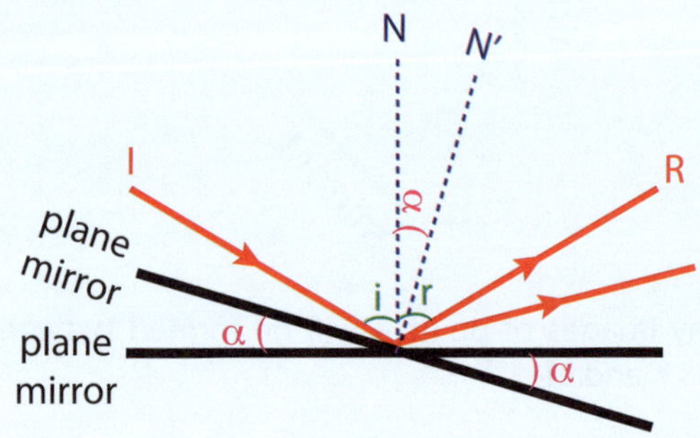

I is the incident and R is the reflected ray. When the plane mirror is rotated by α, the normal to the surface also rotates by α in the direction of rotation. R' is the reflected ray after the rotation.

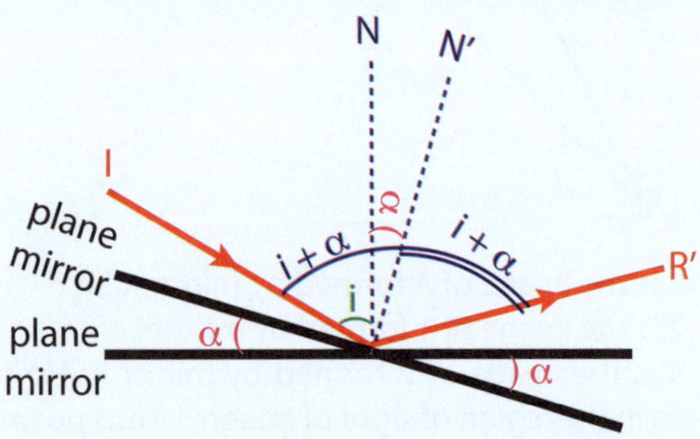

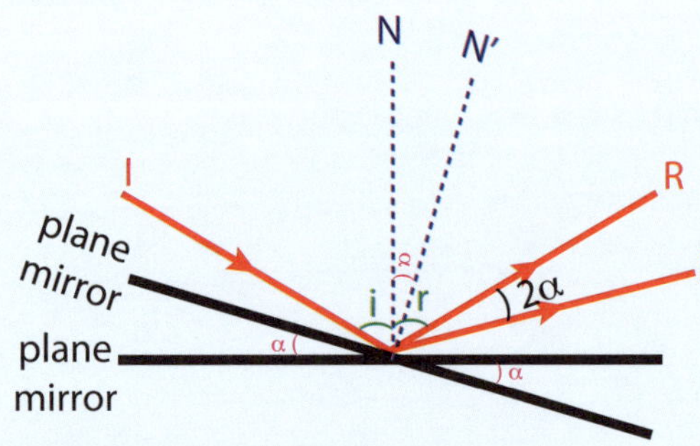

Before the rotation the angle between the incident and the reflected rays is: $i + r = 2i = 2r$

After the rotation the angle between the incident and the reflected rays is: $i' + r' = i + \alpha + i + \alpha = 2i + 2\alpha$

Difference as a result of rotation is: $2i + 2\alpha - 2i = 2\alpha$

If the plane mirror is rotated by α without a change in the incident light ray, the reflected ray rotates by 2α.

Physics Extra

If an object approaches to a plane mirror with a velocity of v, than the image of the object also approaches to the mirror with the same magnitude of velocity, v.

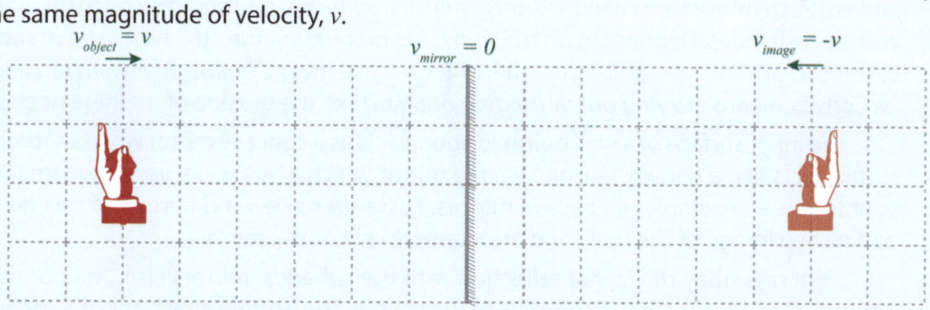

What if the mirror is moving also? Study the following cases:

- The object moves with a velocity of $2v$ and the mirror moves with a velocity of v in the same direction.

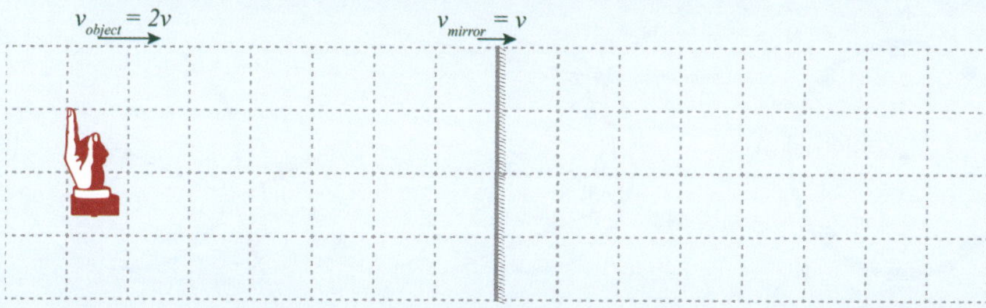

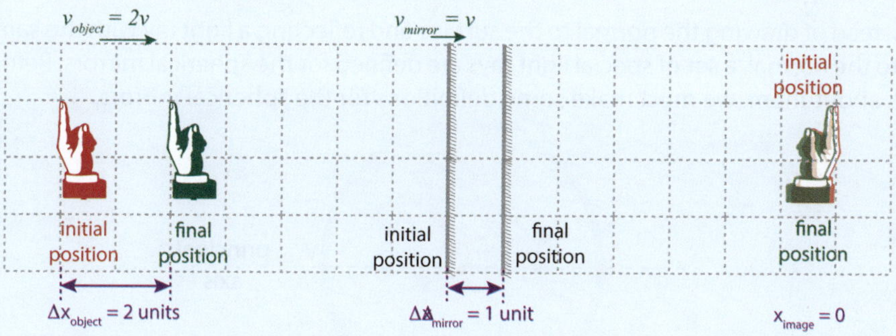

The object is 7 units in front of the mirror initially. So its first image will be formed 7 units behind the mirror.

Let's say the object changes its position by 2 units in unit time. Since the mirror has half the velocity of the object, the mirror will move 1 unit towards right. Now the object is 6 units in front of the mirror so the image will also be formed 6 units behind the mirror.

The initial and final images of the object are formed at the same position, the displacement of the image is zero. We can conclude that the velocity of the image with respect to the ground is zero.

$$v_{image} = 0$$

- The object moves with a velocity of $2v$ and the mirror moves with a velocity of v in the opposite direction.

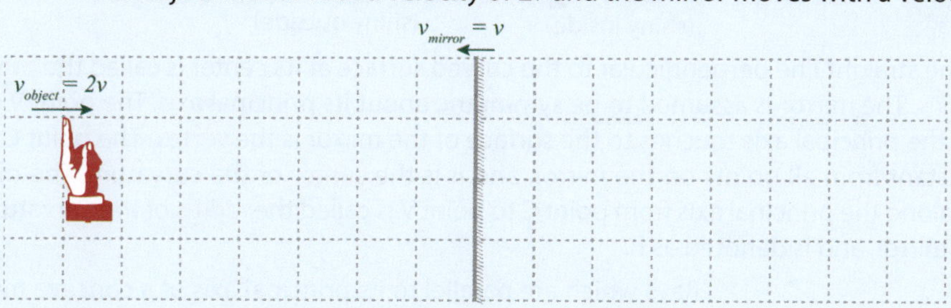

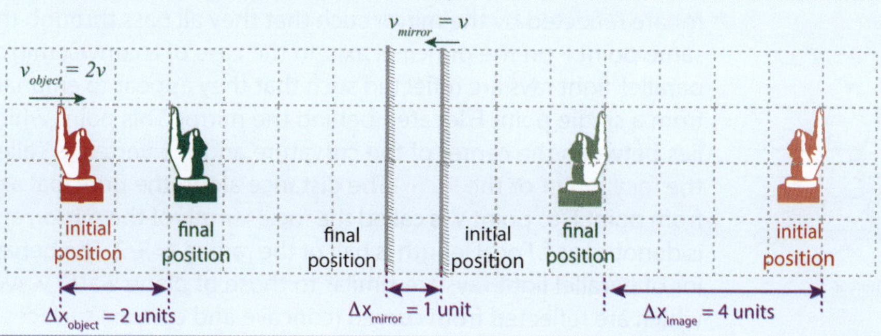

The object is 7 units in front of the mirror initially. So its first image will be formed 7 units behind the mirror.

Let's say the object changes its position by 2 units in unit time. Since the mirror has half the velocity of the object, the mirror will move 1 unit towards left. Now the object is 4 units in front of the mirror so the image will also be formed 4 units behind the mirror.

The distance between initial and final images of the object is 4 units, the displacement of the image is 4 units. We can conclude that the velocity of the image with respect to the ground is $4v$.

$$v_{image} = 4v$$

4.5. Spherical Mirrors

4.5.1 Fundamental Concepts about Spherical Mirrors

Not all mirrors have a flat surface like a plane mirror. Surface of some mirrors are curved. Such mirrors are called spherical mirrors. Spherical mirrors are used to focus light and create images. Depending on the shiny side of the curvature, the two types of spherical mirrors are concave and convex mirrors. Concave means *hollowed inward* or *curving in*. Convex means *curving out* or *bulging outward* like the outside of a sphere or circle.

Image formed on the inner and outer curvatures of a spoon

Magnifying mirrors are concave mirrors

Convex mirrors have a large region of sight, they are very useful in seeing blind spots at roads

The inner surface of a well polished spoon acts as a concave mirror whereas its outer surface acts like a convex mirror. Shaving mirror which gives you a very large image of your face is an example of concave mirrors. Passenger side wing mirrors of cars gives a diminished image of the road, and are examples of convex mirrors.

Light rays obey the law of reflection with the spherical mirrors also. The normal to a spherical surface is along the radius of the sphere. The normal directions of a spherical surface are not parallel to each other as in planar surfaces. A light ray, incident upon a spherical surface makes the same angle with the line joining the center of the curvature and the point of contact as it is being reflected.

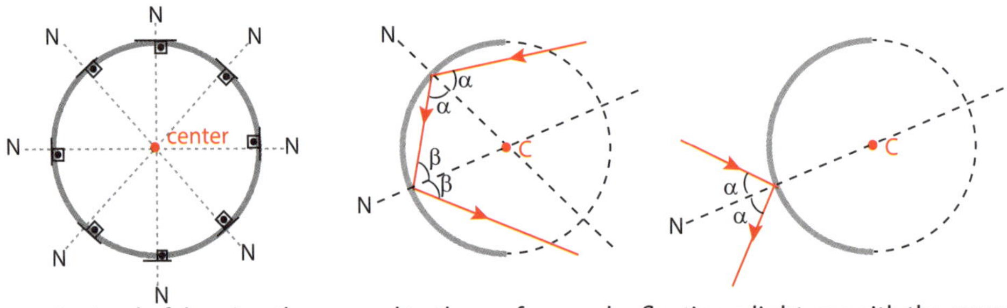

Instead of drawing the normal to the surface and reflecting a light ray with the same angle to the normal, a set of special light rays are defined for the spherical mirrors. Before we talk about them, we must make some definitions for the spherical mirrors.

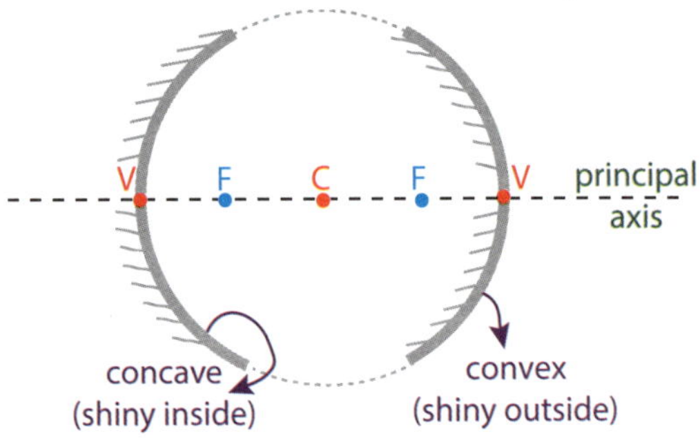

The straight line perpendicular to the curved surface at its center is called the principal axis. The mirror is assumed to be symmetric about its principal axis. The point V at which the principal axis touches to the surface of the mirror is the vertex. The point C is equidistant from all points on the mirror, and it is the center of the curvature. The distance along the principal axis from point C to point V is called the radius of the curvature of the mirror, and is denoted as R.

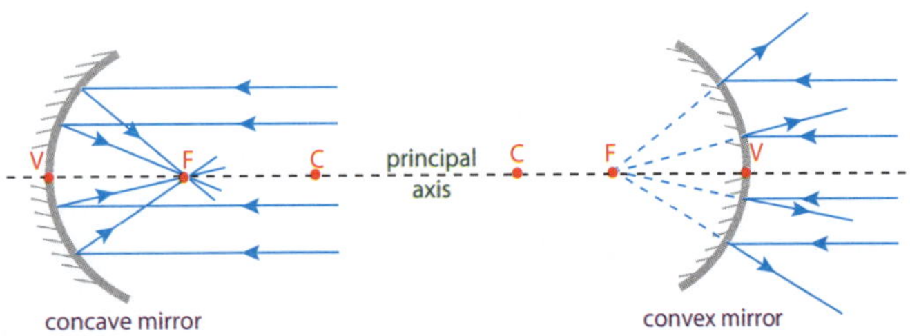

Rays which are parallel to its principal axis of a concave mirror are reflected by the mirror such that they all pass through the same point F on the principal axis. In the case of a convex mirror, parallel light rays are reflected such that they appear to emanate from a single point F located behind the mirror. This point which lies between the center of the curvature and the vertex, is called the focal point or the focus. The distance along the principal axis from point F to point V is called the focal length of the mirror, and is denoted as f. Focal length is half of the radius, f=R/2. The behavior of parallel light rays are similar to those of plane water waves which are reflected from curved (concave and convex) surfaces.

Now we can draw the special light rays for concave and convex mirrors:

Concave mirror			Convex mirror		
Incident light ray	Drawing	Reflected light ray	Incident light ray	Drawing	Reflected light ray
Parallel to the principal axis	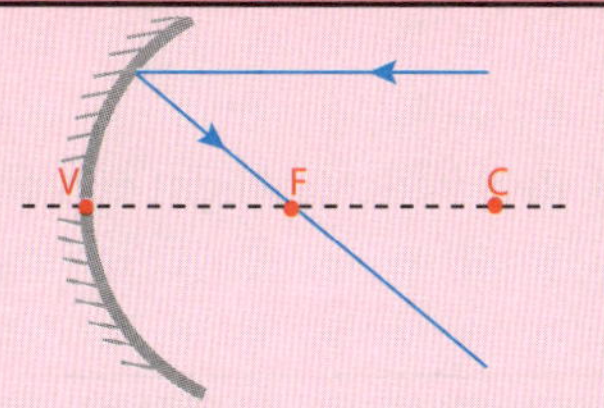	Passes through the focus	Parallel to the principal axis	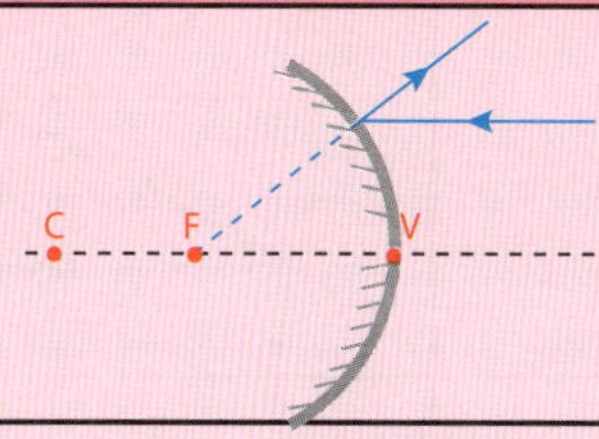	Extension passes through the focus
Passing through the focus	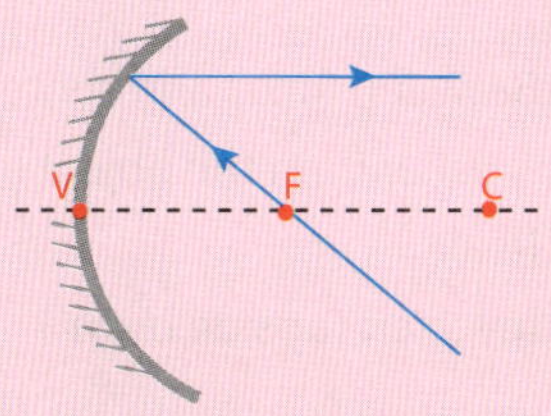	Parallel to the principal axis	Towards the focus	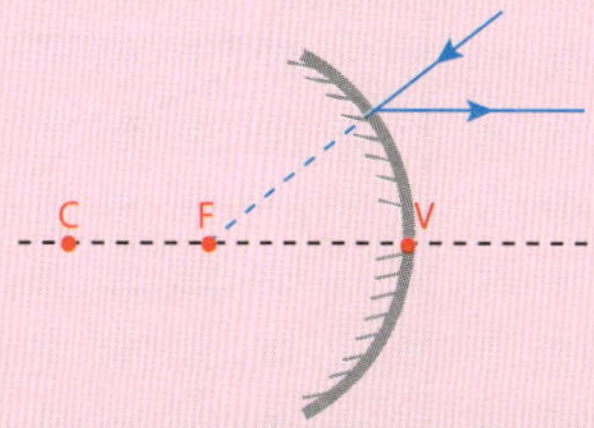	Parallel to the principal axis
Passing through the center	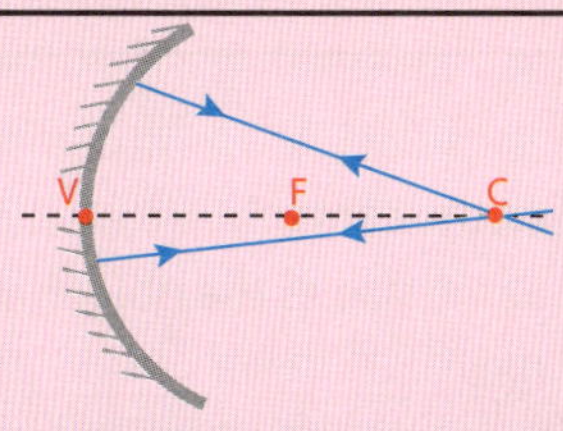	Reflected back on itself (normal to the surface)	Towards the center	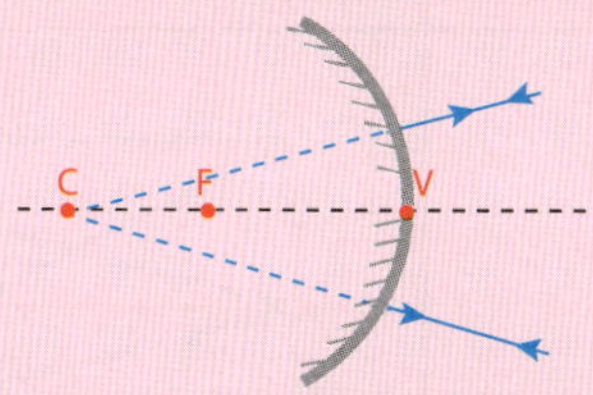	Reflected back on itself (normal to the surface)
Towards the vertex	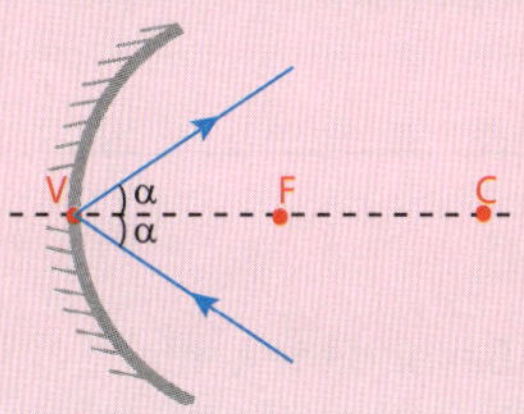	Makes the same angle with the principal axis	Towards the vertex	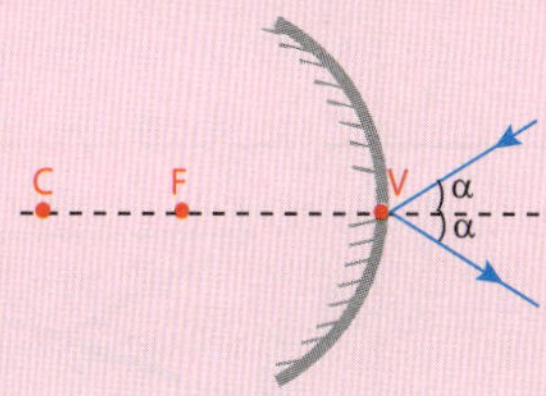	Makes the same angle with the principal axis

The first and second light rays for both mirrors display the reversibility of light rays. Any light ray will follow back the same route if it was coming from the direction of the reflected light ray. It should be kept in mind that, these special light rays still obey the law of reflection. We will prefer to use them in drawing the image of an object formed by spherical mirrors.

4.5.1 Images Formed by Spherical Mirrors

The special light rays we had learnt will be helpful to us as we are drawing the image of an object formed by spherical mirrors. Just as we did for finding the image of an object with a plane mirror, we will find the image of top and bottom points of the object with respect to the mirror. As we are forming the image of a point, we will not work on points on the principal axis. The image of a point on the principal axis will be formed on the principal axis. For other points we will use any of two special light rays to find the image of that point.

For the plane mirror, the reflected light rays were not intersecting. The image formed was a virtual image. If the reflected light rays are intersecting, the images formed are real images. Real images can be projected upon a screen.

For a concave mirror, we will begin with finding the image of an object at infinity. Infinity to a mirror is a very large distance when compared to the focal length of the mirror. Sun is an object at infinity for a spherical mirror. Then we will get the object closer to the mirror step by step.

Position of the object	Drawing	Properties of the image
At infinity	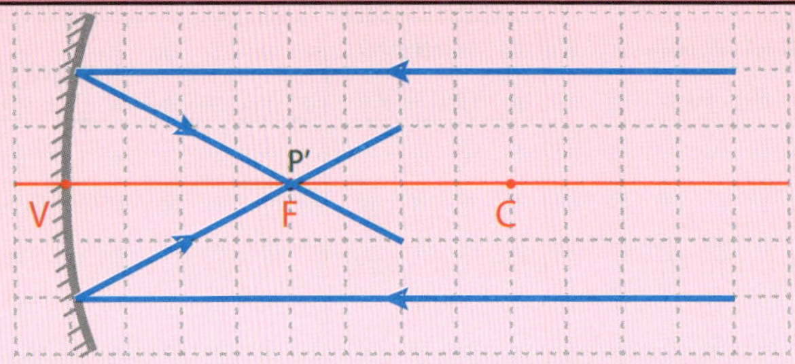	1. At focus 2. A point 3. Real (reflected light rays are intersecting
Beyond the center	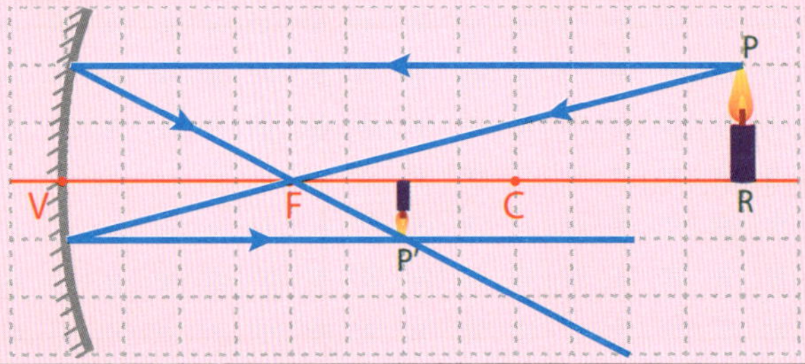	1. Between focus and center 2. Smaller than the object (diminished) 3. Real 4. Inverted (upside down)
At center	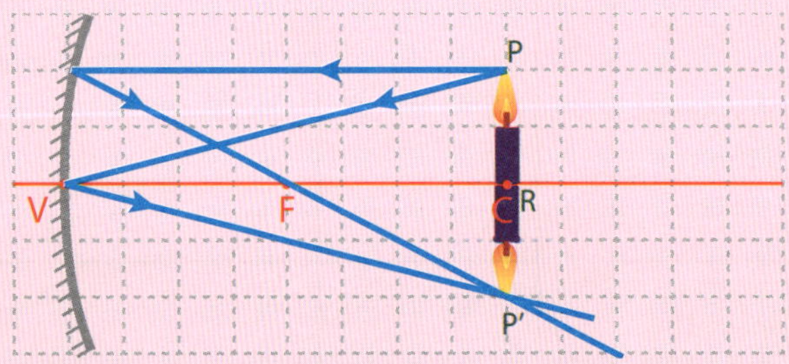	1. At the center 2. Same size as the object 3. Real 4. Inverted
Between focus and center	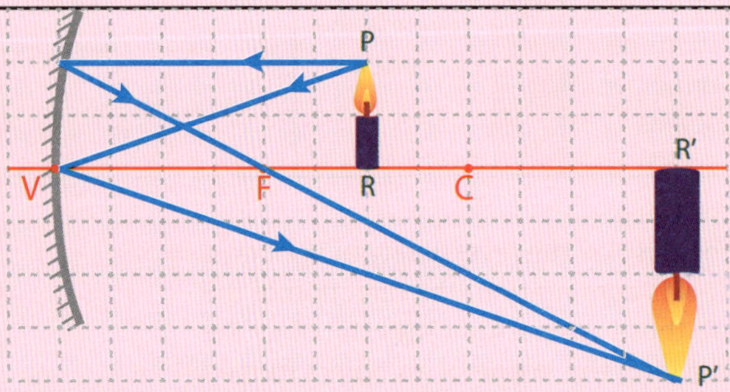	1. Beyond the center 2. Larger than the object (magnified) 3. Real 4. Inverted
At focus	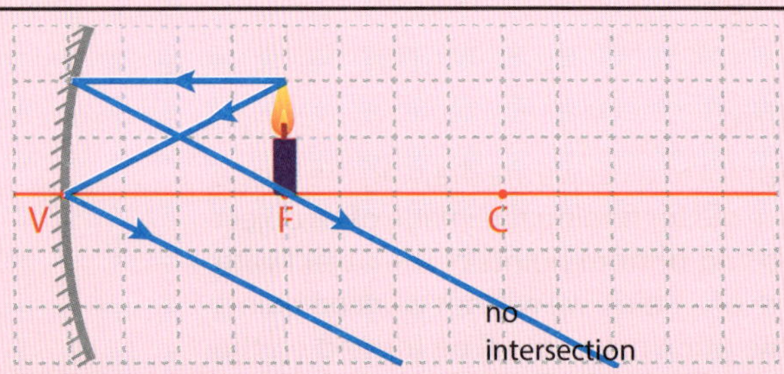	1. At infinity
Between focus and vertex	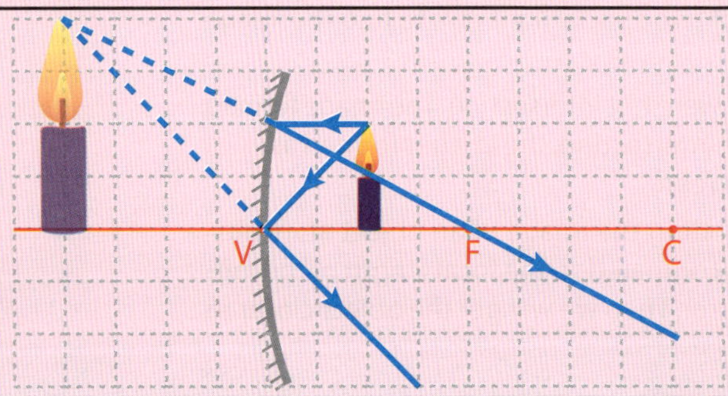	1. Behind the mirror 2. Larger than the object (magnified) 3. Virtual 4. Upright

As the object gets closer to the mirror, the image gets away from the mirror and magnifies. The relationship between the distance of the image to the mirror and the distance of the object to the mirror is not a constant one such as we had for a plane mirror. Let's analyze the geometry of the image formed below:

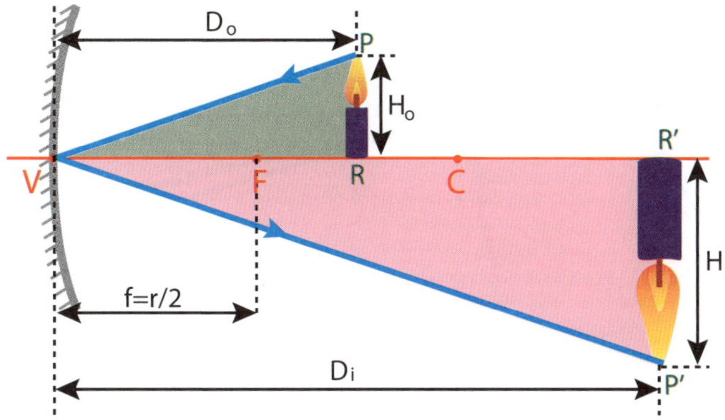

$$\Delta PVR \approx \Delta P'VR'$$

$$\frac{P'R'}{PR} = \frac{P'V}{PV}$$

$$\frac{D_i}{D_o} = \frac{H_i}{H_o} \quad (magnification)$$

Example

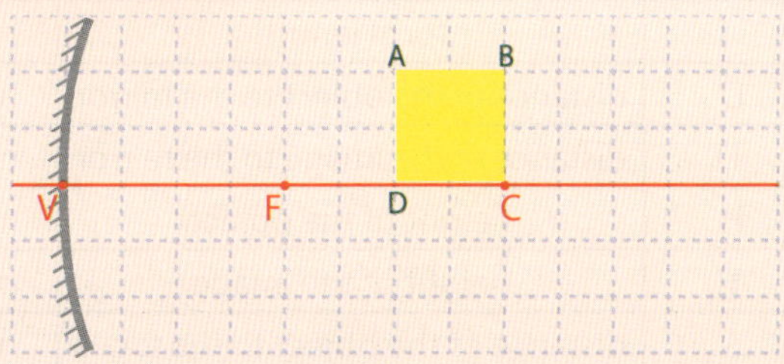

Draw the image of the object ABCD formed by the concave mirror.

Solution

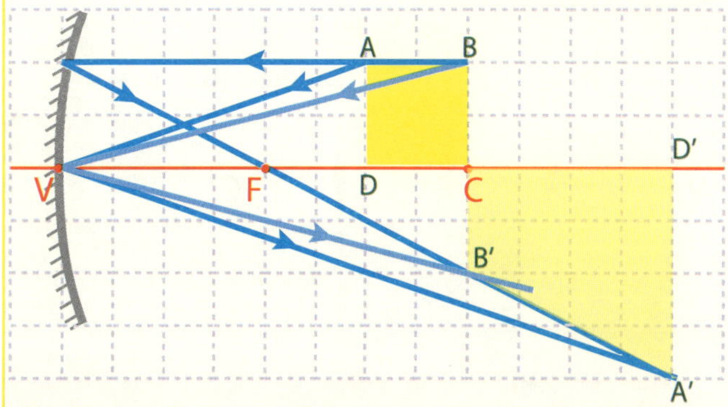

The properties of an image formed by a convex mirror are as follows:

Position of the object	Drawing	Properties of the image
At infinity	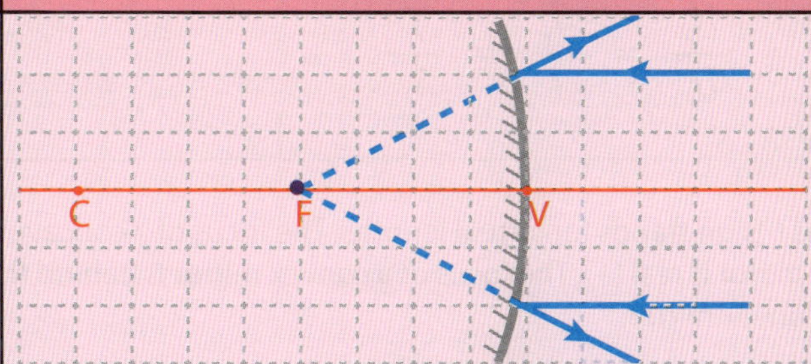	1. At focus 2. A point 3. Virtual (extensions of the reflected light rays are intersecting)
In front of the mirror	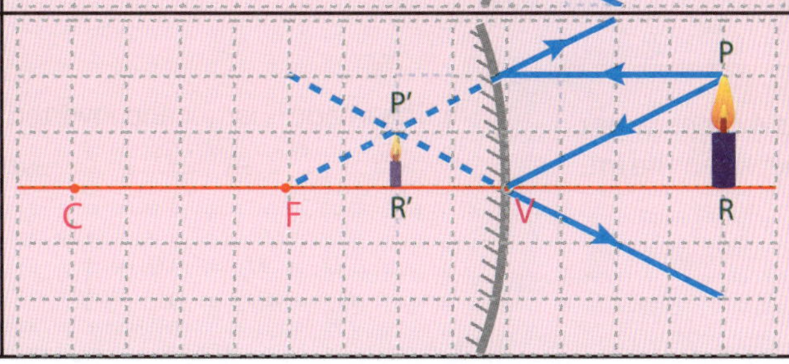	1. Always between focus and vertex 2. Always smaller than the object 3. Always virtual 4. Always upright

As the object gets closer to a convex mirror, the image moves from the focus towards the mirror and gets larger. The size of the image is equal to the size of the object when the object is placed by the vertex.

Example

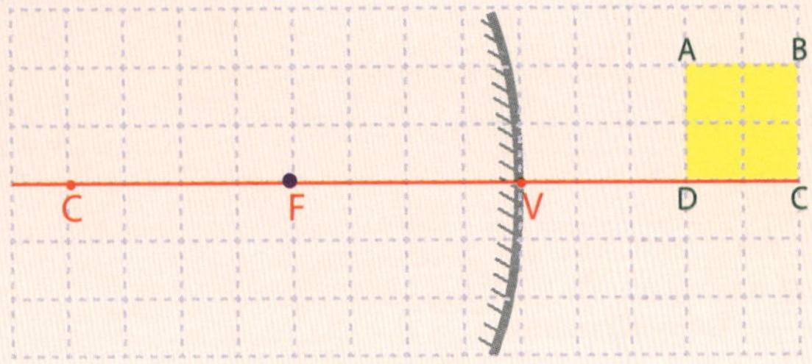

Draw the image of the object ABCD formed by the concave mirror.

Solution

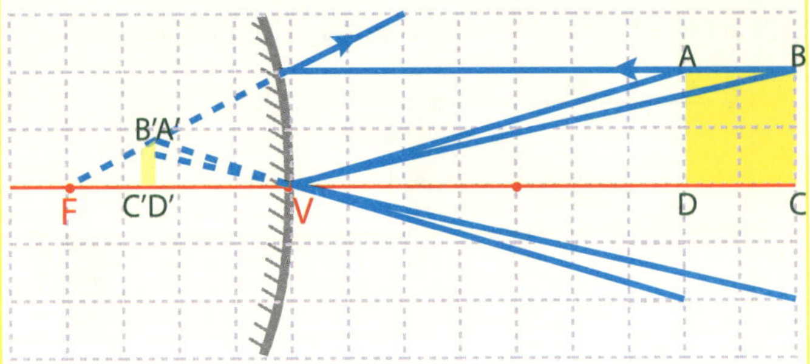

Physics Extra

Equations used in spherical mirrors:

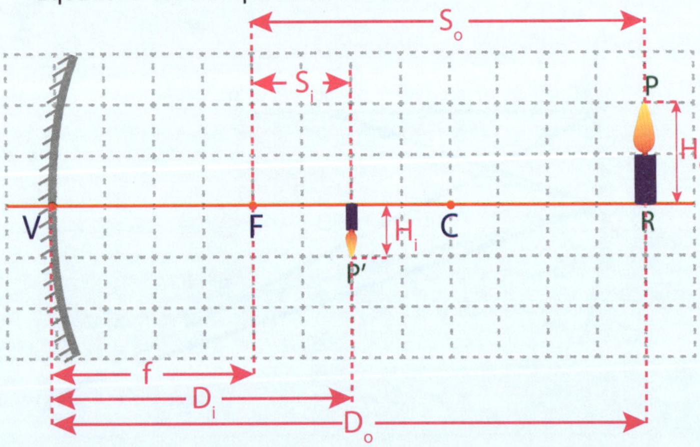

Symbol	Quantity	Unit
f	focal length	m
D_o	distance of the object to the mirror	m
D_i	distance of the image to the mirror	m
H_o	height of the object	m
H_i	height of the image	m
S_o	distance of the object to the focus	m
S_i	distance of the image to the focus	m

Location of the image or the object: $\pm\frac{1}{f} = \pm\frac{1}{D_o} \pm \frac{1}{D_i}$

Magnification: $\frac{D_o}{D_i} = \frac{H_o}{H_i} = \frac{S_o}{f} = \frac{f}{S_i}$

Focal length: $f^2 = S_o.S_i$

The (+) sign stands for real lengths whereas (-)sign stands for virtual lengths. The focus of a concave mirror is at the shiny side of the mirror so the focal length of a concave mirror is real (+). The focus of a convex mirror is behind the shiny side of the mirror this is why the focal length is virtual (-) for a convex mirror.

Example

The image formed by a spherical mirror is 4 times larger than the object and inverted. The distance between the image and the object is 300 cm.

a. What is the type of the mirror?

b. Find the focal length of the mirror.

Solution

a. The image is larger and inverted so it is a concave mirror.

b. $\frac{D_o}{D_i} = \frac{H_o}{H_i} \Rightarrow \frac{D_o}{D_i} = \frac{1}{4} \Rightarrow D_i = 4D_o$

$$\left.\begin{array}{l} D_o + D_i = 300\ cm \\ D_o + 4D_o = 5D_o \end{array}\right\} 5D_o = 300\ cm \left\{ \begin{array}{l} D_o = 60\ cm \\ D_i = 240\ cm \end{array}\right.$$

$$\pm\frac{1}{f} = \pm\frac{1}{D_o} \pm \frac{1}{D_i}$$

$$\frac{1}{f} = \frac{1}{60} + \frac{1}{240} = \frac{4+1}{240} = \frac{5}{240} = \frac{1}{48} \Rightarrow f = 48\ cm$$

4.6. Refraction

A change in the depth of water was causing a change in the velocity and wavelength of water waves. Light which is an electromagnetic wave should also display the same property as it is changing the medium it is travelling through. Indeed, as we had mentioned before; Jean Foucault proved that light changes its speed in different transparent media. Now the question is why does the light change its speed in different media?

A bullet moving with a speed of v_1 in vacuum is approaching to a wooden block. It will slow down due to frictions within the block. If it has a velocity of v_2 as it is leaving the block, it will continue its motion with v_2 in vacuum. But according to the measurements of speed light in different media, if light is moving with speed v_1 in vacuum, its speed decreases to v_2 in a transparent medium such as glass. But as light leaves the glass and re-enters to vacuum, it will again have its initial velocity v_1. What happens inside the glass so that the light slows down to v_2? And what changes as it leaves the glass?

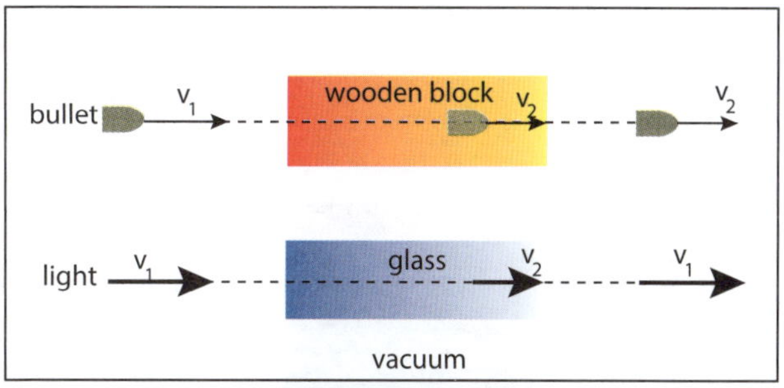

Once the light enters to the glass, it may encounter an electron which is bound to an atom, such as an atom at point K. The atom may absorb the light and this may cause the electron to oscillate. The electron will act as an antenna and radiate the beam of light toward an atom at point L. Again the light is absorbed by that atom at L. The electrons act like the runners in a relay race who are passing the baton to the runners in the same team. Once the light emerges into the vacuum, the absorptions and emissions of light end and the speed of light returns back to its initial value.

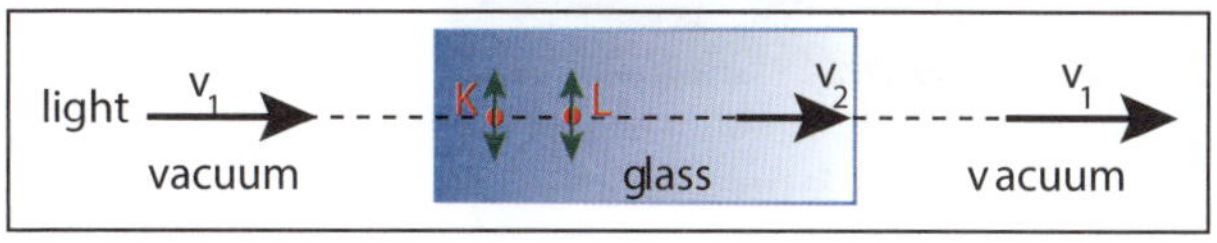

The speed of light in vacuum is 299792.458 km/s which is rounded to 300000 km/s most often. Speed of light in vacuum is commonly denoted by c.

$c = 3 \times 10^5\ km/s = 3 \times 10^8\ m/s$

The absorption and emission of light by the atoms of the medium take time. The speed of light reduces to 2×10^8 m/s in glass, 2.25×10^8 m/s in water, 1.24×10^8 m/s in diamond.

The ratio of the speed of light in vacuum to its speed in a transparent medium is called the refractive index.

$$n = \frac{c}{v}$$

Symbol	Quantity	Unit
c	speed of light in vacuum	m/s
v	speed of light in transparent medium	m/s
n	refractive index	(none)

Refractive index does not have a unit. The speed of light in vacuum is the limit of speed that anything in our physical world can have. This is why the refractive indices of different media will always be greater than 1.

The frequency of a wave is dependent on its source. A light wave entering to a different medium will still have the same frequency as it propagates in that medium.

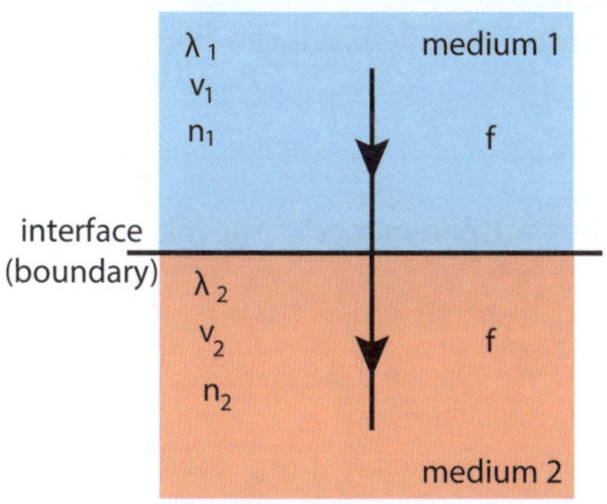

For a ray of light passing from transparent medium 1 to transparent medium 2:

$$\frac{n_1}{n_2} = \frac{c/v_1}{c/v_2} = \frac{v_2}{v_1}$$

The velocity of a wave is dependent on its frequency and also wavelength.

$$v = \lambda . f$$

$$\frac{v_1}{v_2} = \frac{\lambda_1 . f}{\lambda_2 . f} = \frac{\lambda_1}{\lambda_2}$$

$$\frac{n_1}{n_2} = \frac{v_2}{v_1} = \frac{\lambda_2}{\lambda_1}$$

Refractive indices of some media

Medium	n=c/v
Vacuum	1.0000
Air	1.0003
Water	1.33
Crown glass	1.52
Plexiglas	1.51
Sodium chloride	1.53
Diamond	2.42

The refractive index of a medium is inversely proportional to the speed and the wavelength of light in the medium. Recall from water waves, the depth of water was also inversely proportional to the speed and wavelength of water waves.

4.6.1 Law of refraction

When a light ray crosses the boundary between two transparent media, a small part of the light, like 5 % of the light rays, bounces back, in other words is *reflected to the medium of incidence.* As you are looking at a shop window, so often you see your reflection on the window glass. The rest of the light, the remaining 95 % of light rays, continues into the second medium. The transmitted light changes its direction as it crosses the boundary.

Recall that water waves were propagating slower in shallow media with a smaller wavelength. As they were propagating from a deep medium to a shallower one, the transmitted water waves were bending towards the boundary.

A polar bear in water looks fatter in water

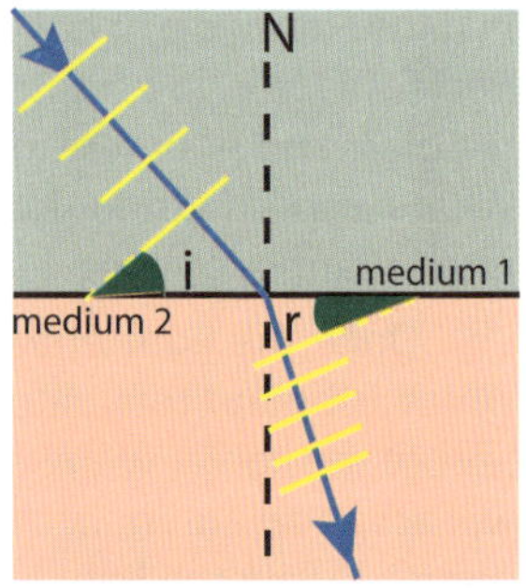

In the aside drawing; the wavelength in medium 1 is greater than the wavelength in medium 2. The speed of light should also be greater in medium 1 since it is directly proportional to the wavelength. In the case of light waves, speed and wavelength was inversely proportional to the refractive index of the medium. If the wavelength and speed are greater in medium 1, then its refractive index should be smaller than medium 2. Moreover, the light waves are bending towards the boundary so that the angle of refraction, r, is smaller than angle of incidence, i.

A pencil bent in a glass filled with water

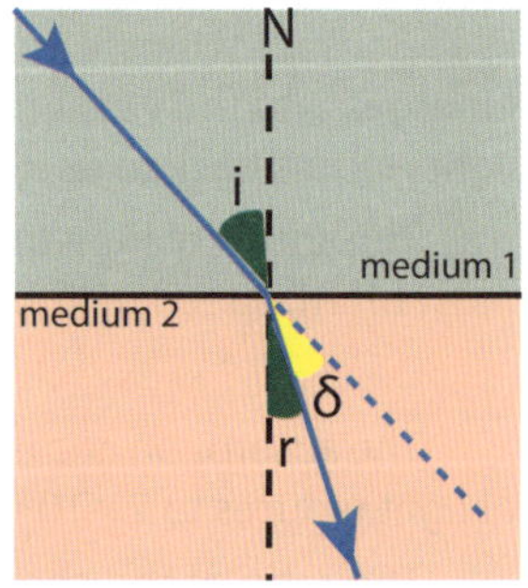

Remember that in optics calculations, the angles were measured with respect to the normal to the surface. If the geometry of the figure is studied, light bends away from the normal as it is passing from a greater refractive index medium to a smaller one. Aside, i is the angle of incidence, r is the angle of refraction and d is the angle of deviation.

$$\delta = |i - r|$$

Quantity	Comparison
Wavelength	$\lambda_1 > \lambda_2$
Speed	$v_1 > v_2$
Angle	$i > r$
Refractive index	$n_1 < n_2$

Are the pens broken or....?

The transmission of light from one transparent medium to another with a change in the propagation direction is called refraction of light. The refraction occurs only at the boundary. Once the light has crossed the boundary between the two media, it continues to travel in a straight line. Only now, the direction of that line is different than it was in the former medium.

A medium with a greater refractive index is called an optically more dense medium. A medium with a smaller refractive index is called an optically less dense medium.

Physics Extra

The angle of refraction depends on the angle of incidence and the refractive indices of the transparent media. The relationship is given by Snell's Law of Refraction.

$$n_i.sin\theta_i = n_r.sin\theta_r$$

Symbol	Quantity	Unit
n_i	refractive index of the incidence medium	-
n_r	refractive index of the refractive medium	-
$sin\theta_i$	sine of the angle of incidence	-
$sin\theta_r$	sine of the angle of refraction	-

Physics Extra

Example

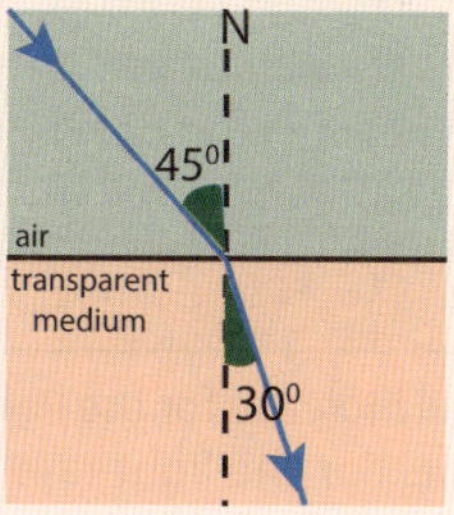

The path followed by a light ray as it enters to a transparent medium from air is given in the figure.

Find the refractive index of the medium.

$$\left(\sin 45 = \frac{1}{\sqrt{2}};\ \sin 30 = \frac{1}{2};\ n_{air} = 1 \right)$$

Solution

According to Snell's law:

$$n_i . \sin\theta_i = n_r . \sin\theta_r$$

$$1 . \sin 45 = n . \sin 30$$

$$1 . \frac{1}{\sqrt{2}} = n . \frac{1}{2}$$

$$n = \sqrt{2} \simeq 1.4$$

4.6.2 Critical Angle and Total Internal reflection

A light wave incident from an optically less dense medium will bend towards the normal as it is passing to an optically more dense medium.

Quantity	Comparison
Wavelength	$\lambda_{less} > \lambda_{more}$
Speed	$v_{less} > v_{more}$
Angles	$i > r$
	$i = i'$
	$\delta = i - r$

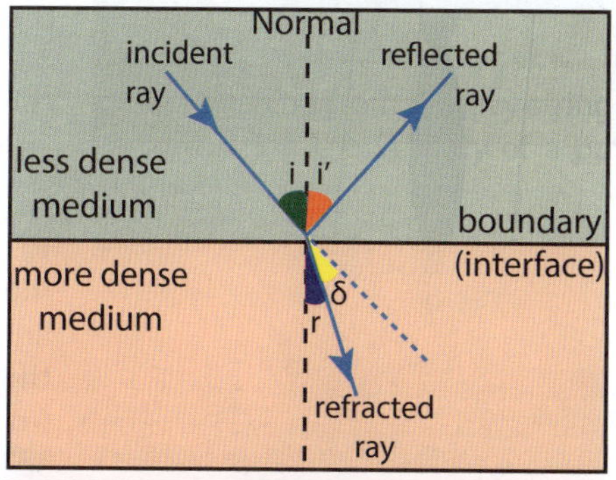

Keeping reversibility of light in our mind, we can expect light to bend away from the normal as it is passing from an optically more dense medium to a less dense one.

Quantity	Comparison
Wavelength	$\lambda_{more} < \lambda_{less}$
Speed	$v_{more} < v_{less}$
Angle	$i < r$
	$i = i'$
	$\delta = r - i$

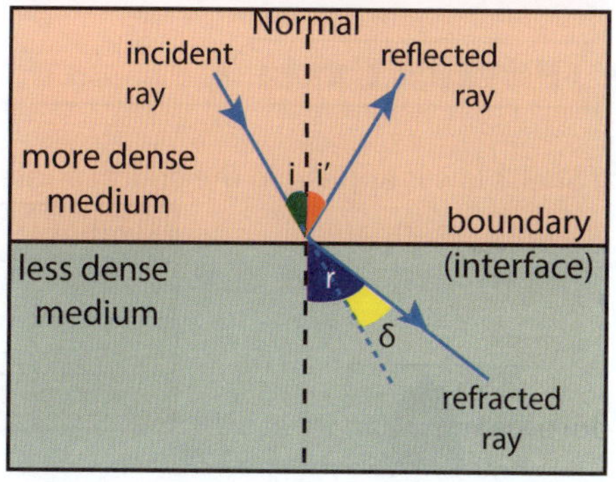

Light is always transmitted from a less dense medium to a more dense one. You can always see the fish in an aquarium if you are in air. What about the fish in aquarium? Can they see us always? Or is there a limit for this?

The interface acts as a mirror on the boundary of water and air as you look from the water to air. Why?

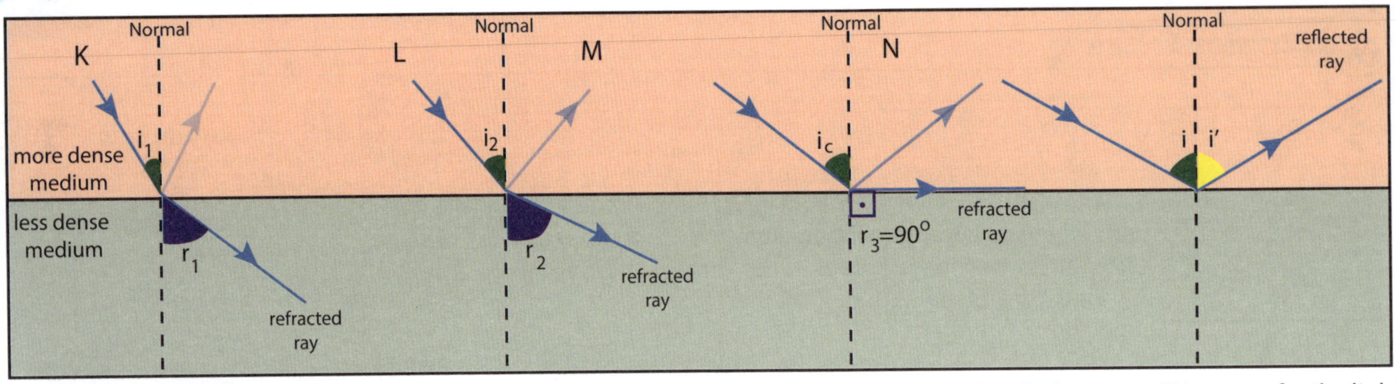

As the angle of incidence increases, the angle of refraction will increase for the light rays transmitted from a more dense medium to a less dense one. For the light rays K and L in the figure above $i_1 < i_2$ then $r_1 < r_2$. If the angle of incidence is increased more, angle of refraction becomes 90° at a certain value. This value of incident angle is called the critical angle. Light rays cannot pass to the less dense medium if angle of incidence is equal to the critical angle, instead they propagate at the interface. Any light ray which has a value of angle of incidence greater than the critical angle will not be transmitted to the less dense medium. The light rays will be reflected back. This is called total internal reflection (TIR).

If we recall the analogy we had for the passage of soldiers from concrete ground to mud, concrete can be referred as the less dense medium and mud can be referred as more dense medium. Soldiers can always walk from concrete ground to mud (transfer of light from less dense medium to a more dense one). But if the soldiers are in a swamp, they may not be able to get out of the mud and pass to concrete ground (transfer of light from more dense medium to a less dense one). Moreover walking on concrete is easier so the soldiers will move faster on concrete ground.

Physics Extra

If Snell's law is applied in the critical angle case:

$$n_i . \sin\theta_i = n_r . \sin\theta_r$$
$$n_i . \sin\theta_c = n_r . \sin 90$$
$$n_i . \sin\theta_c = n_r . 1$$
$$\sin\theta_c = \frac{n_r}{n_i}$$

Symbol	Quantity	Unit
n_i	refractive index of the incidence medium	-
n_r	refractive index of the refractive medium	-
$\sin\theta_c$	sine of the critical angle	-

According to the equation we derived for critical angle, as the difference of refractive indices increase, the critical angle decreases. As a result, total internal reflection occurs easier.

Example

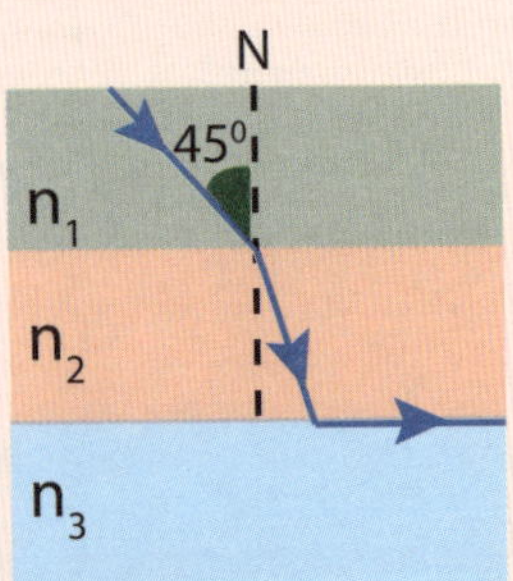

The path followed by a light ray as it enters to a transparent media is given in the figure. The refractive indices of the media are:

$$n_1 = \sqrt{2} \; ; \; n_2 = \sqrt{3}$$

Find the refractive index of the third medium, n_3.

$$\left(\sin 45 = \frac{1}{\sqrt{2}} \; ; \; \sin 90 = 1 \right)$$

Solution

$$n_i . \sin\theta_i = n_r . \sin\theta_r \text{ (Snell's Law)}$$

$$\left.\begin{aligned} \sqrt{2} . \sin 45 &= \sqrt{3} \sin\theta_r \\ \sqrt{3} \sin\theta_r &= n_3 . \sin 90 \end{aligned}\right\} \sqrt{2} . \sin 45 = n_3 . \sin 90$$

$$\sqrt{2} . \sin 45 = n_3 . \sin 90$$

$$\sqrt{2} . \frac{1}{\sqrt{2}} = n_3 . 1$$

$$n_3 = 1$$

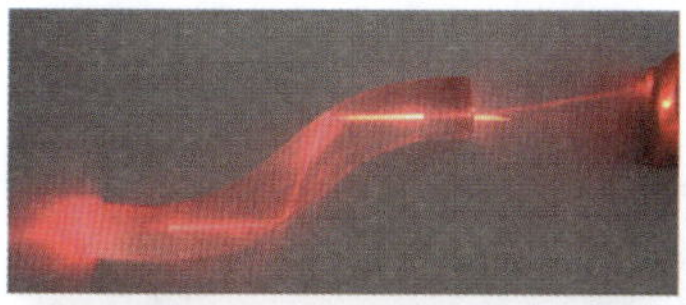

Although light travels in straight paths, using TIR; the propagation direction of light can be changed. Light is *confined* to travel within these glass optical fibers even on a curved path. A thin glass strand will be flexible when compared to a thick glass rod.

When a light ray is shone at one end of the fiber in such a way that it strikes the glass surface with an angle of incidence greater than the critical angle of the glass, TIR occurs. If the fiber were straight, the light ray will undergo multiple reflections and only emerge at the other end of it. The fiber usually has an outer plastic coating which helps to prevent the glass surface from being scratched. A damaged glass surface may allow light to escape through its side rather than at its end. There is very little loss of light intensity within optical fibers. This is due to reflections from the two ends of the fiber and the absorptions by the fiber material. Optical fibers are used in endoscopes to see the image of internal organs. They can carry much higher volume of telephone calls or other forms of communication than the electrical wires.

Refraction occurs even if the change of refractive index is gradual. Although the refractive index of air is approximately equal to 1; it is slightly larger than 1 ($n_{air} = 1.0003$). Its value depends on the density, hence the temperature of air. On a hot summer day, the layer of air just above the black asphalt road surface is considerably hotter than the air which is several centimeters above the road. A ray of light approaching to the road is refracted away from the normal as it propagates in the hot layer of air. The light ray may re-enter the upper layer of air as though it had been reflected by the road. This effect stimulates a presence of layer of water on the road some distance away from the car. Thus the driver of the car will observe a mirage on the road.

Mirages formed on the dessert

Mirages formed on the road

As you are looking out from the window, the light rays are refracted on the closer side of the glass boundary as they are passing from air to glass. As they are leaving the window glass, a second refraction takes place on the opposite boundary of the glass. The glass pane used in a window is relatively thin. Our eye cannot distinguish the difference of path for the transmitted light.

But if the glass slab is wider so that the light has a longer path to pass, the shift in the path of light can easily be observed. If the geometry of the incident ray and the refracted rays is analyzed, it can easily be seen that the angle of incidence in the first boundary is equal to the angle of refraction in the second boundary. We can conclude that the direction of light is not changed in passing from air to (glass to) air. The second refracted light ray in air is parallel to the incident light ray from air.

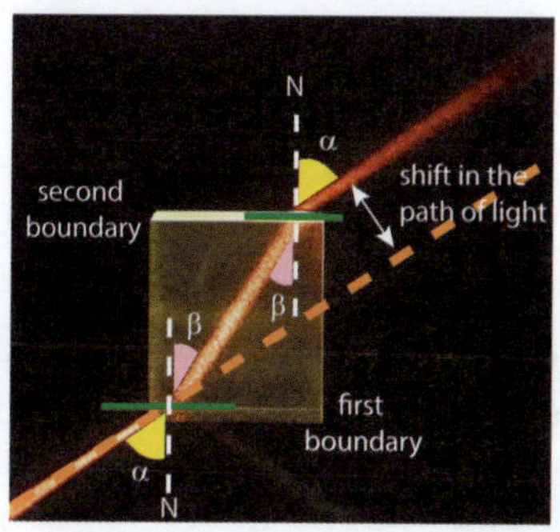

The first photo is the image of a pencil placed in water and the second photo is for the same pencil but placed in an emulsion of oil and water. In the third photo, the part of pen in water is cropped and pasted over the pencil in the second photo (the red shaded part). Although there is a slight deviation in oil (the slight blue parts in the oil layer), the red shaded part completely covers the portion of pencil in water. Removing the oil layer wouldn't effect the path followed by light as it is passing from air to water. As we are comparing the propagation of light in three different media, a comparison between the first and third layers can be made.

As a mathematical proof, remember the ratio we had written for the refractive indices and velocities of water in different media.

$$\left.\begin{aligned}\frac{n_1}{n_2} = \frac{v_2}{v_1} \Rightarrow n_2 = \frac{v_1}{v_2}n_1 \\ \frac{n_2}{n_3} = \frac{v_3}{v_2} \Rightarrow n_2 = \frac{v_3}{v_2}n_3\end{aligned}\right\} \frac{v_1}{v_2}n_1 = \frac{v_3}{v_2}n_3 \Rightarrow \frac{n_1}{n_3} = \frac{v_3}{v_1}$$

Example

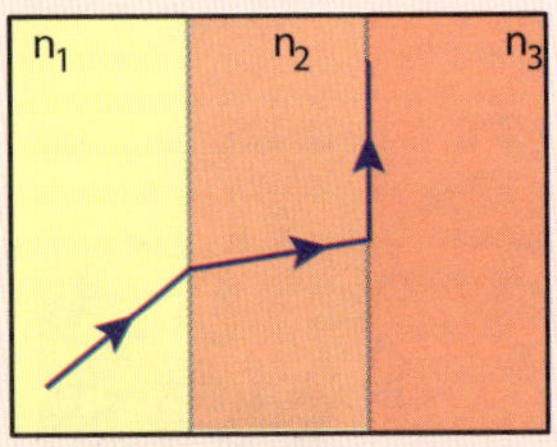

The path followed by a monochromatic light ray in different transparent media is given in the figure.

a. What is the relationship between the refractive indices?

b. What is the relationship between the speed of light in these media?

Solution

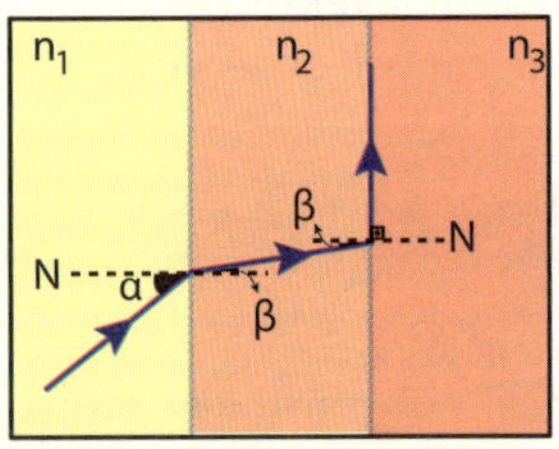

Since $\alpha > \beta$; $n_1 < n_2$ and $90 > \beta$; $n_3 < n_2$

A comparison can be made between first and third media as if the second medium is not present, as explained above.

$\alpha < 90$ so $n_1 < n_3$

$n_3 < n_1 < n_2$

Refractive indices are inversely proportional to the speeds:

$v_2 < v_1 < v_3$

4.6.4 Apparent Depth

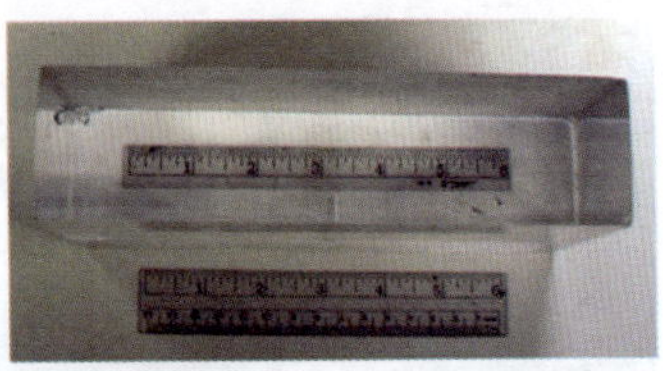

Even though both rulers are placed on the table, and photographed from the same height, the ruler below the glass slab observed closer to the eye.

As you are looking down into the water in a pool or sea (or even a glass full of water) it is usually hard to decide on the depth of water. The bottom seems closer to you. But usually picking that beautiful sea shell is harder once you stretch out to take it out of water. This is another result of refraction of light that we observe in our daily life.

As we have mentioned, light rays emerging from an object should reach to our eye to be seen. In the case of objects in a different medium; the light rays refract as they are transmitted from one medium to the other. The image of the object is formed at the intersection of the refracted light rays reaching to our eye.

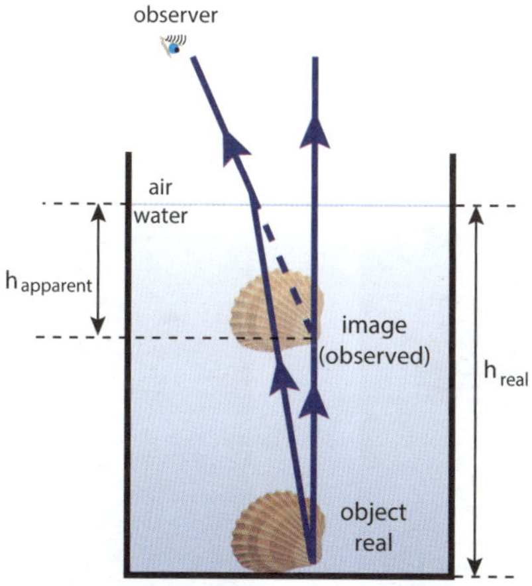

For an observer *in a less dense medium and looking at a more dense medium from normal direction,* the objects will appear *closer* to the observer. The light rays from the object to the observer bend away from the normal to the surface. The object would be observed *closer and larger* if the observer looks from *a greater angle.*

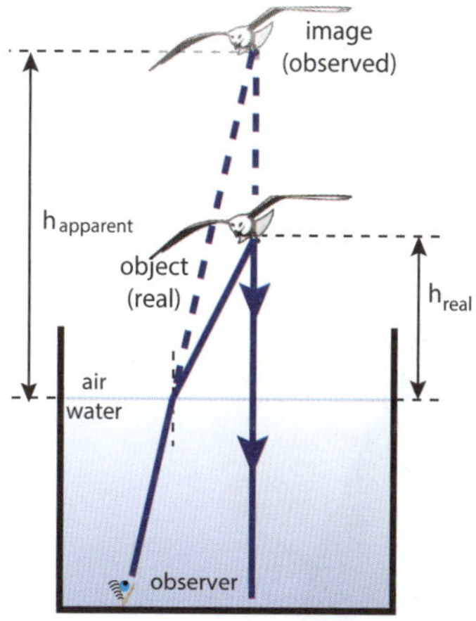

For an observer *in a more dense medium and looking at a less dense medium from normal direction,* the objects will appear *farther* from the observer. The light rays from the object to the observer bend closer to the normal to the surface. The object would be observed *farther and smaller* if the observer looks from a *greater angle.*

Physics Extra

In order to find the apparent height of the object:

$$h_{apparent} = h_{real} \cdot \frac{n_{observer}}{n_{object}}$$

Symbol	Quantity	Unit
$n_{observer}$	refractive index of the medium where the observer is	-
n_{object}	refractive index of the medium where the object is	-
h_{real}	the real height of the object	m
$h_{apparent}$	the observed height of the object	m

Example

A girl is observing an object through a glass pane with a thickness of 6 cm. If the object is 30 cm away from the pane and the girl is 16 cm away from the glass pane.

How far away will she see the object?

($n_{air} = 1$, $n_{glass} = 1.5$)

Solution

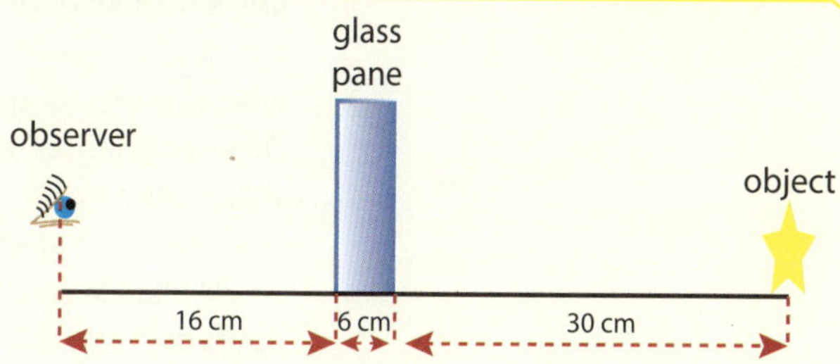

$$h_{apparent} = h_{real} \cdot \frac{n_{observer}}{n_{object}} = 6 \cdot \frac{1}{1.5} = 4 \text{ cm}$$

$$x_{total} = 30 + 4 + 16 = 50 \text{ cm}$$

4.7. Lenses

The photos are examples of images formed by transparent media. With these refractive media a daisy can fit in a raindrop, the sky any the trees can fit in a glass of water, a house can fit into a drop of water carried by an ant.

If a piece of glass or other transparent material takes the appropriate shape, it becomes capable either converging incident parallel rays to a point or appear to diverge them from a point. Such a piece of glass is referred to as a lens. Lenses are transparent materials that uses refraction of light to form the image of an object.

Being able to converge the light rays is a useful property. However, the trash that people leave behind, after spending a beautiful day under the trees, may result forest fires along with causing environmental pollution. The glass bottles may converge the light rays from the sun and cause them to focus at some point. If the light rays focus for enough time, the dry leaves on the forest ground can reach to the flammability temperature. This will result a forest fire. Collecting your trash as you are leaving your picnic area is the easiest way to prevent a forest fire which may be caused by the glass bottles.

4.7.1 Properties and Types of Lenses

There are a variety of types of lenses. Lenses differ from one another in terms of their shape and the materials from which they are made.

A converging lens converges rays of light which are traveling parallel to its principal axis. Converging lenses can be identified by their shape; they are thicker across their middle and thinner at their upper and lower edges.

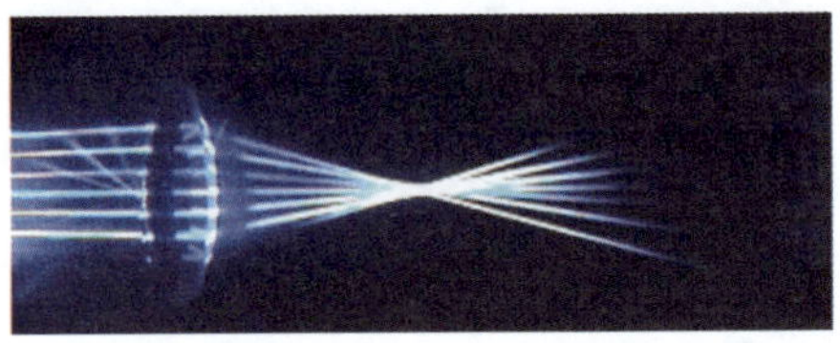

A double convex lens is symmetrical across both its horizontal and vertical axis. Each face of the lens thought of as being part of a sphere. The fact that a lens is thicker across its middle and thinner at its ends is an indicator that it will converge rays of light which travel parallel to its principal axis. A double convex lens is a converging lens.

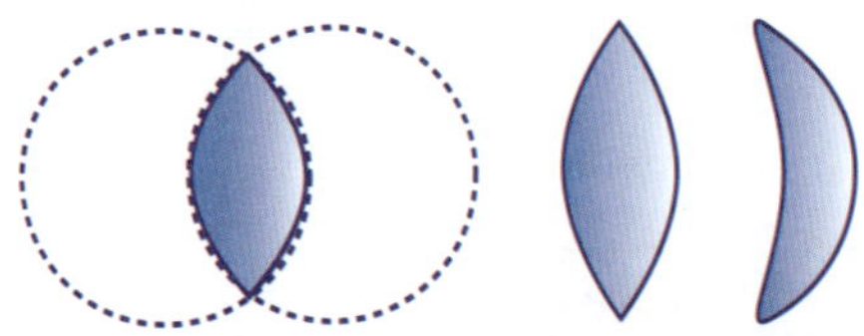

A diverging lens diverges rays of light which are traveling parallel to its principal axis. Diverging lenses can be identified by their shape; they are thinner across their middle and thicker at their upper and lower edges.

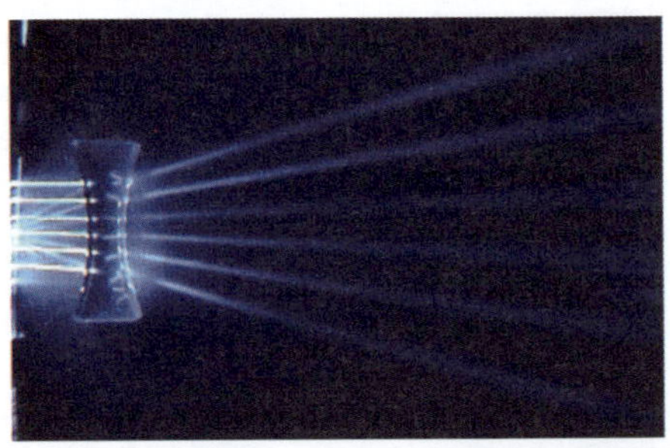

A double concave lens is also symmetrical across both its horizontal and vertical axis. The two faces of a double concave lens can be thought of as originally being part of a sphere. The fact that a lens is thinner across its middle and thicker at its ends is an indicator that it will diverge rays of light which travel parallel to its principal axis. A double concave lens is a diverging lens.

A symmetrical lens can be thought of as being a slice of a sphere. Then the lens will have two symmetry axes. The horizontal symmetry axis of the lens is known as the principal axis of a lens. An imaginary vertical axis bisects the symmetrical lens in two. The intersection of the horizontal and vertical symmetry axes is called the optical center of a lens.

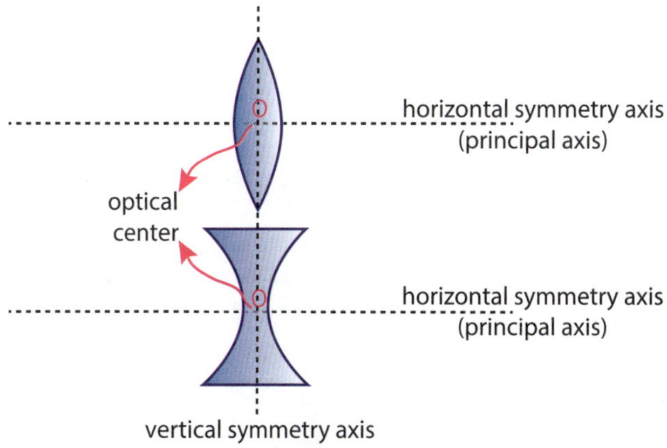

Light rays may be incident towards either face of a lens. A lens may converge or diverge a beam of light which are parallel to the principal axis. The point at which the parallel light rays are converging to is known as the focal point of the converging lens. If the light rays diverge, then the diverging rays can be traced backwards until they intersect at a point. This point is known as the focal point of a diverging lens. The focal point is denoted by the letter F on the diagrams. Each lens has two focal points; one on each side of the lens. Unlike mirrors, lenses can allow light to pass through either face, depending on where the incident rays are coming from. The focal point on the side from which the light rays are incident from the object are approaching the lens is called the near focal point. The focus at the opposite side from which the refracted light rays are approaching is called the far focal point. The distance from the optical center to the focal point of a lens is known as the focal length. Since the refracted light rays themselves are converging to the focus, the focal length of a converging lens is a real length for the lens, and considered as a positive value. However, since the extensions of the light rays appear to diverge from the focus of the diverging lens, its focal length is considered as a virtual length and taken as a negative value. The 2F point is an imaginary point on the principal axis and it is 2f distance from the optical center of the lens.

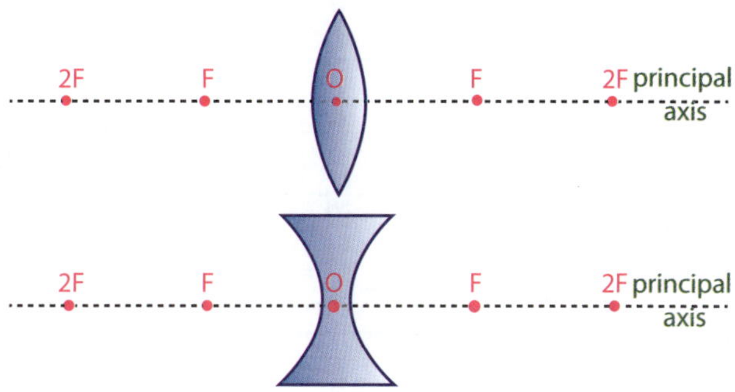

Both surfaces of a lens need not to have the same radius. A lens may have a concave and a convex surface and can still display the properties of a converging lens. The focal length of a lens is dependent on the radius of the curvatures of the spherical surfaces. In the most general case, lenses thinner at the ends and thicker at the middle display properties of converging lenses and the ones thicker at the ends and thinner in the middle display properties of a diverging lenses.

Parallel light rays are expected to be converged by the first three lenses whereas they are expected to be diverged by the final three lenses in the figure aside.

The refractive index of the transparent media plays role in the focal length of a lens also. If a concave lens is placed in a denser medium, it diverges parallel light rays just like a diverging lens. Likewise a diverging lens behave like a converging lens if it were to be placed in a denser medium.

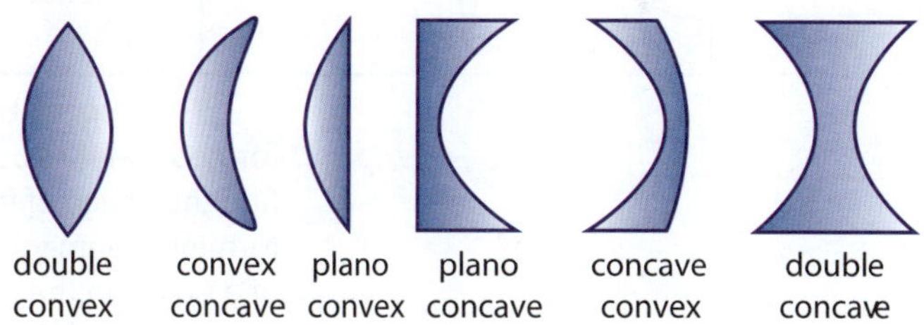

4.7.2. Images Formed by Lenses

Lenses are refractive media. They use refraction of light to form the image of an object.

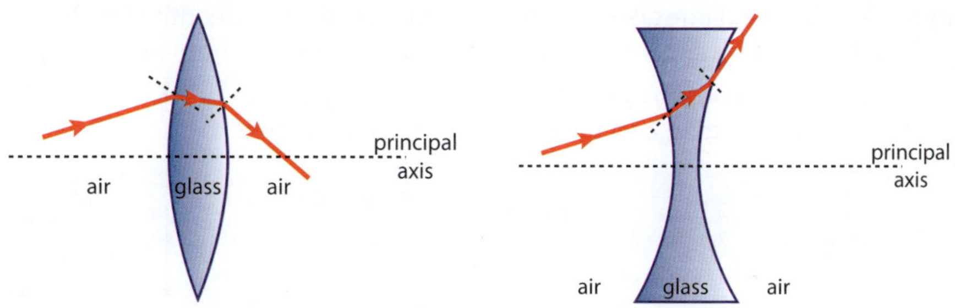

However, just like the special light rays we had in spherical mirrors, there are some special light rays for lenses also. These rays obey the law of refraction as they are being transmitted from one side of the lens to the other. But working with them is easier in drawing the image of an object formed by a lens.

Converging lens			Diverging lens		
Incident light ray	Drawing	Refracted light ray	Incident light ray	Drawing	Refracted light ray
Parallel to the principal axis	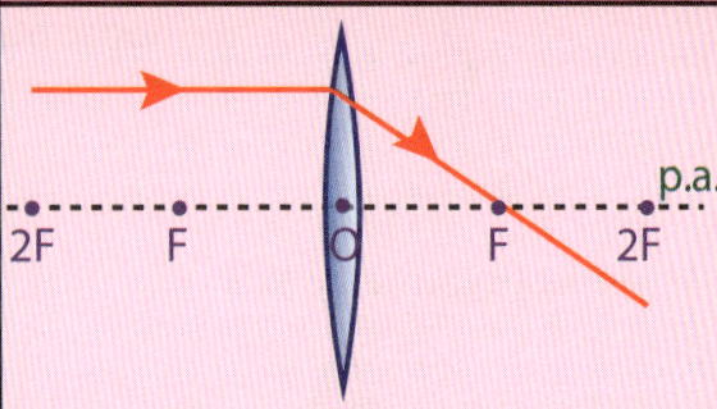	Passes through the focus	Parallel to the principal axis	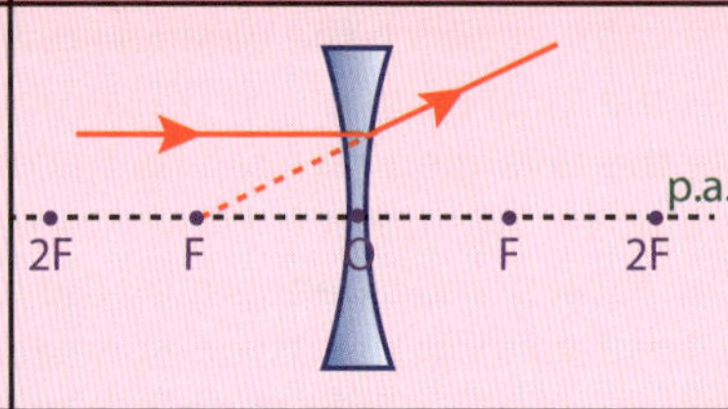	Extensions passes through the focus (diverges)
Coming from the focus	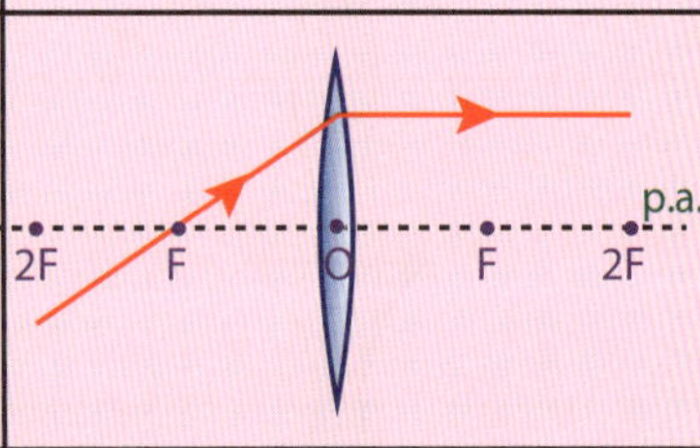	Parallel to the principal axis	Towards the focus	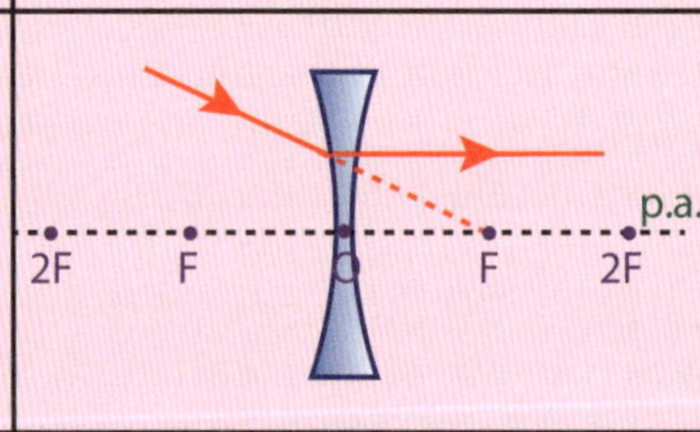	Parallel to the principal axis
Coming from the 2F point	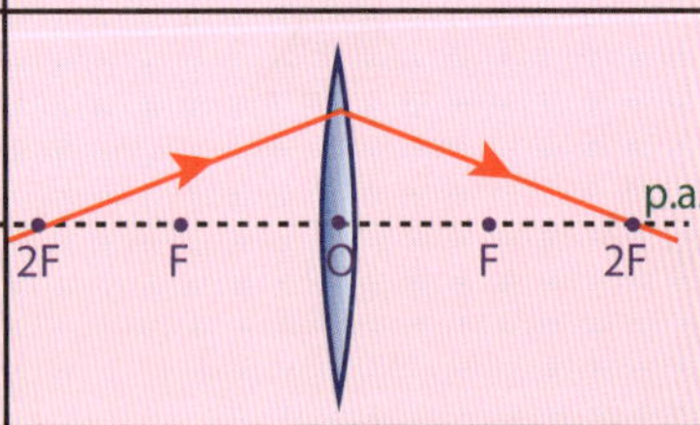	Passes from the 2F point	Coming in the direction of the 2F point	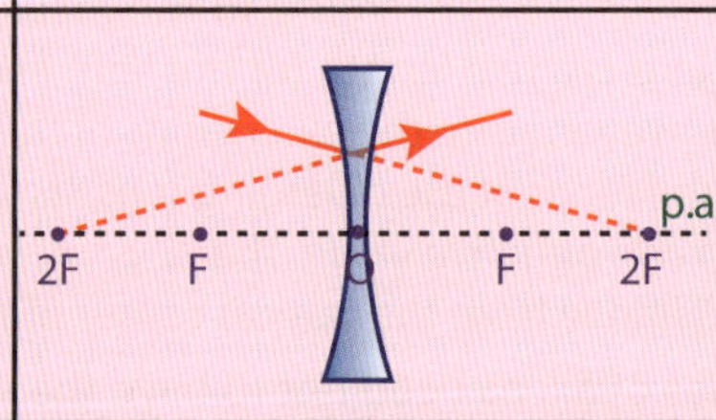	Its extension passes from the 2F point
Towards the optical center	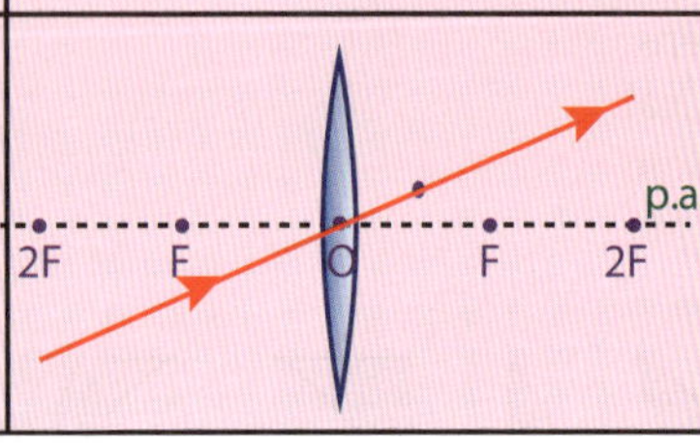	Continues in the same direction, no refraction	Towards the optical center	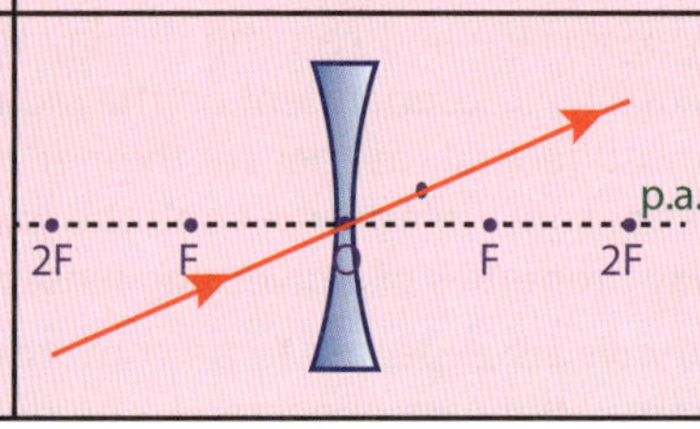	Continues in the same direction, no refraction

These special light rays will be helpful to us as we are drawing the image of an object formed by lenses. Just as we did for finding the image of an object with mirrors, we will find the image of top and bottom points of the object with respect to the lens. As we are forming the image of a point, we will not work on points on the principal axis. The image of a point on the principal axis will be formed on the principal axis. For other points we will use any of two special light rays to find the image of that point.

If the refracted light rays are intersecting, the images formed are real images. Real images can be projected upon a screen. If the refracted light rays are not intersecting, then the image formed is a virtual image.

For a converging lens, we will begin with finding the image of an object at infinity. Infinity to a lens is a very large distance when compared to the focal length of the lens. Sun is an object at infinity for a lens. Then we will get the object closer to the mirror step by step.

As we are drawing the images we will use the aside symbols to represent the converging and diverging lenses for an easier sketching.

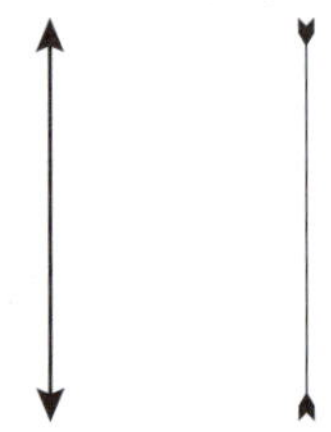

Position of the object	Drawing	Properties of the image
At infinity	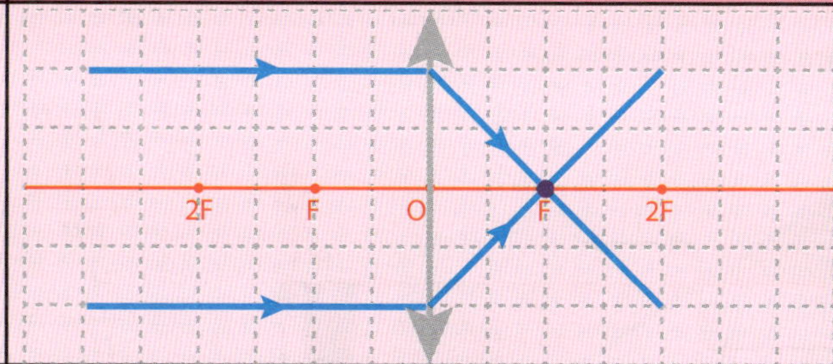	1. At focus 2. A point 3. Real (refracted light rays are intersecting)
Beyond the center	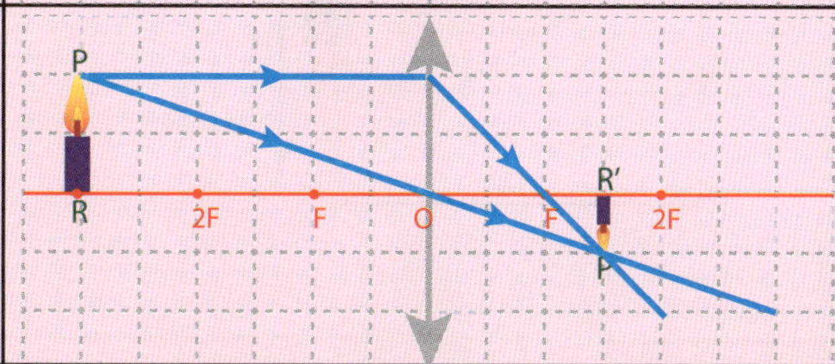	1. Between focus and 2F 2. Smaller than the object (diminished) 3. Real 4. Inverted (upside down)
At center	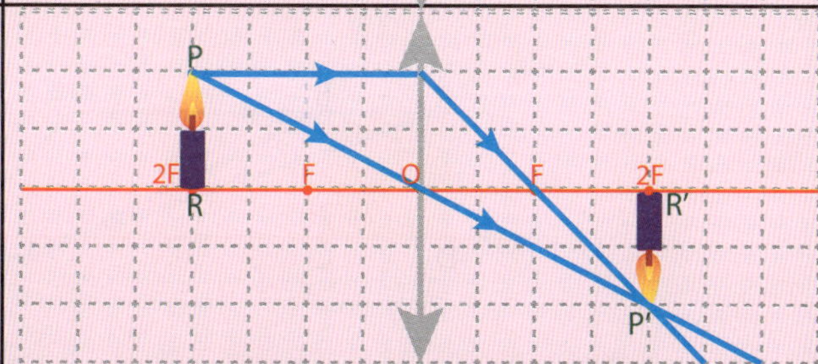	1. At 2F 2. Same size as the object 3. Real 4. Inverted
Between focus and center	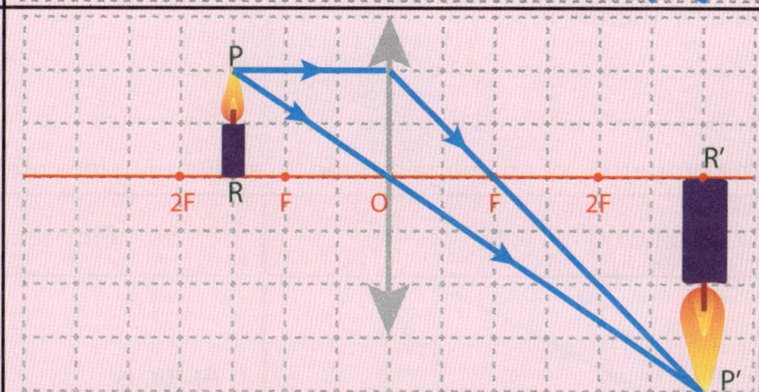	1. Beyond the 2F 2. Larger than the object (magnified) 3. Real 4. Inverted
At focus	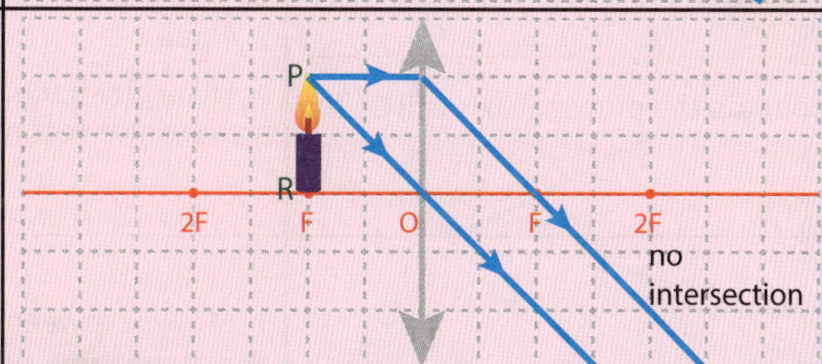	1. At infinity
Between focus and optical center	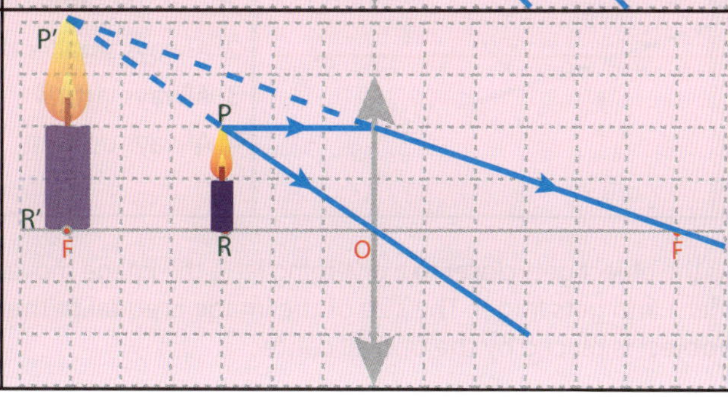	1. Behind the object and at the same side with the object 2. Larger than the object (magnified) 3. Virtual 4. Upright

As the object gets closer to the converging lens, the image gets away from the lens and magnifies. The relationship between the distance of the image to the lens and the distance of the object to the lens is not a constant one. Let's analyze the geometry of the image formed below:

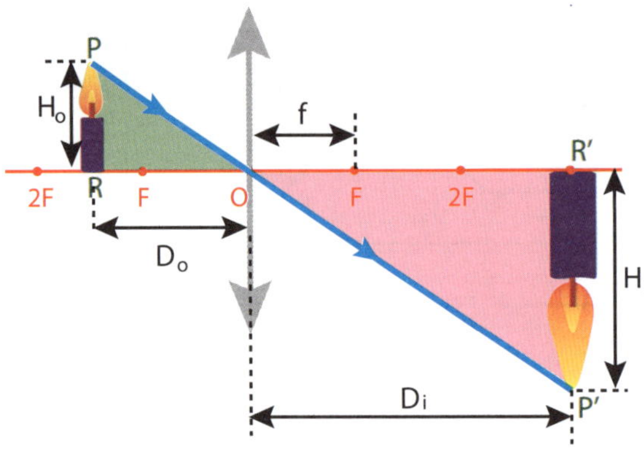

$$\Delta POR \approx \Delta P'O'R'$$

$$\frac{P'R'}{PR} = \frac{O'R'}{OR}$$

$$\frac{H_i}{H_o} = \frac{D_i}{D_o} \quad (magnification)$$

Example

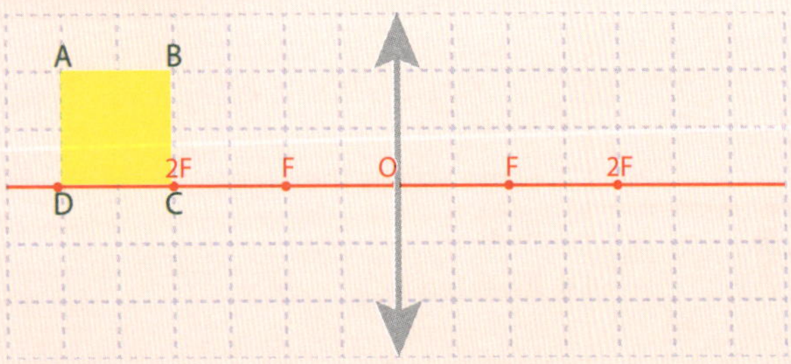

Draw the image of the object ABCD formed by the diverging lens.

Solution

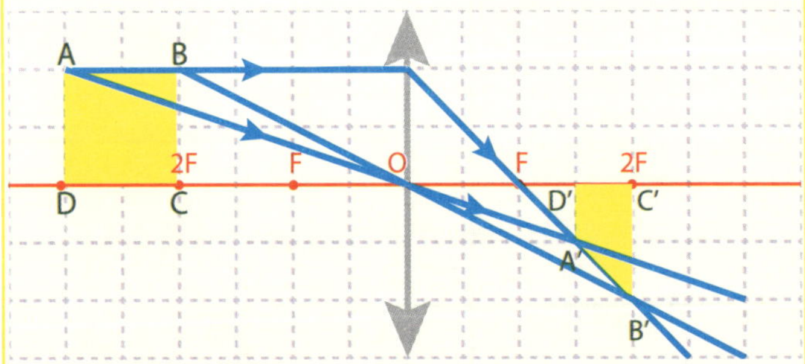

The properties of an image formed by a diverging lens are as follows:

Position of the object	Drawing	Properties of the image
At infinity	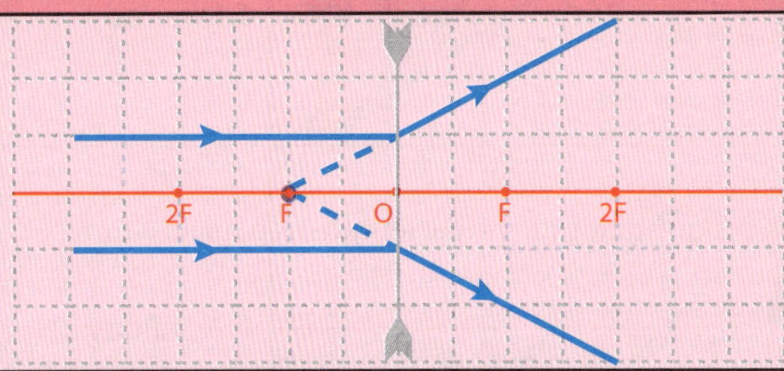	1. At focus 2. A point 3. Virtual (extensions of the refracted light rays are intersecting)
In front of the lens	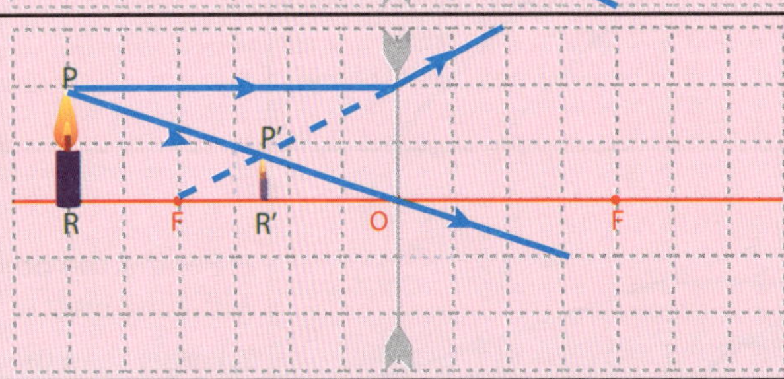	1. Always between focus and optical center and at the same side as the object. 2. Always smaller than the object 3. Always virtual 4. Always upright

As the object gets closer to a diverging lens, the image moves from the focus towards the lens and gets larger. The size of the image is equal to the size of the object when the object is placed by the optical center.

Example

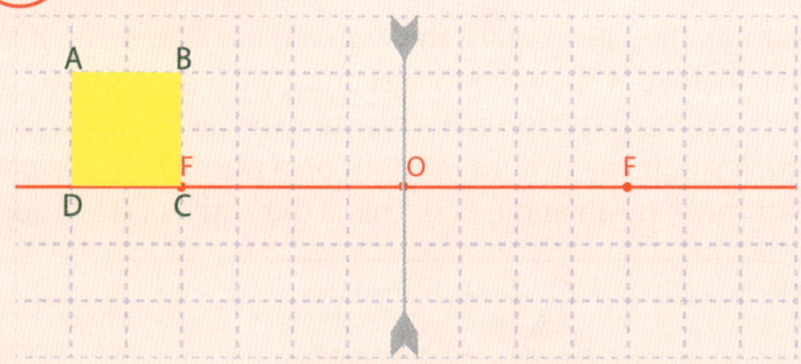

Draw the image of the object ABCD formed by the diverging lens.

Solution

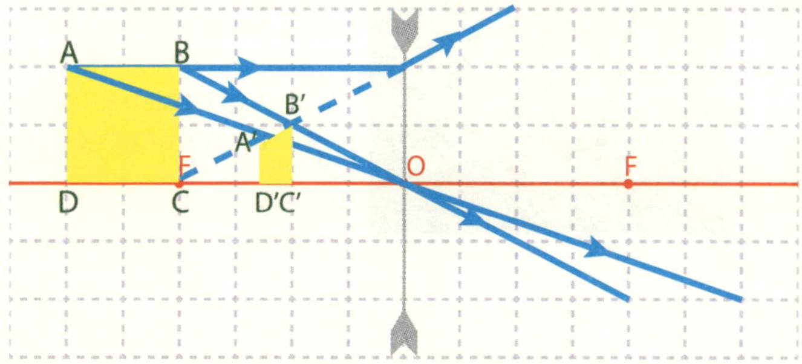

Physics Extra

Equations used in thin lenses:

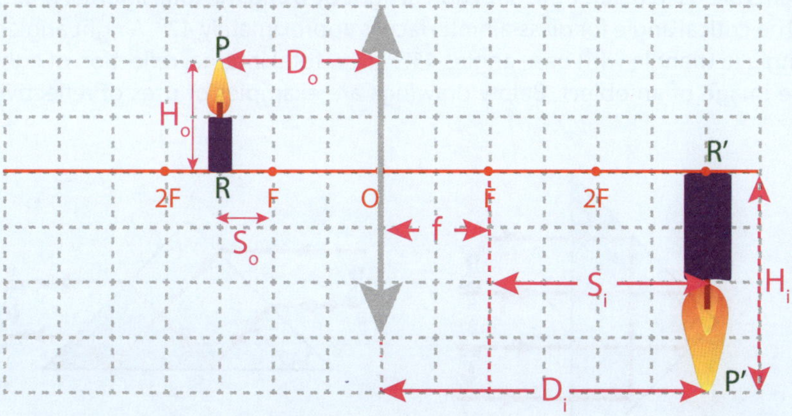

Symbol	Quantity	Unit
f	focal length	m
D_o	distance of the object to the lens	m
D_i	distance of the image to the lens	m
H_o	height of the object	m
H_i	height of the image	m
S_o	distance of the object to the focus	m
S_i	distance of the image to the focus	m

Location of the image or the object: $\pm\frac{1}{f} = \pm\frac{1}{D_o} \pm \frac{1}{D_i}$

Magnification: $\frac{D_o}{D_i} = \frac{H_o}{H_i} = \frac{S_o}{f} = \frac{f}{S_i}$

Focal length: $f^2 = S_o . S_i$

The (+) sign stands for real lengths whereas (-)sign stands for virtual lengths. A parallel beam of light is converged at the far focus of a converging lens so the focal length of a converging lens is real (+). A parallel beam of light is diverged at the near focus of a diverging lens so the focal length of a converging lens is virtual (-) for a diverging lens.

Example

The image formed by a lens is 3 times larger than the object and upright. The distance between the image and the lens is 120 cm.

a. What is the type of the lens?

b. Find the focal length of the lens.

Solution

a. A larger and upright image is a virtual image in a converging lens

b. $\frac{D_o}{D_i} = \frac{H_o}{H_i} \Rightarrow \frac{D_o}{D_i} = \frac{1}{3} \Rightarrow D_i = 3D_o$

$D_i = 3D_o = 120 \ cm \Rightarrow D_o = 40 \ cm$

$\pm\frac{1}{f} = \pm\frac{1}{D_o} \pm \frac{1}{D_i}$

$\frac{1}{f} = \frac{1}{40} - \frac{1}{120} = \frac{3-1}{120} = \frac{2}{120} = \frac{1}{60} \Rightarrow f = 60 \ cm$

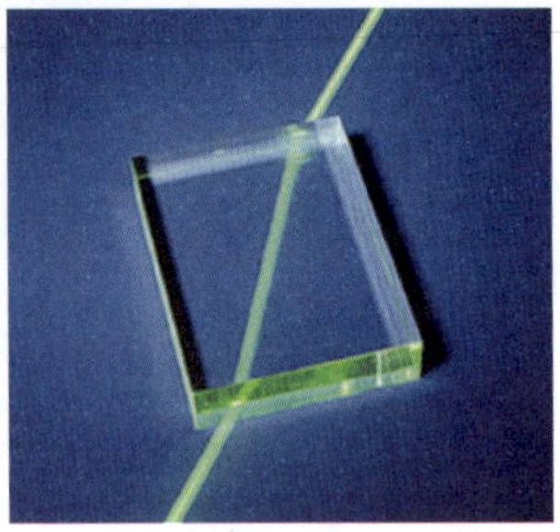

4.8. Prisms

4.8.1 Light Prisms and Their Uses

The light rays shift their path as they propagate through a glass slab in air. If the faces of the slab are not parallel, instead if they are intersecting with each other; the light ray cannot propagate with the same direction as it re-enters the air medium. Such a glass slab is called a light prism. A single ray of light incident on a prism emerges bent away from its original direction of travel by an angle of d. This is the angle of deviation.

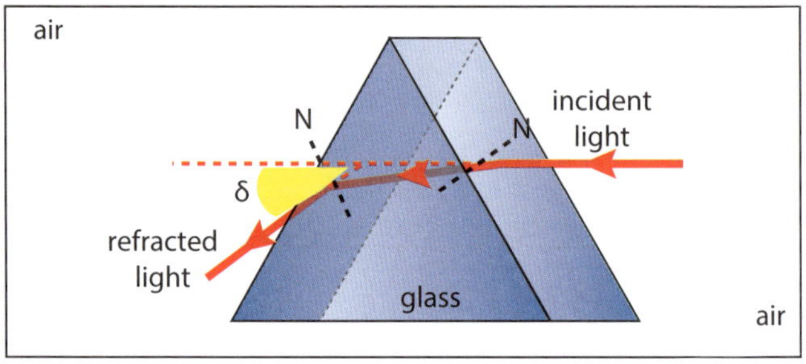

Light prisms are used to alter the direction of light by using the total internal reflection of light. The critical angle for glass-air interface is approximately 42°. A right angled isosceles triangle (a triangle with base angles 45°) allow total internal reflection of light and alters the image of an object. Below drawings are examples of uses of reflective prisms.

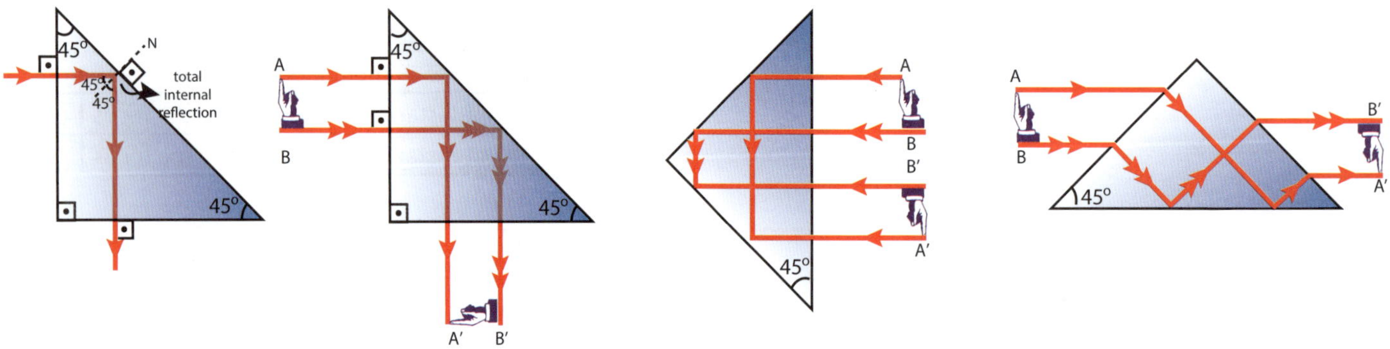

A submarine periscope, a binocular uses total internal reflective light prisms to form the image of an object.

In binoculars, the lenses are much more apart from each other than our eyes. The light rays are needed to be brought together before exiting the eyepieces. Rather than using a mirror which get dirty and easily scratched and require alignment, binoculars use a pair of prisms on each side. The light undergoes two total internal reflections and emerges from the eyepiece.

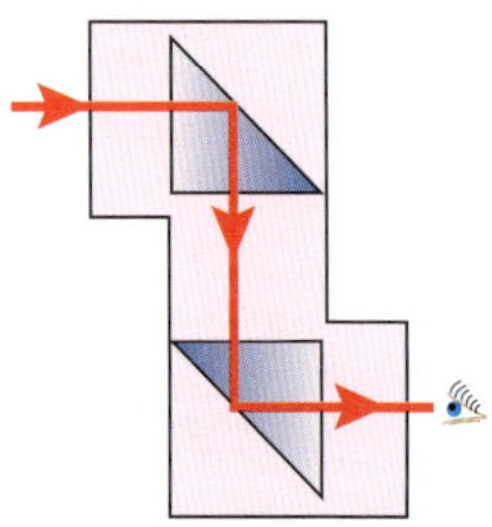

Schematic drawing of periscope

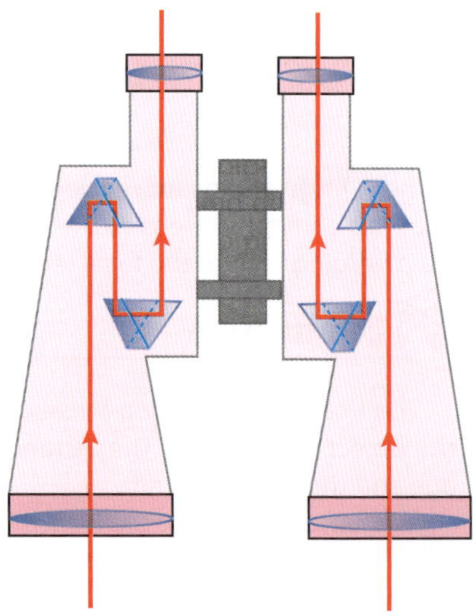

Binoculars
(Schematic drawing)

It was readily known by people that if light is passed through a triangular glass slab (a glass prism), white light would disperse into the colors of the rainbow. Before Newton's studies, people were thinking that the prism was giving color to light.

Newton passed a beam of sunlight through a prism first. Then he used a second prism inverted with respect to the first one. What he had observed was a beam of pure white light on the other side of the second prism. He also observed another color of light on the other side of the second prism if he let a portion of light to pass through the second prism. Only if he let all the colors of the rainbow to pass through the second prism, he had white light on the other side of the second prism. He concluded that color is associated with the light itself. The prisms had nothing to do with the colors.

Different colors experience different refractive indices as they are propagating through a prism. This results different refraction angles for each color of light. The refractive index for violet is greater than that of red. This is why, red appears at the uppermost part of the spectrum as the light rays are leaving the glass prism. Red refracts the least since the refractive index for it is the smallest. Violet refracts the most and observed at the lowermost part of the spectrum.

The slight variation of refractive index with varying wavelength, in other words, color is known as dispersion. For a greater wavelength a smaller refractive index will be experienced.

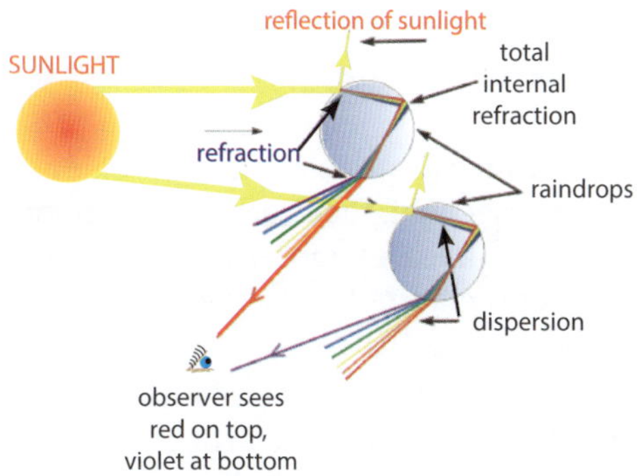

Rainbows are spectacular displays of dispersion. The raindrops acts like prisms and disperse the sunlight. Light experiences total internal reflection within the raindrop as depicted in the figure aside. As a result, red appears at top and violet appears at the bottom of a rainbow. An observer sees only one color from one raindrop depending on his or her position.

To see a rainbow, sun must be shining on one part of the sky and, rain should be falling on the opposite side of the sky. If you turn your back to the sun, you can observe the *bow of spectrum* in sky.

A secondary rainbow with reversed colors is usually observed at a greater angle around the primary bow. It is a result of double reflections within the raindrops. Because of extra reflections the secondary bow is usually much dimmer than the primary bow.

If a rainbow is observed high enough from an airplane a complete circle rather than a bow will be observed. If there weren't the ground all the rainbows would appear circular.

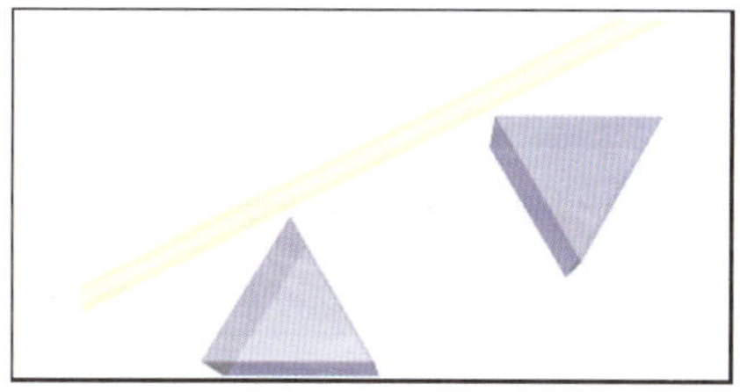

Two glass prisms and white light.

White light passes through the first prism and gives the colors of rainbow.

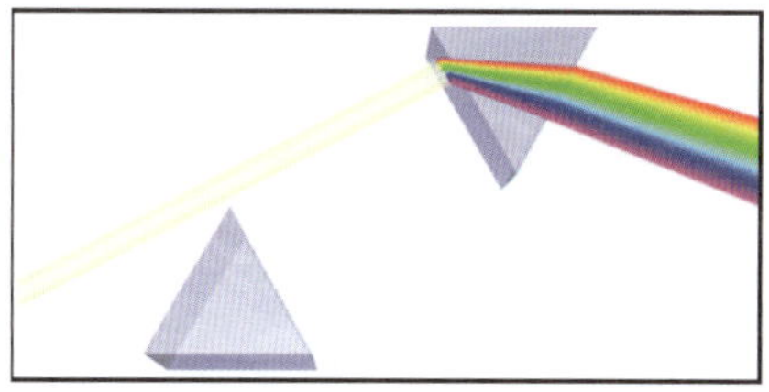

White light passes through the second prism and gives the colors of rainbow.

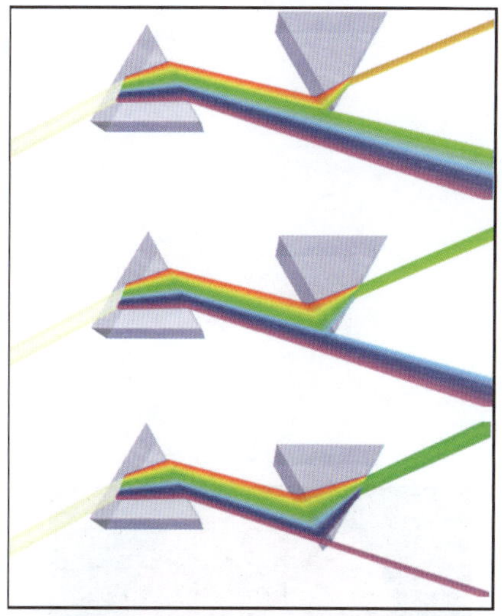

If some part of the rainbow passes through the second prism, colors other than white is observed on the other side of the second prism.

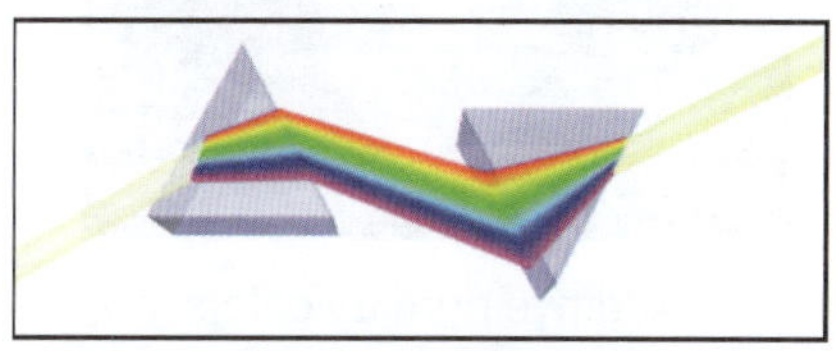

If whole rainbow is passed through the second prism, white light is observed on the other side of the second prism.

4.9. Color

Wavelengths and frequencies of visible light colors

Color	Wavelength ($nm=10^{-9}$ m)	Frequency ($THz=10^{12}$ Hz)
Violet	400	749
Blue	475	631
Green	510	588
Yellow	570	526
Orange	590	508
Red	650	461

Color is a perception, not a physical quantity. It arises from the responses of our visual system to the environment. Colors are associated with the wavelength of light. The wavelength of visible light varies between 400 nm to 650 nm. Seeing light with wavelength 400 nm as violet is the response by our visual system to an electromagnetic wave of this wavelength. There is nothing about violet associated with the electromagnetic wave itself.

Within the visible wavelength spectrum, violet and blue wavelengths are scattered by the molecules in the atmosphere more than the other wavelengths. The sky looks blue, not violet, because our eyes are more sensitive to blue light. At sunrise and sunset, red or orange colors are present because the wavelengths associated with these colors are less scattered by the atmosphere than the shorter wavelength colors like blue. A large amount of blue and violet light has been removed as a result of scattering and the long wave colors, such as red and orange, are more readily seen.

4.9.1 Colored Vision

Atoms vibrate with their own natural frequencies. The electrons of one kind of atom can be set into vibrational frequencies which are different from the frequencies of other atoms. If the electrons are vibrating at the resonant frequency, they have a large amplitude and they absorb the light falling upon them. For frequencies below and above the resonant frequency, light falling upon the substance is re-emitted.

For both opaque and transparent materials, the re-emitted light is the reflected light. Opaque objects appear colored as a result of the pigments that absorb light of some wavelength and reflect the rest of light at other wavelength. A carrot would appear orange since it mainly reflects light with the wavelength of orange. The neighboring colors (wavelengths) of orange (red and yellow) are reflected weakly also. However, our eyes are not capable of detecting those colors.

The chemical reactions that take during photosynthesis are able to use red, blue and violet light. Chlorophyll absorbs the red, blue and violet light. Green and yellow are not used in photosynthesis so they are reflected. As a result leaves appear green. Any substance reflecting all colors appears white. If no color is reflected back, than the material appears black.

An object can only reflect frequencies that are present in the illuminating light. If red light is incident upon a red object it appears red. But if blue light is incident on the same red object, since it absorbs all the light incident on it, it will appear black.

Under the bright spotlights of shopping centers, it is usually difficult to differentiate some colors. The appearance of a colored object depends on the colored light used. Colors appear different in daylight from how they appear when illuminated by fluorescent (blues are enhanced) and incandescent (reds are enhanced) lamps.

Sunlight is a mixture of different wavelengths of light. Although we have a perception that sunlight is white, it actually is transparent.

We can have white light by mixing different light colors also. If red, blue and green lights are incident upon a surface, they will result white. Red, blue and green are the *additive primary light colors.*

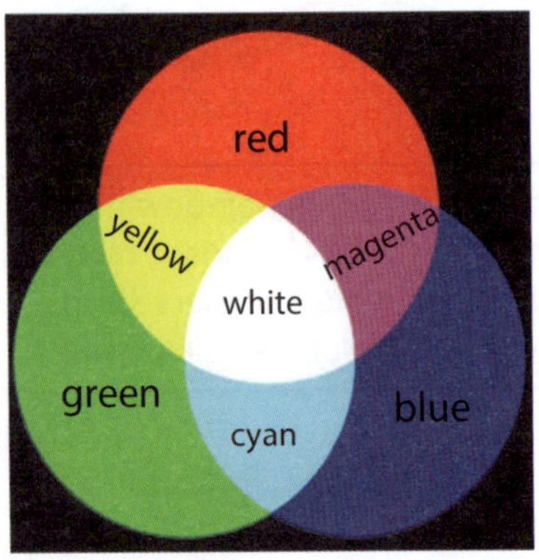

Additive primary colors

BLUE + GREEN = CYAN
RED + BLUE = MAGENTA
RED + GREEN = YELLOW
RED + BLUE + GREEN = WHITE

CYAN, **MAGENTA** and **YELLOW** are the secondary light colors since they are obtained by combinations (additions) of two additive primary colors. The yellow obtained by mixing red and green is referred as *artificial yellow.* The yellow as a component of sunlight has its own frequency and it is referred as the *natural yellow.*

Any two colors that add together to produce white are *complementary colors.*

BLUE + YELLOW = WHITE
RED + CYAN = WHITE
GREEN + MAGENTA = WHITE

The secondary colors of light are the *subtractive primary painting colors.* The primary colors of light can be obtained by combinations (subtractions) of two subtractive primary colors.

YELLOW + CYAN = GREEN
MAGENTA + YELLOW = RED
MAGENTA + CYAN = BLUE
MAGENTA + YELLOW + CYAN = BLACK

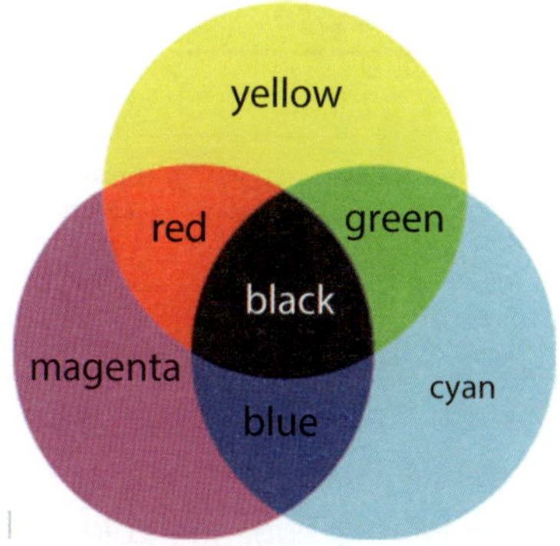

Subtractive primary colors

Unlike additive primary colors, the combination of three subtractive primary colors gives black.

Window glasses let all colors of light to pass through them if they don't have their own color. Any colored transparent material will let the components of light in its color to pass through it. Such materials *filter* the colors of light.

If sunlight which we consider as white light is incident on a light filter, what really happens is the wavelength of light in the color of the filter and the neighboring wavelengths of the visible spectrum is allowed to pass through it. The colors of the rainbow are in the order of visible spectrum of light. The order of the colors is **RED** / **ORANGE** / **YELLOW** / **GREEN** / **BLUE** / **VIOLET**. For a blue filter if sunlight is incident on it, a high intensity of blue and lower intensities of green and violet lights passes through it. Since intensities of green and violet are small; the color observed on a screen will be blue. In the aside picture, red piece of glass lets high intensity of red light and lower intensity of orange to pass through it. Green glass, lets high intensity of green and lower intensities of blue and yellow lights to pass through it. But for yellow, there will be a slight difference. Yellow filter lets yellow, green and red to pass through it. The combination of primary colors red and green gives the artificial yellow as a secondary color of light. This is why red, even though it is not a neighboring color, can pass through a yellow filter. In a similar manner, a magenta filter would pass its components of primary colors; blue and red whereas a cyan filter lets blue and green lights to pass through it. A screen behind a filter would appear black if no light rays are incident upon it.

Glass filters

RED filter

GREEN filter

BLUE filter

MAGENTA filter

YELLOW filter

CYAN filter

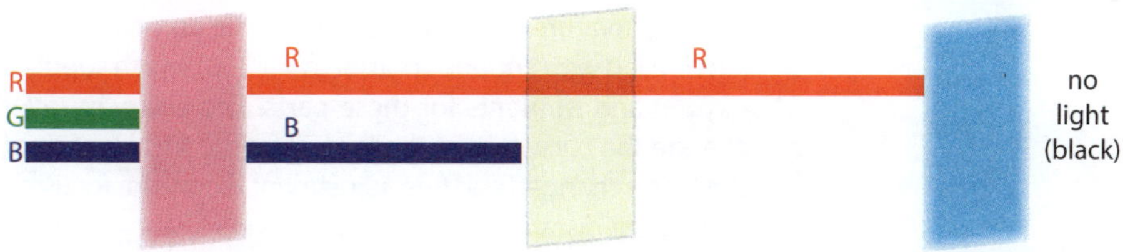

MAGENTA filter YELLOW filter CYAN filter

Example

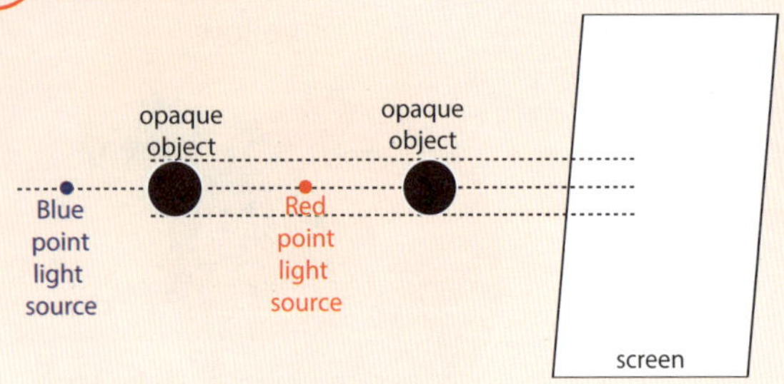

Two identical opaque objects, blue and red point light sources and a screen are placed as shown in the figure.

Draw the shadow that will be formed on the screen.

Solution

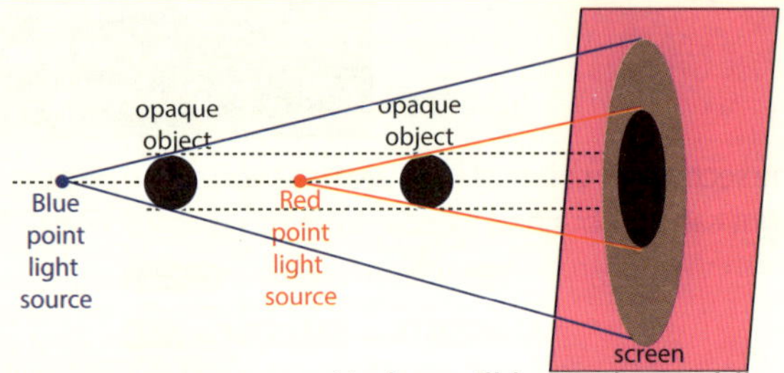

Umbra will be in the middle, penumbra will be illuminated by red light. The rest of the screen will be illuminated by both red and blue lights, so magenta will be observed on the screen.

Example

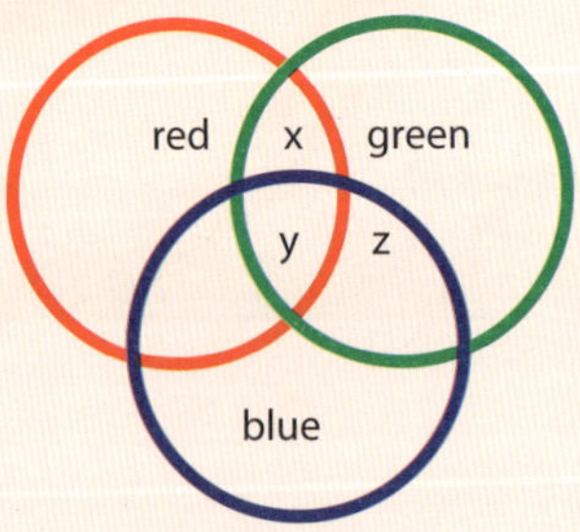

A white table is illuminated by red, blue and green lights as given in the figure.

What are the colors of regions x, y, z in the figure.

Solution

x = red + green = yellow

y = red + green + blue = white

z = blue + green = cyan

Physics Extra

4.10. The Eye and The Optical Devices

4.10.1 Image formation in the eye

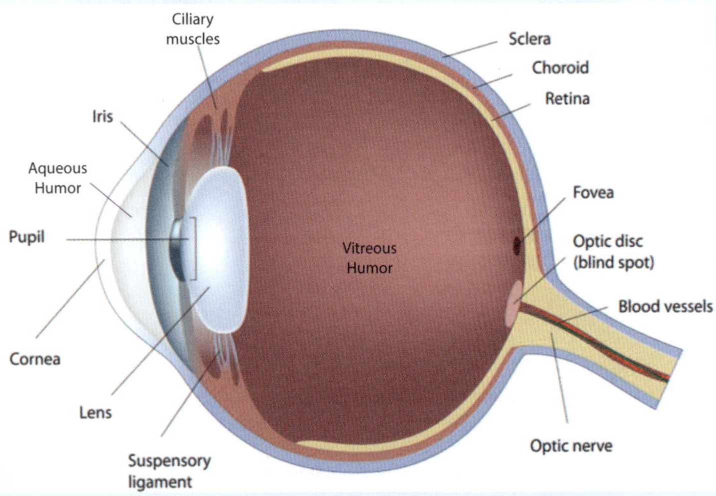

Now, its time to learn about the working principles of our unique sense organ of sight; the eye.

The functioning of the eye and the camera are very similar to each other. Such as a camera, the eye has an optical system to focus the incoming light, a diaphragm to adjust the amount of light and a screen to project the image.

Cornea, aqueous humor and the *lens* are responsible for refracting the incoming light rays. The lens of the eye (n = 1.386 to 1.406) does little of the bending of the light rays. Most of the refraction is done at the front surface of the cornea (n =1.376), at the interface of air and cornea due to its strong curvature. The space between the cornea and the lens is filled with aqueous humor (n = 1.336) which supplies oxygen and nutrients for these parts. The *ciliary muscles* change the curvature of the lens. Thus the focal length of the lens is changed and fine adjustment is done for focusing at different distances.

Physics Extra

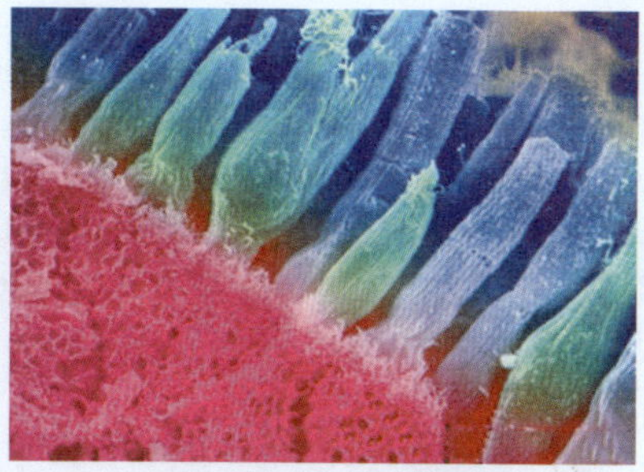

Rod (blue), cone (green) cells of the retina

The colored part of the eye, the *iris* acts like the diaphragm and controls the amount of light entering the eye. *Pupil* is the hole in the middle of the iris and light passes through it. The pupil appears black because no light is reflected from it and very little light is reflected back out from the interior of the eye. The interior of the eye is filled with gel-like transparent liquid, the *vitreous humor* (n = 1.337) which is filling the space between the lens and the retina. The curved rear surface of the eye is the *retina*. Retina is the biological equivalent of the CCD of a digital camera and it consists of a complex array of the nerves and receptors. These receptors are known as the *rods* and the *cones*. Rods are responsible for our night vision whereas the cones provide the daytime and colored vision. They change the light energy to electrical signals. These signals travel along the nerves and reach to the seeing center of the brain. *Fovea* is a small area found at the center of the retina. The cones are closely packed at the fovea. The sharpest image and the best discrimination are found at the fovea.

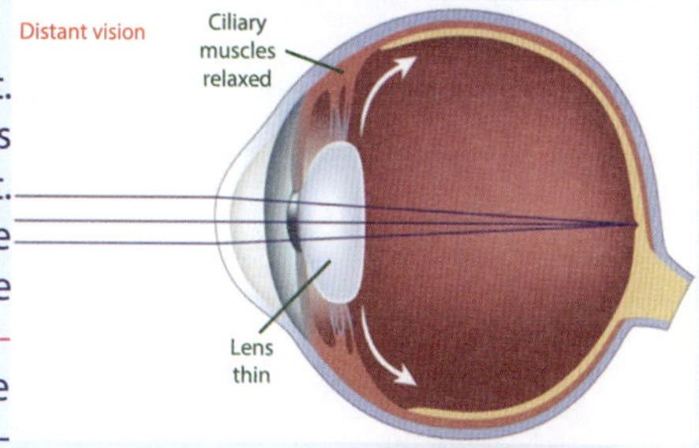

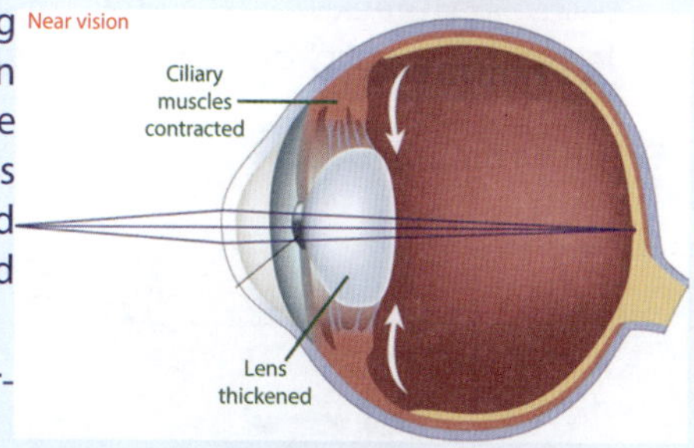

The ciliary muscles are relaxed and the lens is thin in order to focus on a distant object. Parallel light rays emerging from a distant object are focused on the focal point which is located on the fovea of the retina. To focus on a nearby object, the ciliary muscles contract. The lens thickens at the center so the focal length is shortened in order to have the image of the object projected on the fovea again. The change in the shape in other words the focal length of the lens with the contradiction of ciliary muscles is called the accommodation of the eye. For normal vision, the far point is the infinity (any object 6 m away from the eye is assumed to be at infinity). The closest point on which the eye can focus with the ciliary muscles fully contracted is called the near point of the eye. The eye is not able to bring into the sharp images of objects closer than the near point. The near point of the human eye is conventionally taken as 25 cm. Younger people has a closer near point. As people get older, the near point moves away from the eye. With the aging, the lens becomes less elastic and more difficult to curve. By the age of 50, the near point receded to 50 cm and by the age of 70, to 1 m. To check your near point, place this book close to your eye and slowly move it away until you have a clear vision of the letters on the page.

The above values are given for a normal eye. However, not all the people have a normal eye. There are some eye defects which result a blurry vision in different occasions.

Myopia (Greek words "myops" meaning near-sighted and "ops" meaning eye) occurs when the lens or the cornea is too strongly curved or the eyeball is longer than the average. The rays from a distant object focus in front of the retina when the eye is relaxed. The rays begun to diverge as they are propagating towards retina. This is a why myopic person sees distant objects blurry. If the object is brought near, the image distance increases and light rays can focus on the retina. A myopic person can clearly see close objects with relaxed eye. Their far point is closer than the infinity and they can focus objects much closer than 25 cm, the near point distance of a normal eye. Myopia is sometimes referred as nearsightedness which refers to the ability to see only nearby things.

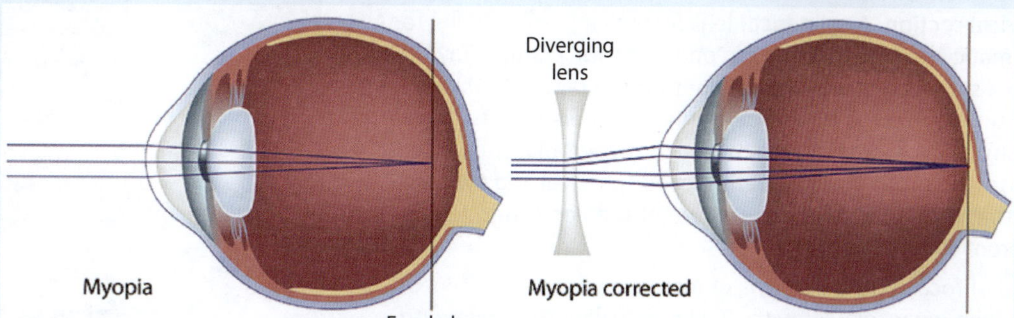

For a diverging lens, light rays from infinity diverge from the lens. The extensions of the light rays intersect and create an image at the focus. The image is as the same side as the object. Thus an object at infinity is brought closer. If a diverging lens is placed in front of a myopic eye and an appropriate correction will be provided. If the focal length of a diverging lens matches to the far point of the eye, the lens creates a virtual image of the object at the far point of the eye. This is the point where the relaxed eye naturally focuses. A diverging lens brings distant objects to a close point where the eye can focus.

Physics Extra

Hyperopia (Greek words "hyper" meaning over or beyond and "ops" meaning eye)occurs when the lens is so weak, the cornea is not sufficiently curved or the eyeball is shorter than the average. Light rays from a distant object will focus behind the retina. If the eye is relaxed, the image of a distant object cannot come to focus on the retina. The eye can focus distant objects if only the ciliary muscles are contracted. But no amount of accommodation is sufficient to focus the light rays from a near object. A hypermetropic person sees the distant objects with effort but not nearby objects. The near point of a hyperopic eye is more than 25 cm. A properly chosen converging lens creates a virtual image of the object at a farther distance of the eye's near point so that the eye can focus on the image there.

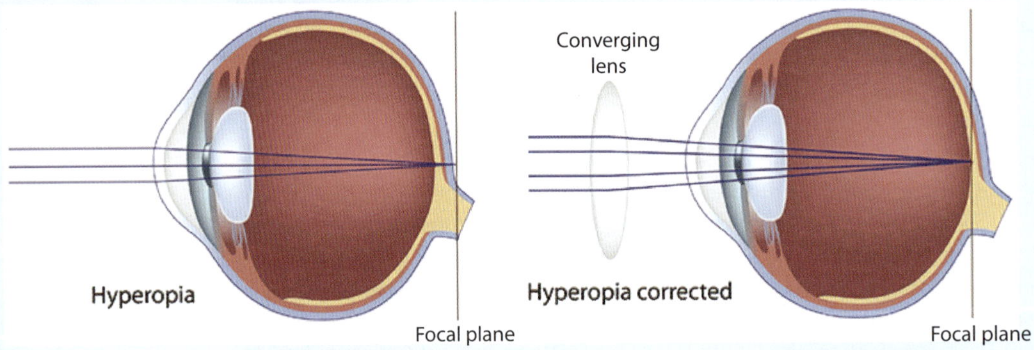

Presbyopia (Greek words "presbys" meaning old person and "ops" meaning eye) is too often misnamed as hyperopia which is the lessening of ability of the eye to accommodate as a person gets older and the near point moves out.

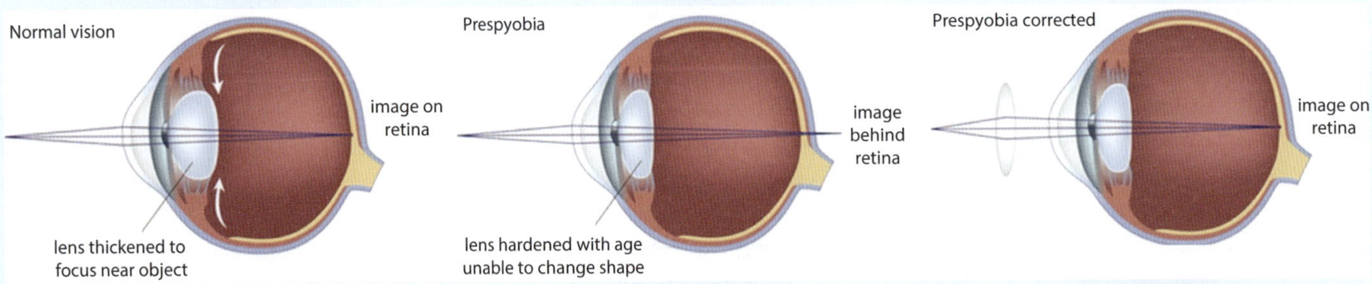

Astigmatism (Greek words "a" meaning without, "stigma" meaning a point, spot or mark and "ops" meaning eye) is caused by an out of round cornea or lens. The point objects are focused as short lines. This result a blurry image. Astigmatism is as if the cornea were spherical with a superimposed cylindrical section. A cylindrical lens focuses a point parallel to its axis. An astigmatic eye may focus rays in one plane for example rays in vertical plane to a shorter distance than it does for the rays in horizontal plane. Astigmatism is corrected by compensating cylindrical lenses. Most often astigmatism is accompanied with myopia or hyperopia. The lenses used for such cases are ground with superimposed spherical and cylindrical surfaces so that the radius of the curvature of the correcting lens is different in different shapes.

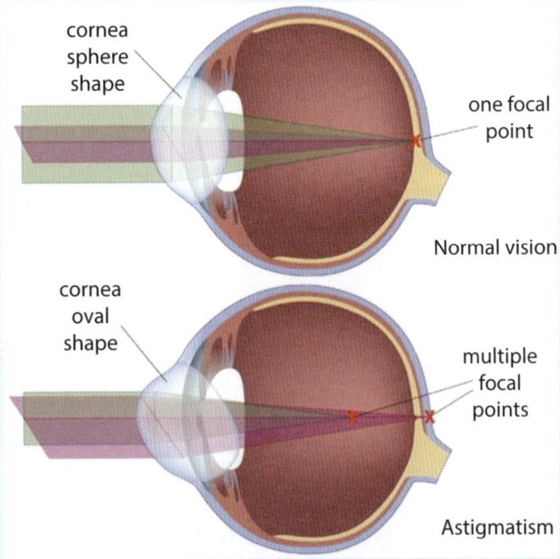

The reciprocal of focal length in terms of meters gives the diopters of a lens. A person who wears glasses with -2.00 is actually using diverging lenses (focal length negative) with a focal length of 50 cm. If the diopters of the glasses is +4.00 the person is hyperopic (+ sign indicates a converging lens) and the focal length is 25 cm.

Distant objects underwater look blurry because at the interface of water (n = 1.33) and cornea (n = 1.376) the difference of the refractive indices is very small. Light rays bent so small so they are focused far behind the retina. But if you wear goggles under water, you can restore the air (n=1.003) and cornea interface. Thus the eye can focus again and you see sharp images.

Physics Extra

4.10.2 Optical Devices

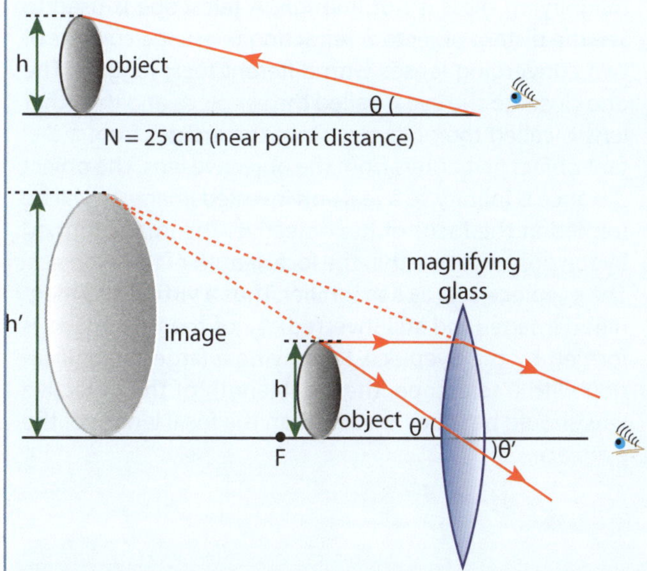

Magnifying glass: So often you tend to bring a object closer to your eye, to see the details on it. As you move an object closer to your eye, the size of the object doesn't change, but its apparent size increases. By bringing the object closer to your eye the object subtends a greater angle. This is just like covering a full moon with a coin. Full moon is a distant object and the coin close to you appears larger that it usually is to your eye.

Our eyes can accommodate only up to near point which is 25 cm for a normal eye. Using a magnifying glass, allows us to place the object closer to our eye so that a larger angle is subtended. If the object is placed at just within the focal point of it, a virtual and larger image is produced by the lens. This virtual image should be at least 25 cm away from the eye so that it can focus.

Camera: It is a successful reproduction of the human eye. A camera consists of a light-tight case and a converging lens. An inverted real image of the object is formed by converging lens. Since the image is real, it can be projected on a screen. A light sensitive surface, a film or an electronic detector, is used as the screen to record the image. If the film is placed behind or in front of the sharp image, then a sharp image cannot be located on the film, and a blur image will be recorded. The focusing rings allows to change the position of the lens in order to project a sharp image of the object on the film. The shutter blocks the light from reaching to the film. As you press the button of your camera, you cause the shutter to move out of the way. The light is allowed to enter into the light-tight box and reach to the film. If the shutter is open for a longer time, more light reaches to the film to expose it. A larger lens will collect more light rays and make a brighter image. If you increase the diameter of the lens more light can enter to the light-tight box. But the lens of a camera cannot be changed repeatedly (at least for amateur cameras). Instead a variable aperture, an iris diaphragm, is placed behind the lens. A diaphragm controls the amount of light entering the light-tight case.

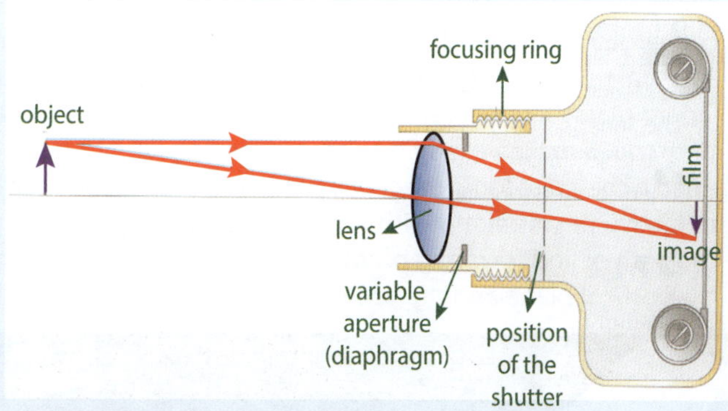

Traditional cameras use the film as the light sensitive surface. Light striking on silver halide crystals creates small particles of silver metal on the crystals and the crystals are converted to dark metallic silver. The parts of the film where light strikes appears dark and a negative image is formed on it. This negative image is projected onto a paper with its photosensitive coating and a positive image, the print, is obtained.

Old fashioned cameras

The electronic light sensitive surface used in digital cameras is called a charged-coupled device; a CCD. A CCD consists of a rectangular array of many millions of small detectors called pixels. When light hits on one of these pixels an electric charge proportional to light intensity is produced. An image is recorded on the CCD in terms of little packets of charge. After the CCD has been exposed, the charges are read out, signal levels are digitized and the picture is stored in the digital memory of the camera.

Digital cameras

Physics Extra

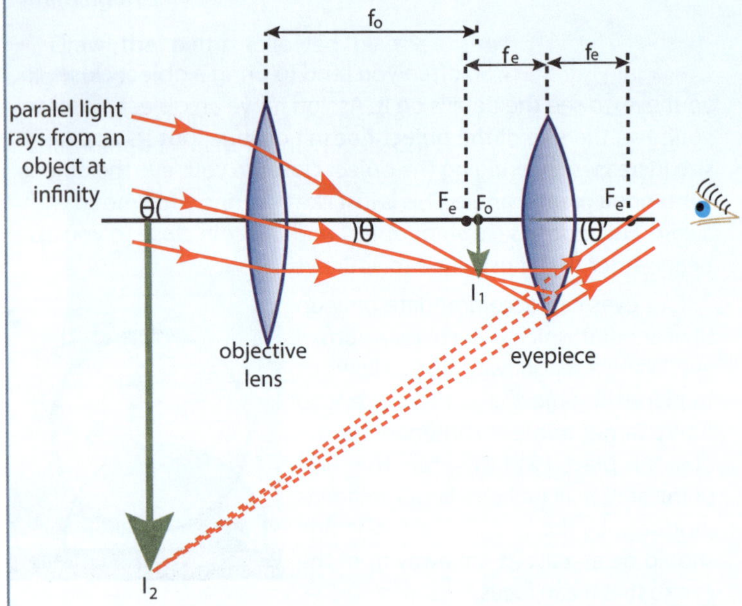

Telescope: To see the distant objects such as stars, a magnifying glass is not enough. A telescope is used to see the distant objects. A refracting telescope consists of two converging lenses with different focal lengths. The one closer to the eye is called the eyepiece and the other lens is called the objective since the images from a distant object first enters from the objective lens. The object distance is infinity so a real and inverted image, I_1, will be formed at the focus of the objective. The image formed by the objective is within the focal length of the eyepiece. The eyepiece acts as a magnifier. Thus a virtual and magnified image (and still inverted), I_2, of the first image is formed by the eyepiece. For having a larger magnification with a telescope, the focal length of the objective lens should be much greater than the focal length of the eyepiece.

Telescopes used to view distant objects on Earth are called terrestrial telescopes. These telescopes have a diverging lens in their eyepiece. The eyepiece intercepts converging rays from the objective lens before they reach to a focus and acts to form a virtual upright image. This type of telescopes are known as the *Galilean type* also. Opera glasses with a reasonably short tube can be given as an example. *Spyglass telescopes* makes use of a third converging lens to have an upright image. They need to be quite long. *Binoculars* which we had mentioned before has total internal reflective prisms along with the objective and eyepiece lenses. The prisms shorten the path of light rays and produce an upright image. One prism inverts the image in the vertical plane and the other in the horizontal plane.

Hubble space telescope

Construction and grinding of large lenses is a very difficult process. A mirror has only one surface to be ground. It can be supported along its entire surface. A large lens supported at its ends would sag under its own weight. This is why largest astronomical telescopes have a concave mirror as their objective. Such type of telescopes are r*eflective telescopes.* The size of a lens should be large in order to collect more light from a distant object. Lenses have a tendency to split the light into its colors as their size gets larger. Mirrors are do not have a tendency to split the light. This is another advantage of reflective telescopes over refractive ones. For most astronomical telescopes, the objective lens is removed from the system so that the real image can be projected on a film or CCD digital sensor.

For small reflective telescopes the image is formed in front of the mirror. Magnifying such an image without getting your head in front of the eyepiece is very hard to obtain.

Physics Extra

Microscope: A microscope is used to see very close and small objects. Unlike the telescope the object distance is very small, and the focal length of the objective is relatively short. The object is placed just beyond the objectives' focal length. The image, I_1, formed by the objective lens is real, quite far from the lens and magnified. The eyepiece enlarges this image a second time to a very large and virtual image, I_2. This image seen by the eye is an inverted image because I_1 acting as an object to form this image is an inverted one.

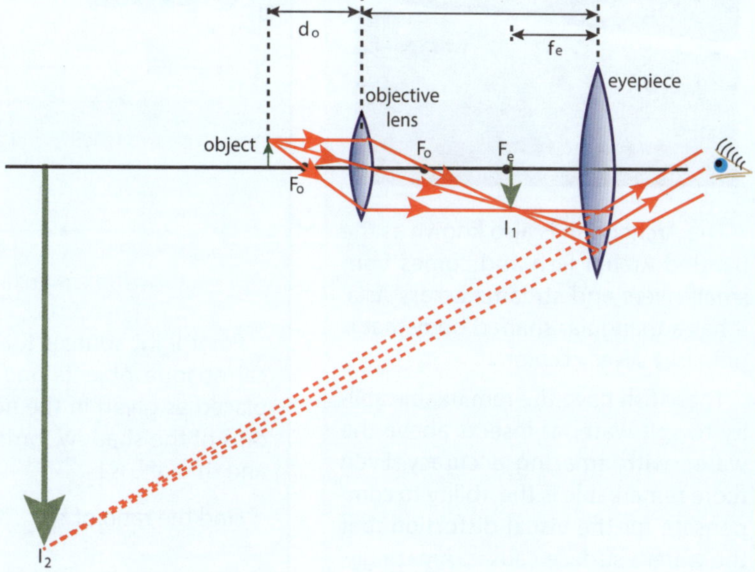

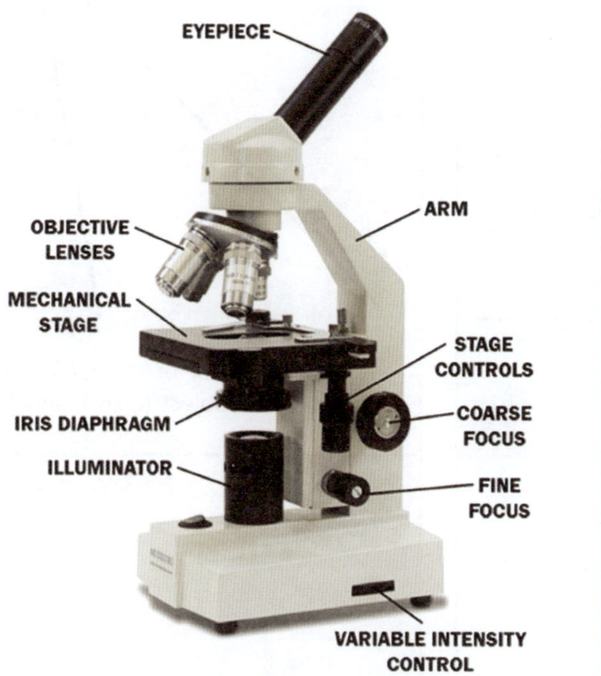

Parts of a compound microscope

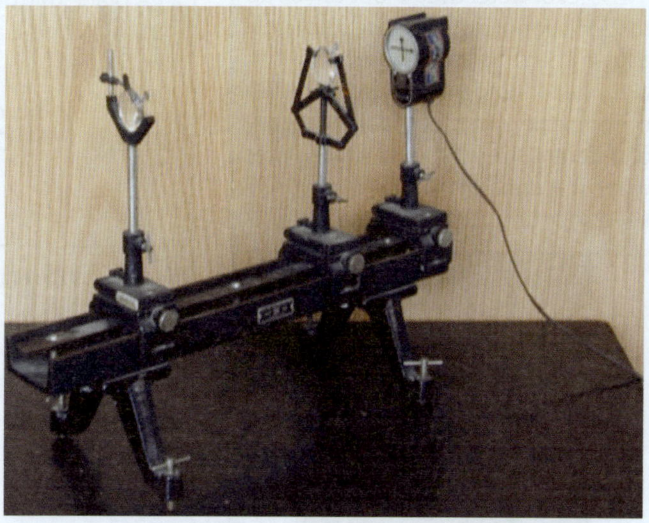

A rough design of a microscope in the lab

ARCHER FISH

The Archer Fish is also known as the Banded Archer Fish and comes from small rivers and streams across Asia. It has a triangular shaped body that is primarily silver in color.

These fish have the remarkable ability to spit water at insects above the water, with amazing accuracy. Even more remarkable is the ability to compensate for the visual distortion that the water's surface causes. Amazingly, adult fish almost always hit the target on the first shot.

The jets of water are produced when archer fish press their tongues against a groove in their mouths to form a gun-barrel-like shape, and close their gills to force out a spurt of water. The water spurts are up to six times stronger than the fish's muscular power - strong enough to abruptly knock even firmly anchored prey into the water.

The jet of water they shoot can be up to 2–5 m long, but their accuracy only allows them to shoot insects 1–2 m away depending on body size. The fish can alter the power of the shot for prey of different sizes. If the fish doesn't get what it wants on the first try, it will keep shooting till it does.

Solved Questions

Question

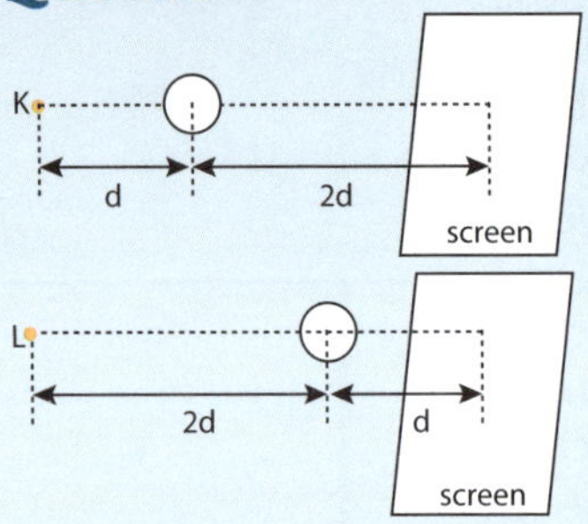

Point light sources K and L, identical opaque objects and screens are placed as given in the figure. The radius of the shadow formed by K is r_1 and that of L is r_2.

Find the ratio of r_1/r_2.

Solution

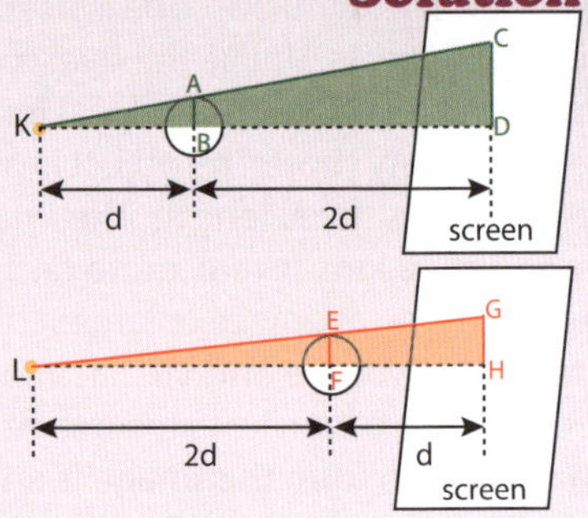

$\Delta AKB \approx \Delta CKD$

$\frac{KB}{KD}=\frac{AB}{CD} \Rightarrow \frac{d}{3d}=\frac{r}{r_1} \Rightarrow r_1=3r$

$\Delta ELF \approx \Delta GLH$

$\frac{LF}{LH}=\frac{EF}{GH} \Rightarrow \frac{2d}{3d}=\frac{r}{r_2} \Rightarrow r_2=\frac{3}{2}r$

$\frac{r_1}{r_2}=\frac{3r}{\frac{3}{2}r}=2$

Question

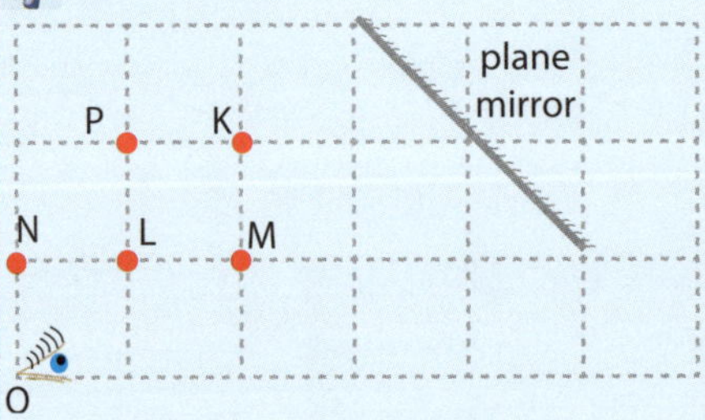

A plane mirror and opaque objects K, L, M, N and P are placed as given in the figure.

Which points can be seen by an observer at point O?

Solution

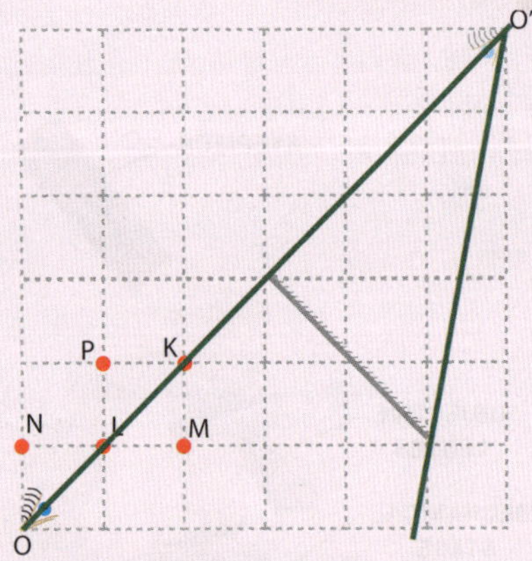

N and P are out of the region of sight. L blocks object K, so L and M can be seen.

Question

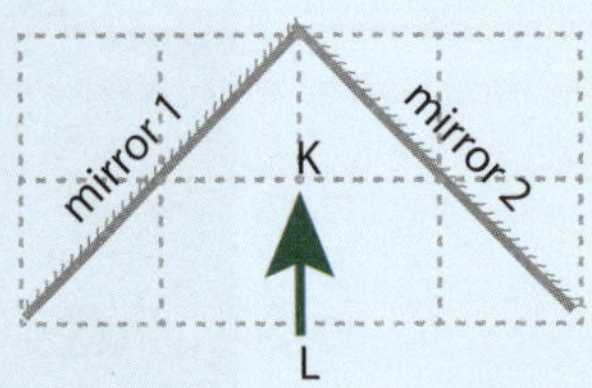

The image of the object KL is first formed by mirror 1, then by mirror 2.

Draw the image of the object.

Solution

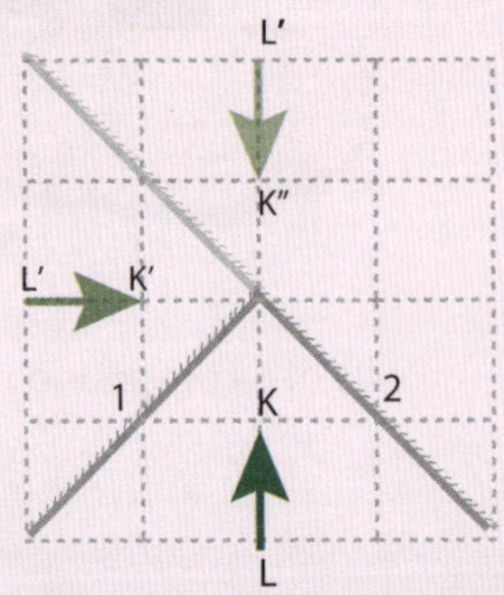

A mirror can be extended in order to draw the image of an object.

Question

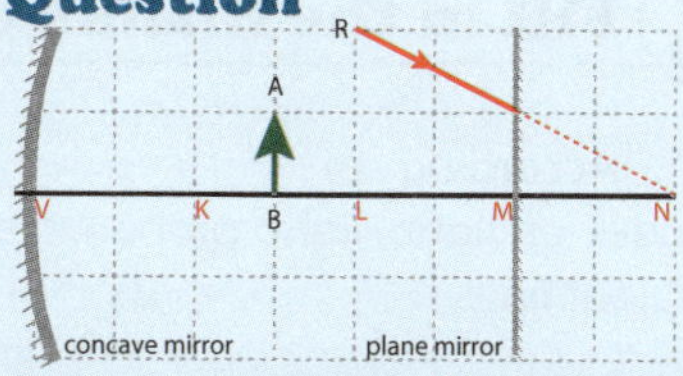

Light ray R is reflected from the plane mirror first and then the concave mirror. It is reflected back on itself from the concave mirror.

Draw the image of the object AB which is placed to midpoint between K and L.

(squares are identical to each other)

Question

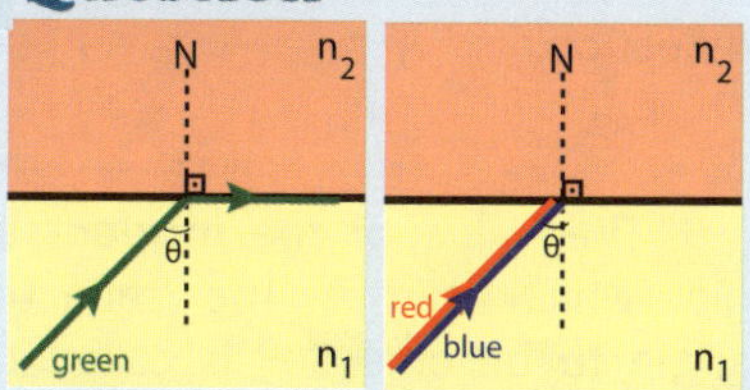

The path followed by a green light ray is given in the figure.

a. What is the relationship between the refractive indices of media n_1 and n_2 and the velocities in these media v_1 and v_2?

b. Draw the paths followed by red and blue light rays if they were sent with the same angle of incidence for the same media.

Question

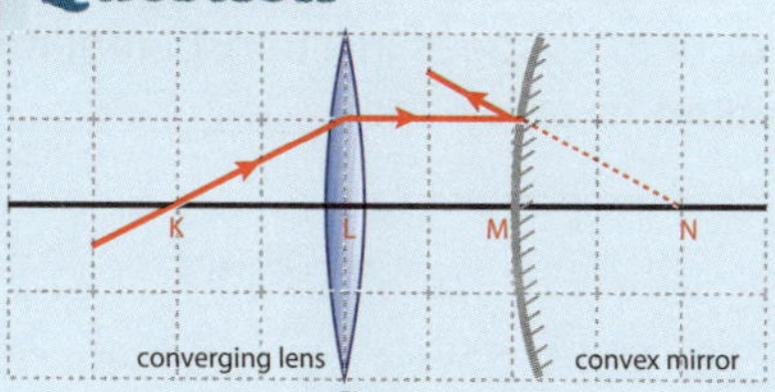

Figure 1

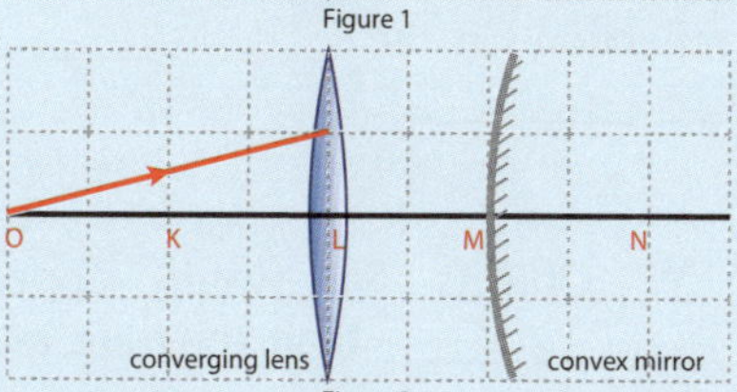

Figure 2

The path followed by a light ray as it is passing through a converging lens is given in figure 1.

Draw the path followed by the light ray in figure 2.

Solution

A light ray coming from the center of a concave mirror will be reflected back on itself. So L is the center, K is the focus of the concave mirror.

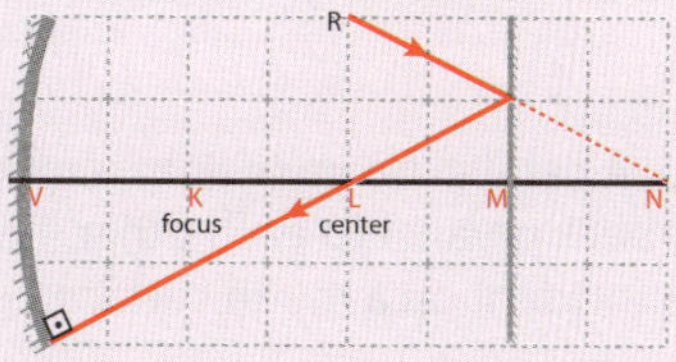

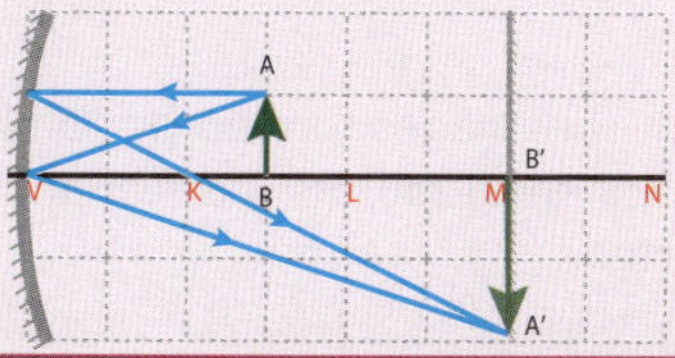

Solution

a. q is the critical angle for these two media. This occurs only when light propagates from a dense medium to a less dense one. Thus $n_1 > n_2$.

The velocity is inversely proportional to the refractive index, thus $v_1 < v_2$.

Red is refracted less than green whereas blue is refracted more than green. Red passes to n_1 whereas blue experiences TIR.

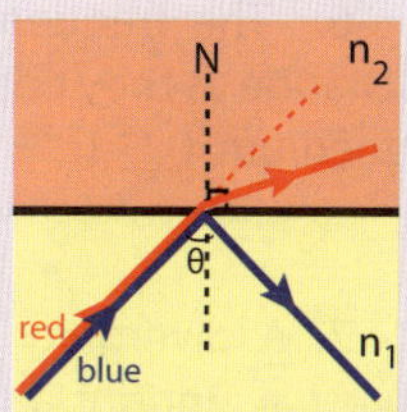

Solution

Light ray is parallel to principal axis as it passes through the lens, so K is the focal point of the lens. The parallel light ray is reflected as if it is coming from point N, so N is the focus of the convex mirror. So O is the 2F point of the converging lens.

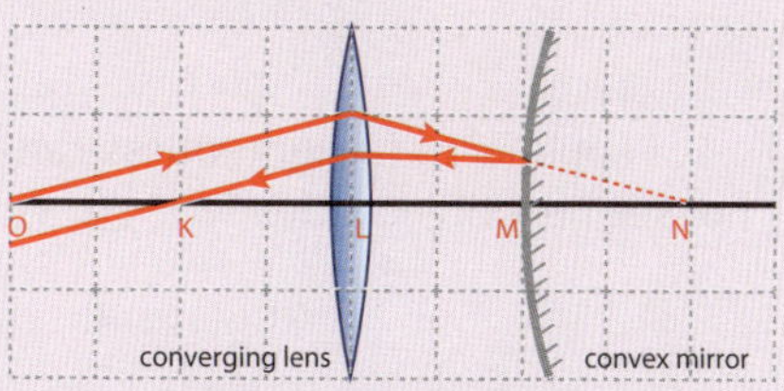

N is the far 2F point of the lens also, so the light will bend towards N. Convex mirror will reflect it parallel to the principal axis and the lens will refract it towards its focus.

INTERESTING FACTS ABOUT THE EYES OF THE ANIMAL KINGDOM

Eyes of an owl are located together in the front of their face, which allows them excellent depth perception during their hunting expeditions –particularly in low-light situations. These massive eyes are fixed in their sockets and can barely move. This is why the owl can turn its head so far.

Nocturnal geckos have to be able to block out the bright sun during the day while still retaining excellent night vision, which is why they have long zigzagged pupils that can tightly constrict to let in only pinpoints of light. While humans cannot see colors in dim moonlight, these animals can discriminate between colors and their eyes are calculated to be almost 350 times stronger when it comes to seeing color.

Butterflies have compound eyes, which are made of hundreds of microscopic, six-sided lenses that allow them to see in every direction simultaneously. While this type of vision prevents the bugs from seeing things in sharp focus, butterflies can see ultraviolet light, which is invisible to the human eye. This aspect of their vision helps to lead them to flowers filled with delicious nectar.

Goats have square pupils. The width of the pupils allows the animals to see at a 330 degree angle, which is a lot larger than for the human eye (average of 185° angle)

Conceptual Questions

1. Some optical instruments, such as periscopes and binoculars use trigonal prisms instead of mirrors to reflect light around corners. Light typically enters perpendicular to the face of the prism, undergoes TIR off the opposite face and then exits out the third face. Why do you suppose the manufacturer prefers the use of prisms instead of mirrors?

2. How and why do diamonds shines with the colors of the rainbow?

3. What are the differences between real and virtual images for the mirrors and lenses?

True - False

1. The eclipses occur if celestial bodies become aligned. (___)

2. A parallel beam of light is reflected back parallel to each other if they are incident on a rough surface.(___)

3. The images formed by all kinds of spherical mirrors are real and inverted.(___)

4. A light ray bends towards the normal as it passes from a more dense medium to a less dense one. (___)

5. As light waves propagate from a more dense medium to a less dense one, they speed up.(___)

6. The frequency of light waves change as they enter to a more dense medium. (___)

7. A diverging lens constitutes a virtual, upright and a smaller image. (___)

8. The shape of eye lens can be changed by the contractions of ciliary muscles. (___)

9. People with farsightedness cannot see near objects clearly. (___)

10. Astronomical telescopes use concave mirrors because construction and grinding of large lenses is very hard. (___)

Fill in the Blanks

1. According to ____________ model of light; light propagates in straight lines.

2. Light cannot propagate through ____________ objects.

3. A filament lamp is an example of ____________ light sources.

4. Angle of incidence is ____________ to the angle of reflection.

5. The value of the incident angle which causes the refraction angle to become 90° is called the ____________ angle.

6. The point at which parallel light rays appear to emanate from a single point is called the ____________ point of a convex mirror.

7. The ratio of speed of light in vacuum to its speed in a transparent medium is called the ____________/ ____________.

8. Light travels via total internal reflection in an ____________/ ____________.

9. Objects in water appear ____________ than they are as a result of ____________.

10. The working principles of ____________ is similar to the human eye.

Conceptual Questions

1. Some optical instruments, such as periscopes and binoculars use trigonal prisms instead of mirrors to reflect light around corners. Light typically enters perpendicular to the face of the prism, undergoes TIR of the opposite face and then exits out the third face. Why do you suppose the manufacturer prefers the use of prisms instead of mirrors?

A prism will allow light to undergo total internal reflection whereas a mirror allows light to both reflect and refract. So for a prism, 100 percent of the light is reflected. But for a mirror, only about 95 percent of the light is reflected. For these reasons, a prism will produce a brighter image due to the greater percent of light being reflected.

2. How and why do diamonds shines with the colors of the rainbow?

The refractive index of diamond is 2.417 and the critical angle for diamond-air interface is 24.5°. As a light ray enters through the surface of a diamond, it bends towards the normal to the surface. White light is strongly dispersed within the diamond. For a properly cut diamond, the light rays within the diamond goes through total internal reflection several times. As the dispersed white light leaves the diamond, it shines.

3. What are the differences between real and virtual images for the mirrors and lenses?

If the reflected light rays themselves are intersecting(refracted rays for lenses) the image formed is a real one. Real images can be projected upon a screen. If the extensions of the reflected light rays themselves are intersecting(refracted rays for lenses) the image formed is a virtual one. It can only be observed but cannot be projected upon a screen.

True - False

1. The eclipses occur if celestial bodies become aligned. (_T_)

2. A parallel beam of light is reflected back parallel to each other if they are incident on a rough surface.(_F_)

3. The images formed by all kinds of spherical mirrors are real and inverted.(_F_)

4. A light ray bends towards the normal as it passes from a more dense medium to a less dense one. (_F_)

5. As light waves propagate from a more dense medium to a less dense one, they speed up.(_T_)

6. The frequency of light waves change as they enter to a more dense medium. (_F_)

7. A diverging lens constitutes a virtual, upright and a smaller image. (_T_)

8. The shape of eye lens can be changed by the contractions of ciliary muscles. (_T_)

9. People with farsightedness cannot see near objects clearly. (_F_)

10. Astronomical telescopes use concave mirrors because construction and grinding of large lenses is very hard. (_T_)

Fill in the Blanks

1. According to ray model of light; light propagates in straight lines.

2. Light cannot propagate through opaque objects.

3. A filament lamp is an example of incandescent light sources.

4. Angle of incidence is equal to the angle of reflection.

5. The value of the incident angle which causes the refraction angle to become 90° is called the critical angle.

6. The point at which parallel light rays appear to emanate from a single point is called the focal point of a convex mirror.

7. The ratio of speed of light in vacuum to its speed in a transparent medium is called the refractive / index.

8. Light travels via total internal reflection in an optical / fiber.

9. Objects in water appear closer than they are as a result of refraction.

10. The working principles of camera is similar to the human eye.

Problems

1.

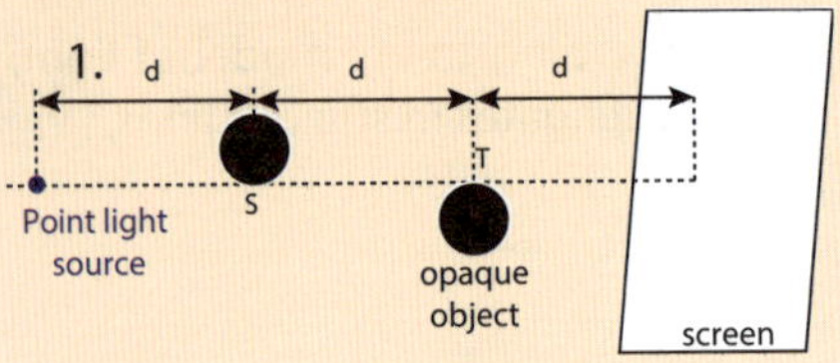

A point light source, two identical opaque objects S and T and a screen are placed as given in the figure.

a. Draw the shadow formed on the screen.

b. Find the ratio of the area of the shadow formed by S to the ratio of shadow formed by T; S/T

a.

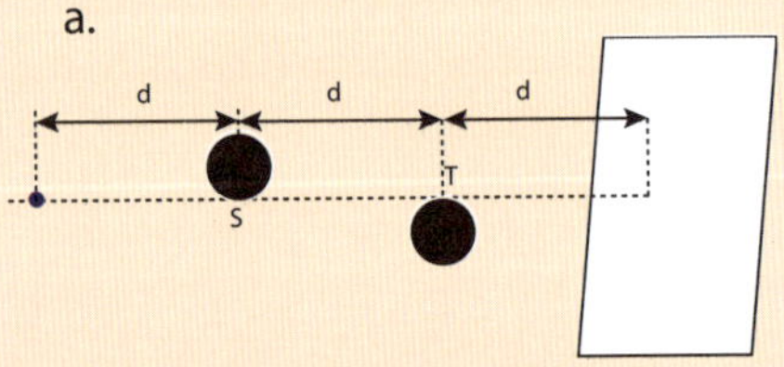

b.

2.

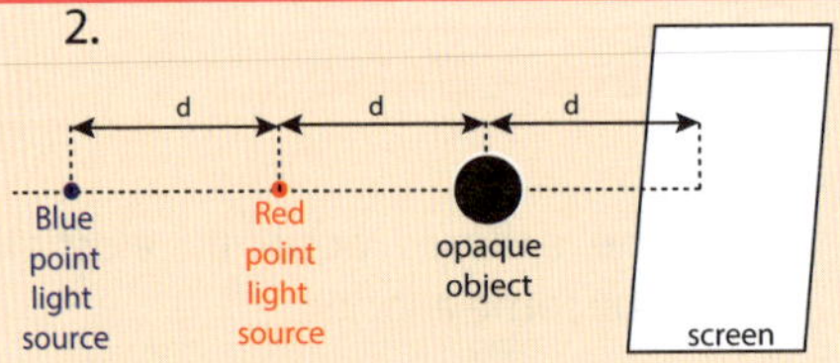

Blue and red light sources, an opaque object and a screen are placed as given in the figure.

a. Draw the shadow formed on the screen.

b. Find the ratio of area of penumbra to umbra; P/U.

a.

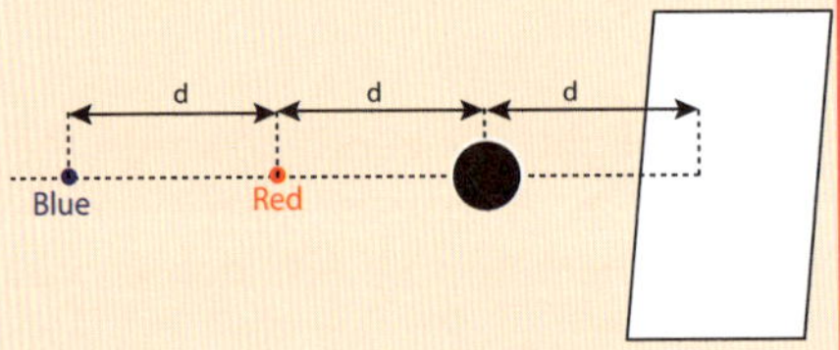

b.

3.

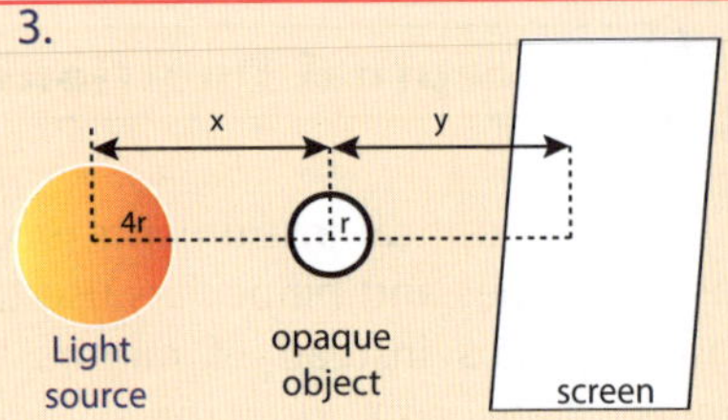

A light source with a radius of 4r, an opaque object with a radius of r and a screen are placed as given in the figure. Only penumbra is observed on the screen. x is the distance between light source and the object, y is the distance between the screen and the object.

a. Find the ratio of x/y.

b. Find the radius of the penumbra in terms of r.

1. a.  ; b. 4 / 2. a. 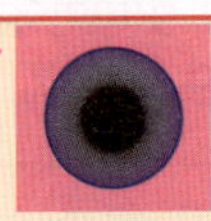 ; b.7/9 / 3. a. 3; b. 8/3

4.

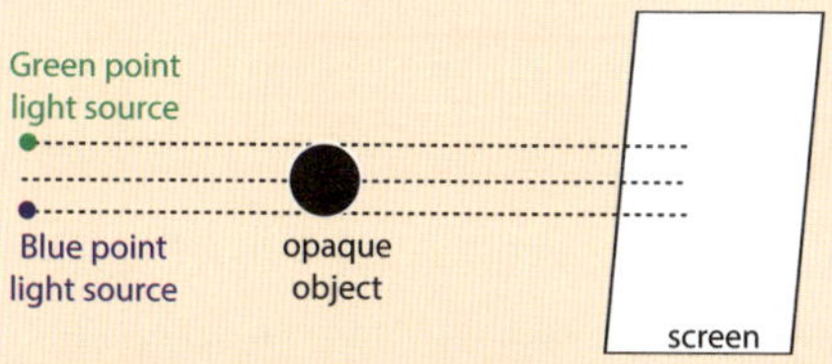

Blue and green light sources, an opaque object and a screen are placed as given in the figure.

Draw the shadow formed on the screen.

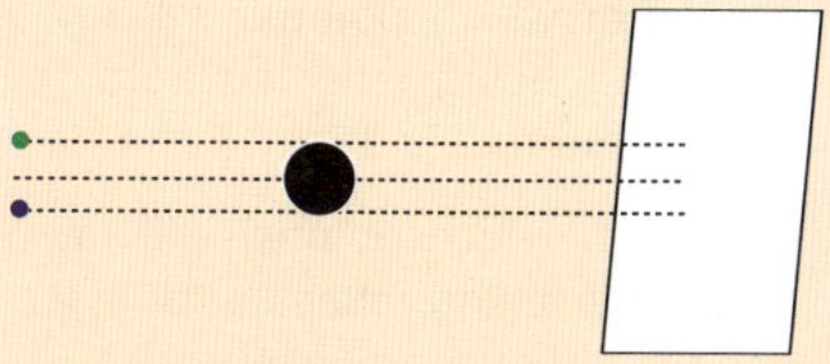

5.

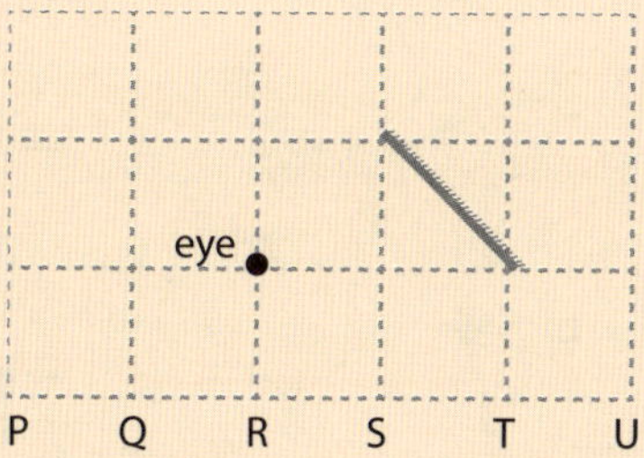

Which points on the wall PU can be observed by the eye looking at the plane mirror given in the figure? Draw the region of sight.

6.

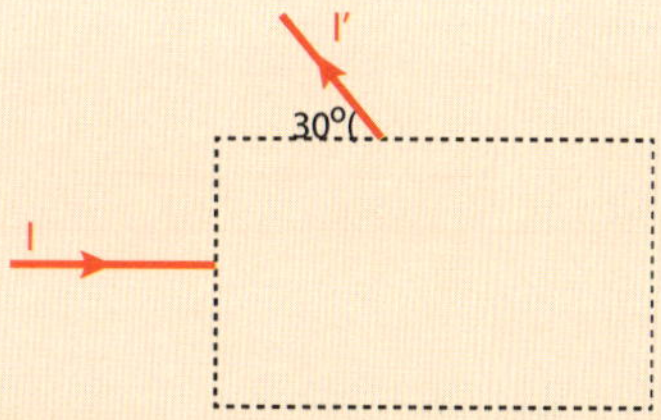

A plane mirror is placed in the rectangular box given in the figure. An incident ray I is reflected as I' as given in the figure.

Find the angle of incidence.

7.

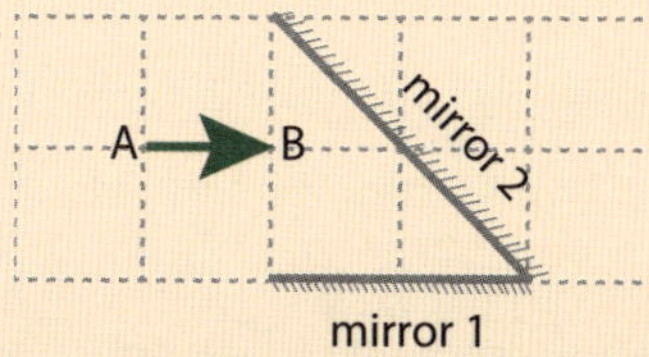

a. Draw the images of the object first formed by mirror 1 then by mirror 2.

b. Draw the images of the object first formed by mirror 2 then by mirror 1.

8.

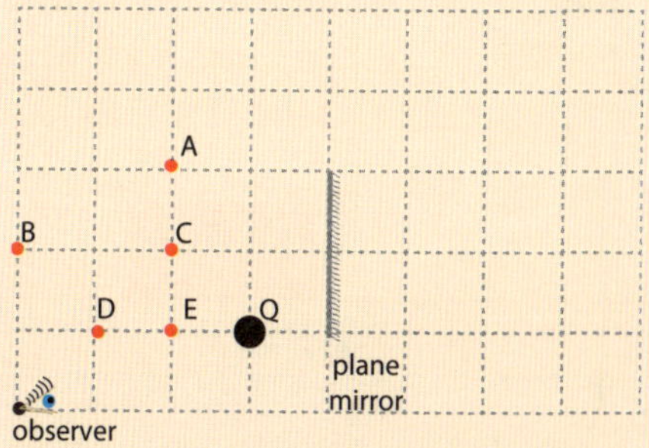

Q is an opaque object.

Which points can be seen by the observer looking into the plane mirror?

9. How many images of the object A will be formed by plane mirrors 1 and 2?

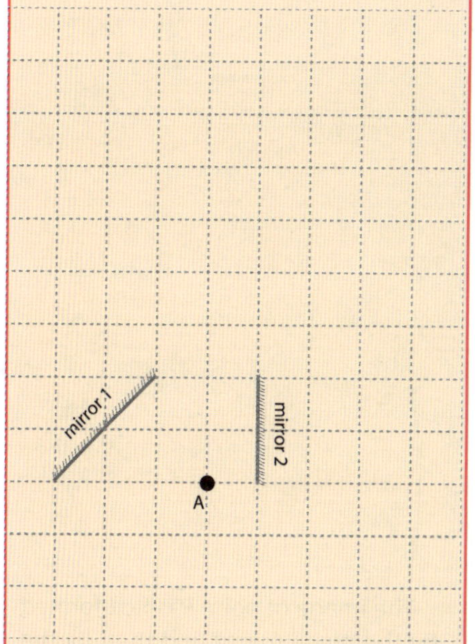

4.  / 5. QT / 6. 15° / 7. a. arrow pointing down; b. arrow pointing up / 8. A, B / 9. 4

10. Draw the image of the objects formed by the concave and convex mirrors.

a.

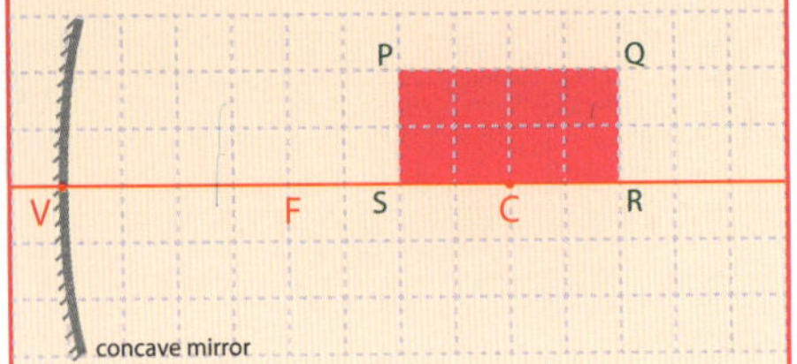

b.

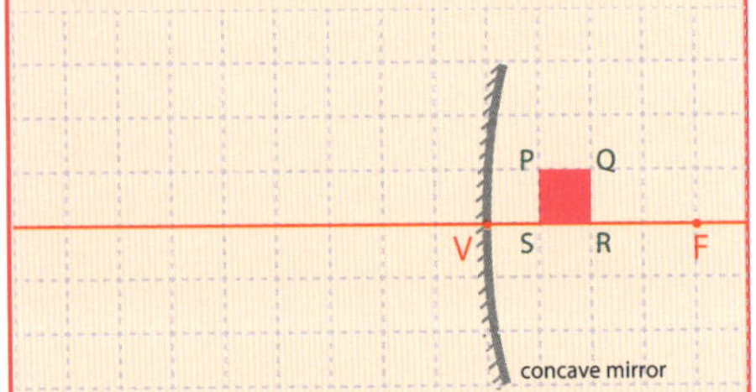

c.

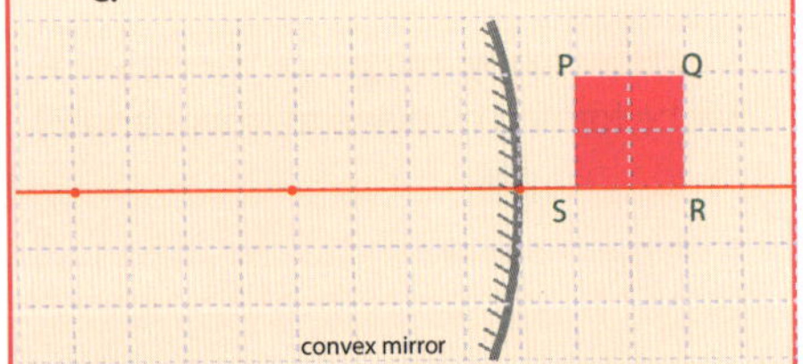

11.

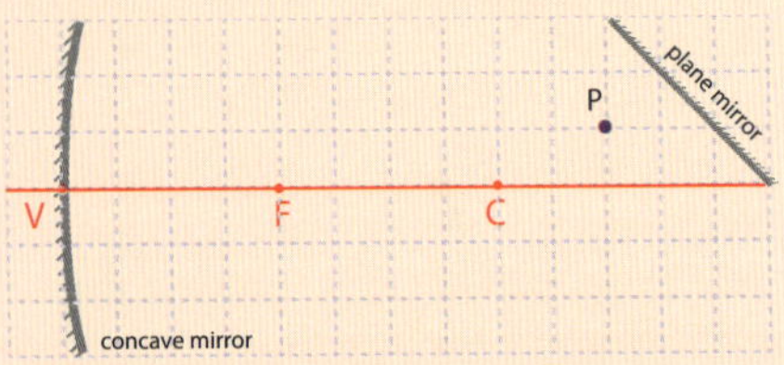

The image of point P is first formed by the plane mirror and then by the concave mirror.

Find the position of the final image formed.

12.

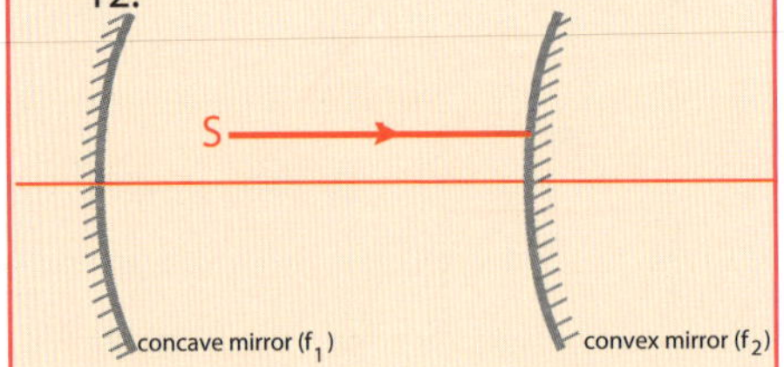

Light ray S is sent parallel to principal axis of the convex mirror. It is reflected *back on itself* from the concave mirror. f_1 is the focal length of the concave mirror and f_2 is the focal length of the convex mirror.

Find the distance between the mirrors in terms of f_1 and f_2.

13.

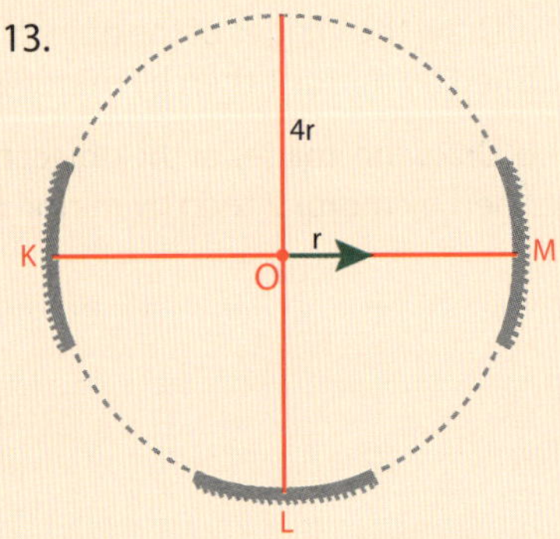

K, L, M are the vertices of the identical concave mirrors. O is the center for these mirrors. 4r is the radius of the spherical surface. Object X with a height of r is placed at the center of these mirrors.

Compare the height of images that will be formed by these mirrors.

14.

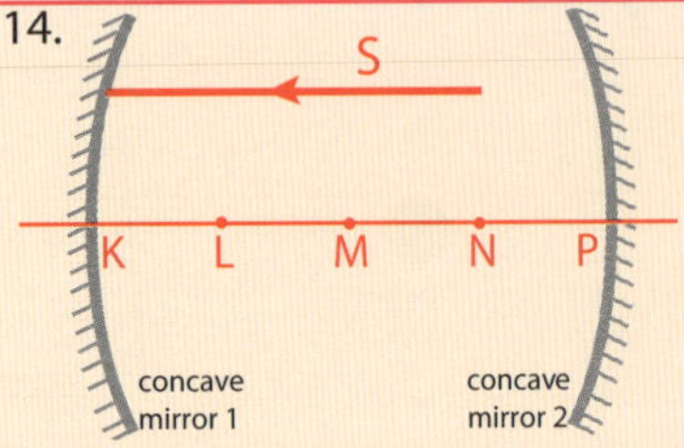

Light ray I is parallel to the common principal axes of the concave mirrors. M is the center for both mirrors. Light ray S is first reflected from concave mirror 1 then from concave mirror 2.

a. From which point(s) will ray S pass after being reflected from the concave mirror 2?

b. From which point(s) will ray S pass after being reflected for a second time from concave mirror 1?

(Distances between points are equal)

a.

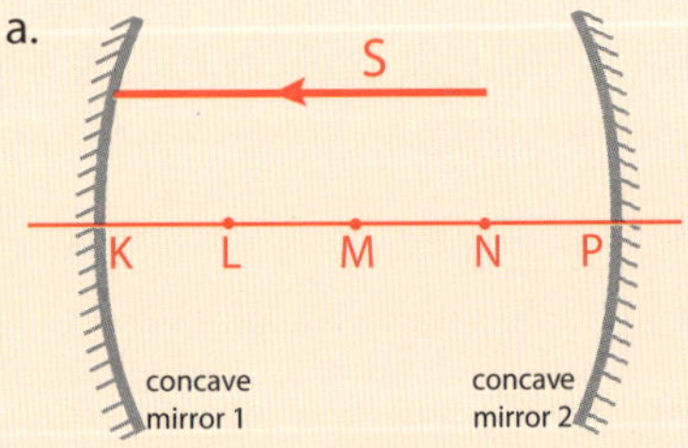

b.

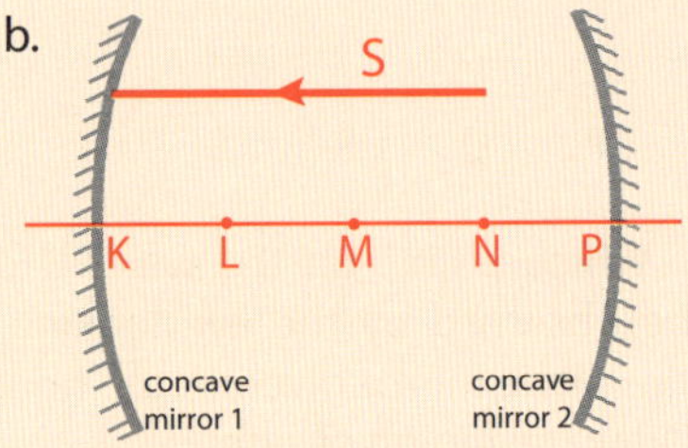

10. a. b. c.

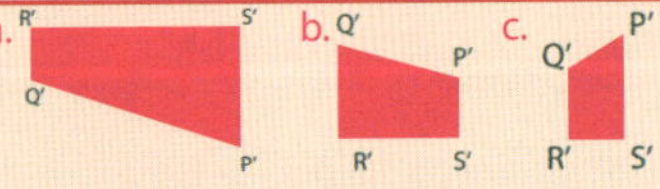

/ 11. Midpoint of FC, 1.5 square below principal axis / 12. $2f_1 - f_2$ /13. M>L>K /14. a. Between MN; b. Between LM

15.

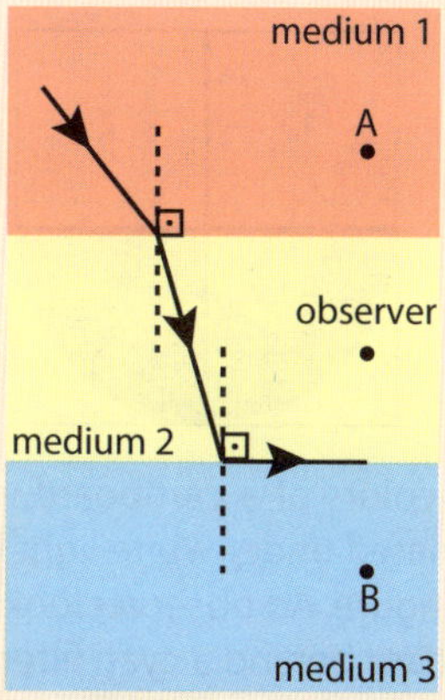

Three transparent media with parallel interfaces are placed as given in the figure. Light ray R follows the path given in the figure.

a. What is the relationship between the refractive indices of the media?

b. What is the relationship between the speed of light in these media?

c. An observer at point O is looking at point A and B respectively.

Will these points appear closer or farther than they really are? Give your reasoning.

16.

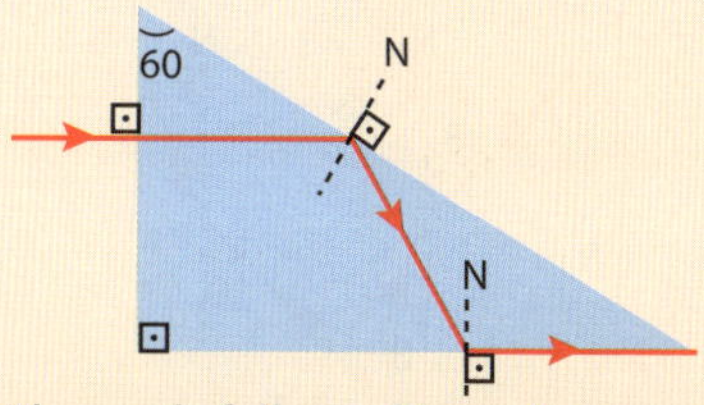

The path followed by a light ray as it propagates in a denser medium is given in the figure.

a. Find the critical angle of the prism.

b. Draw the paths followed in the same medium by the light rays given in the figures.

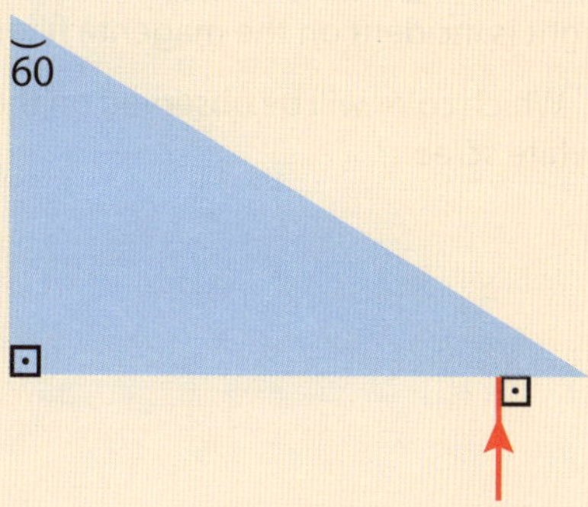

17. Draw the paths followed by the light rays and the images of the objects for the following glass prisms placed in air.

(The critical angle for glass-air interface is 42^{0})

a.

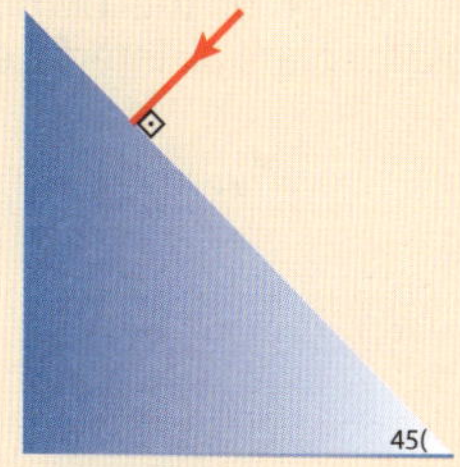

b.

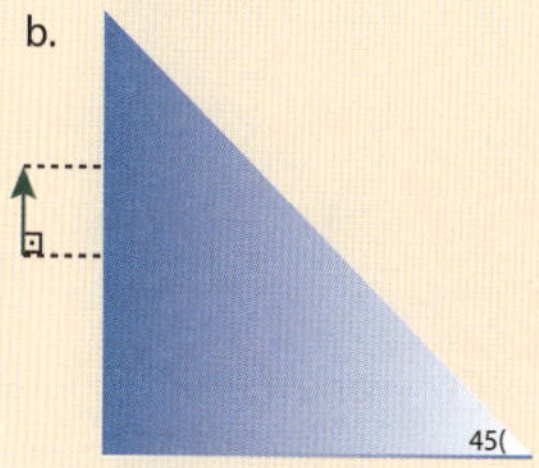

c.

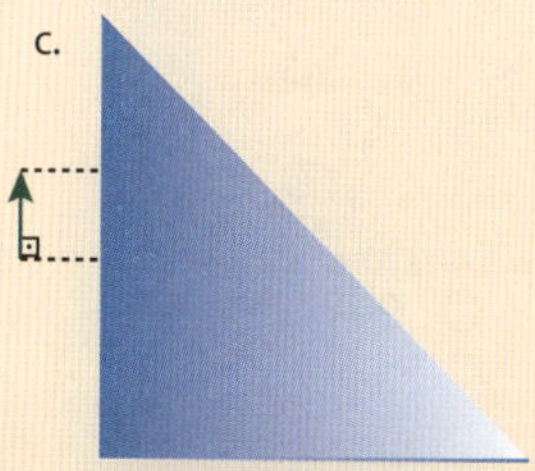

d.

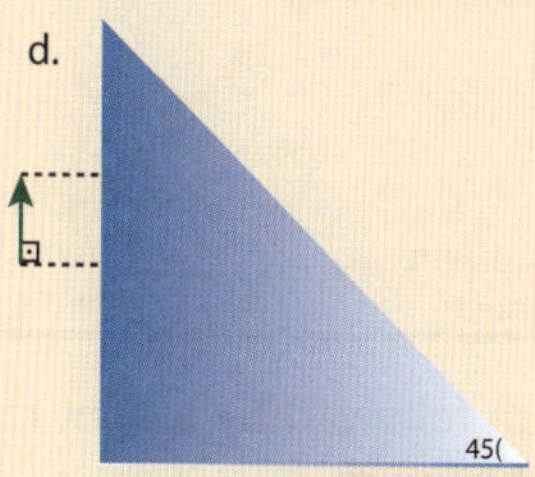

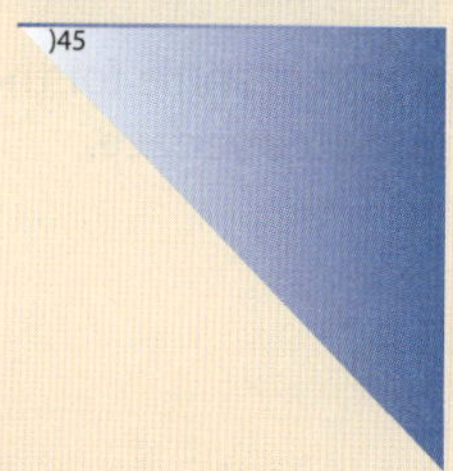

15. a. $3 < 1 < 2$; b. $3 > 1 > 2$; c. both farther because of refraction of light / 16. a. 30^{0}; b.

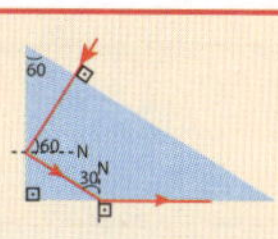

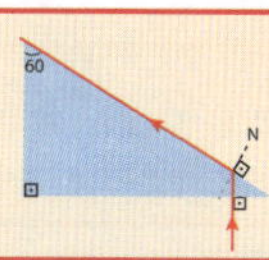

/17.

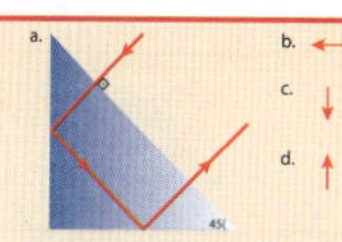

18. A glass hemisphere is placed over a plane mirror the second and third figures.

Draw the paths followed by the light rays.

(O is the center of the hemisphere, critical angle for glass-air interface is 45°)

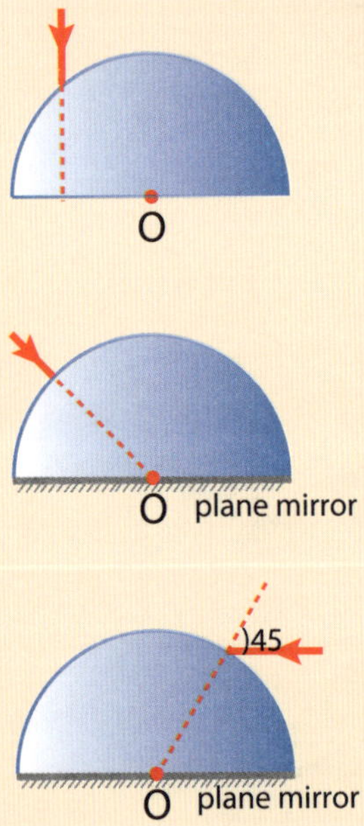

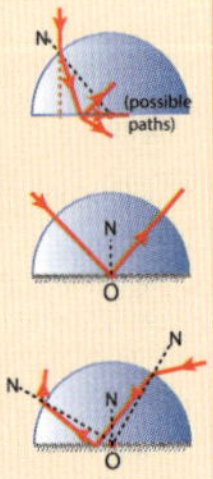

19.

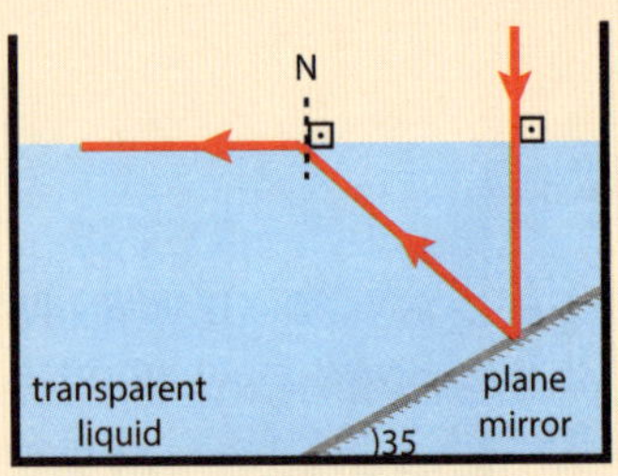

The path followed by a light ray is given in the figure.

Find the critical angle for the transparent liquid-air interface.

20.

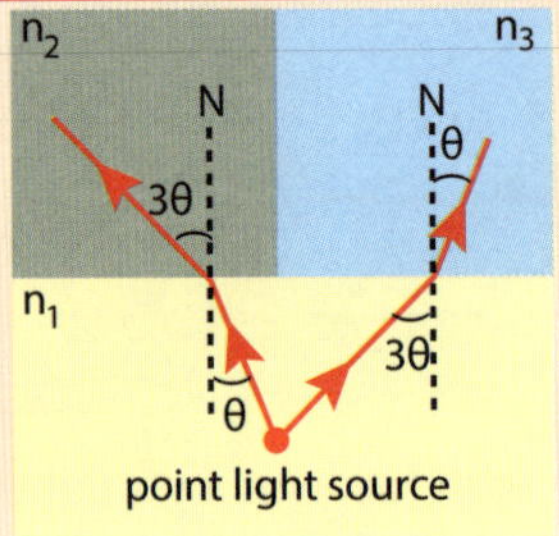

The paths of two different light rays emerging from the same light source is shown in the figure.

What is the relationship between the refractive indices of the media, n_1, n_2 and n_3.

21.

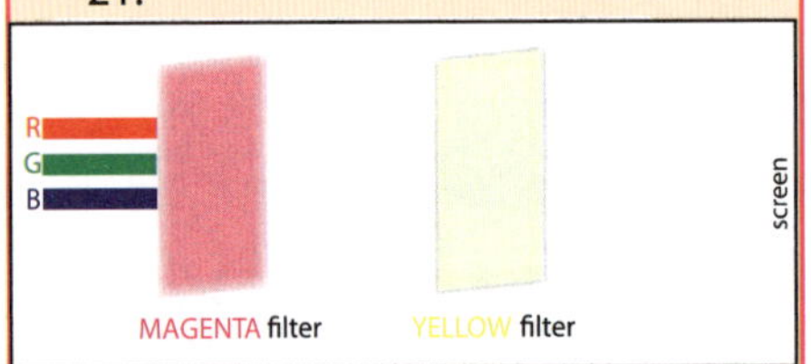

Magenta and yellow filters are placed as given in the figure. White light is incident on the magenta filter.

Which color will be observed on the white screen?

22.

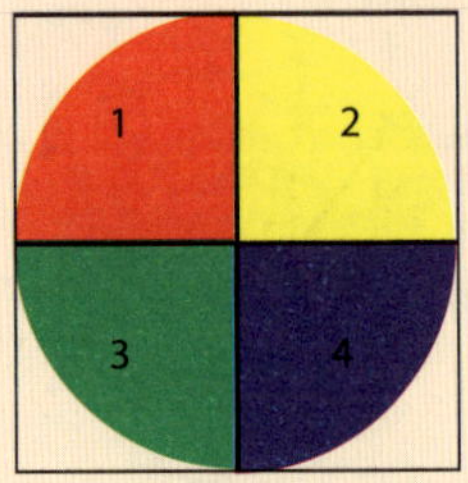

The colors of a cardboard which is illuminated under white light is given in the figure. An observer looks at this cardboard behind a cyan filter.

Which parts will be observed black?

23. Which colors will be obtained as a result of combination of following *light colors*?

a. RED + GREEN =

b. CYAN + RED =

c. BLUE + GREEN =

24. Which colors will be obtained as a result of combination of following *paint colors*?

a. YELLOW + MAGENTA =

b. YELLOW + CYAN =

c. GREEN + MAGENTA =

18.

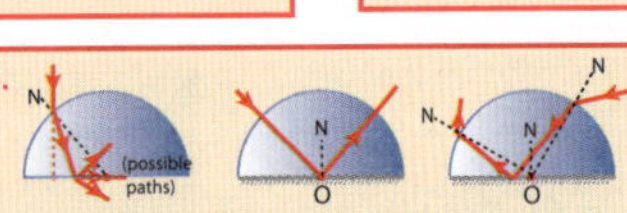

/19. 70° /20. $n_2 < n_1 < n_3$ /21. red /22. 1, 2

/23. a. yellow, b. white, c. cyan /24. a. red, b. green, c. black

25. Draw the image of the objects formed by the converging and diverging lenses.

a.

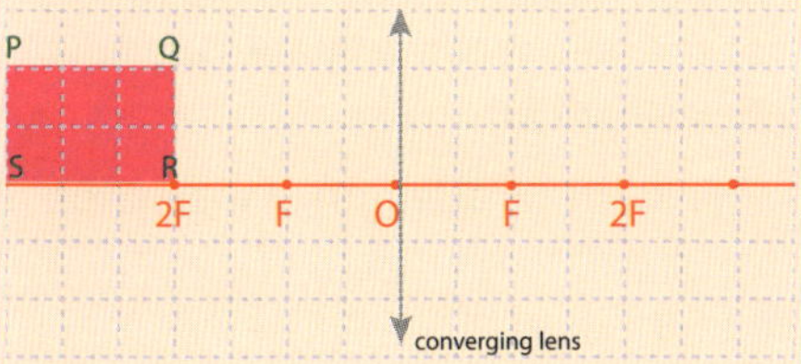

b.

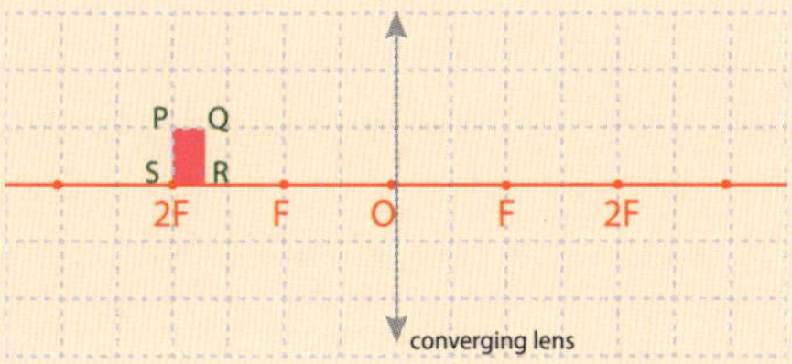

c.

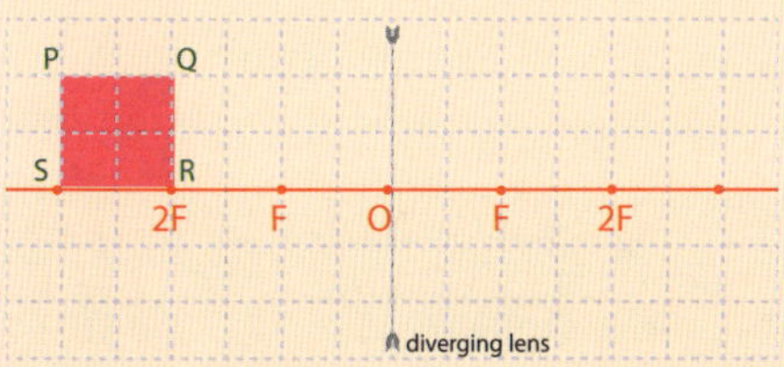

26.

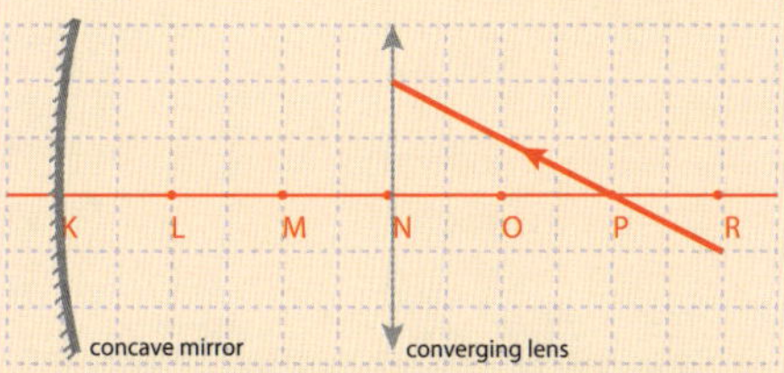

Focal lengths of the concave mirror and the converging lens are equal. P is the 2F point of the converging lens.

Draw the path followed by the light ray.

(Distances between the points are equal)

27.

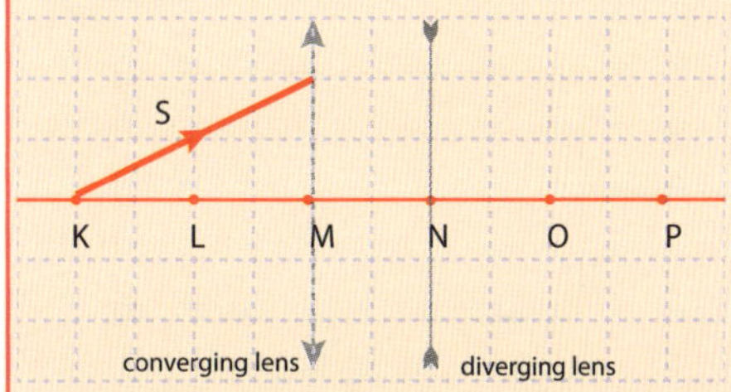

Focal length of the diverging and converging lenses are equal. L is the focus of the converging lens.

Draw the path followed by the light ray S.

(Distances between the points are equal)

28.

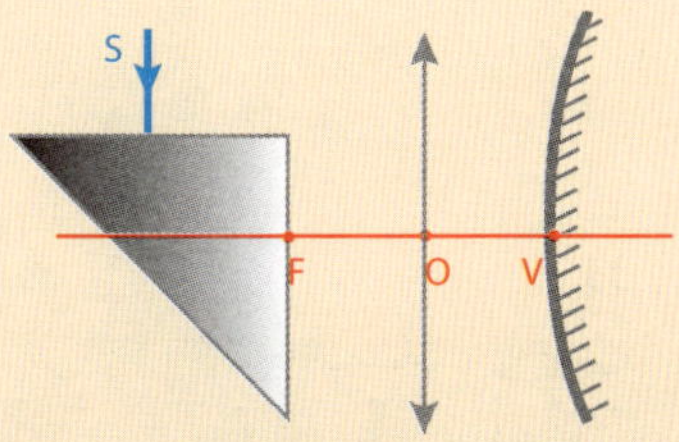

A total internal reflection prism, converging lens and a convex mirror are placed as given in the figure. F is the focus of the converging lens and V is the vertex of the convex mirror.

Draw the path followed by the light ray.

(Distances between the points are equal)

29.

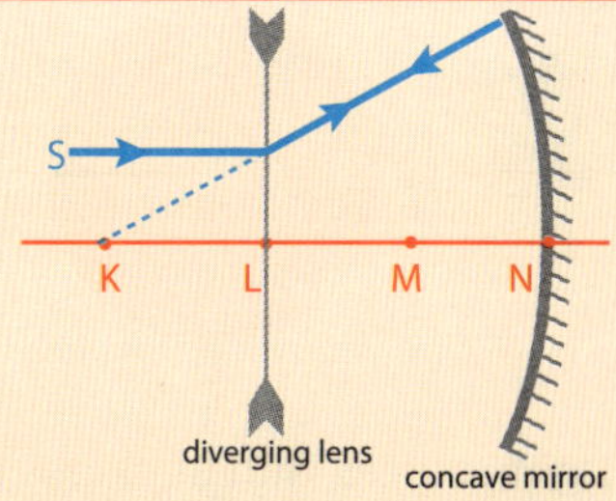

The path followed by a light ray is given in the figure. f_d is the focal length of the diverging lens and f_c is the focal length of the concave mirror.

Find the ratio f_d/f_c.

(Distances between the points are equal)

25. a. b. c.

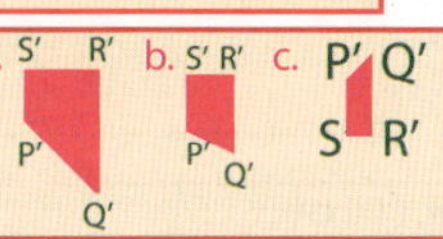

/26. Passes from O / 27. Parallel to principal axis as it leaves the diverging lens / 28. Parallel to the initial light ray as it leaves the system. / 29. 2/3

30.

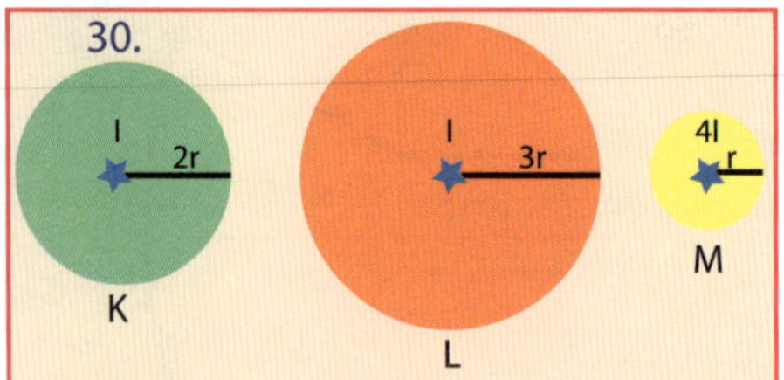

Light sources with luminous intensity I, I and 2I are placed to the center of spheres K, L and M.

a. What is the relationship between the luminous flux at the inner surface of the spheres?

b. What is the relationship between the illuminations at the inner surface of the spheres?

31.

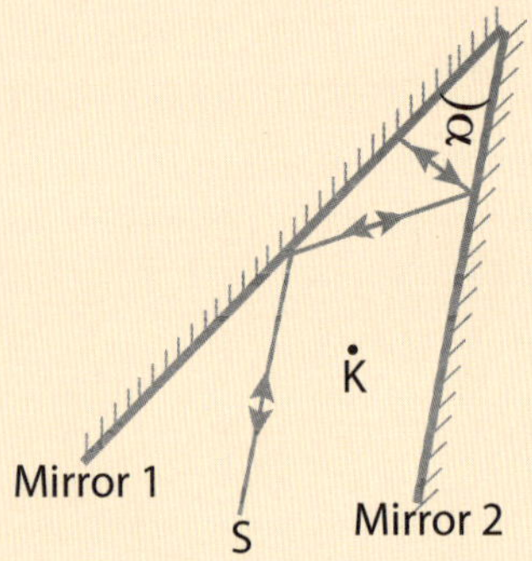

Light ray S is parallel to mirror 2. The path it followed between mirrors 1 and 2 is given in the figure.

a. Find the angle between the mirrors.

b. Find the number of images of point K that will be formed by these mirrors.

32.

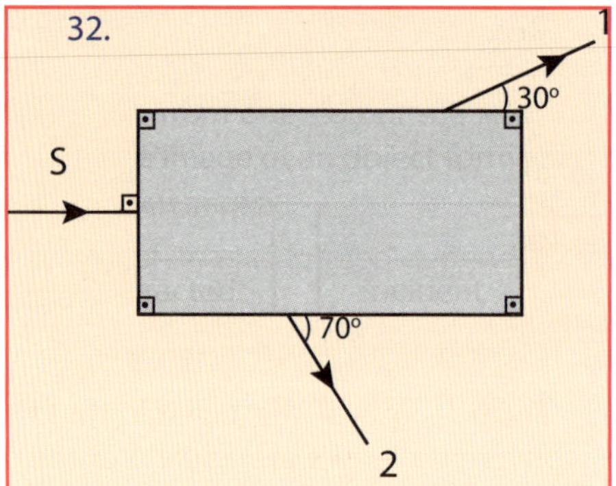

There is a plane mirror in the rectangular box. Light ray S is being reflected as light ray 1. When the mirror is rotated by α, S is being reflected as 2.

Find angle of rotation, α.

33.

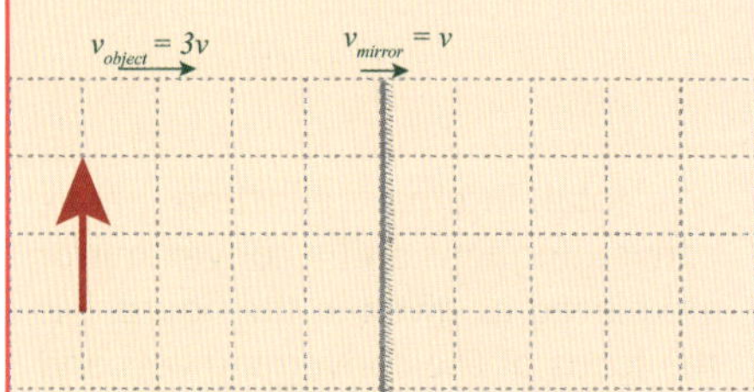

As the object is moving with a velocity of 3v, the plane mirror is moving away from the object with a velocity of v as given in the figure.

What is the velocity of the image of the object in terms of v?

34. When an object is placed 120 cm away from a concave mirror an inverted image is formed. When the same object is placed at 80 cm, an upright image is formed. In both cases the height of the images are equal to each other.

Find the focal length of the mirror.

30. a. $\Phi_K = \Phi_L < \Phi_M$; b. $E_K = E_M > E_L$ / 31. a. 30°; b. 11 / 32. 130° / 33. -v / 34. 100 cm

35. The image formed by a spherical mirror is 3 times smaller than the object and upright. The distance between the image and the object is 480 cm.

a. What is the type of the mirror?

b. Find the focal length of the mirror.

36.

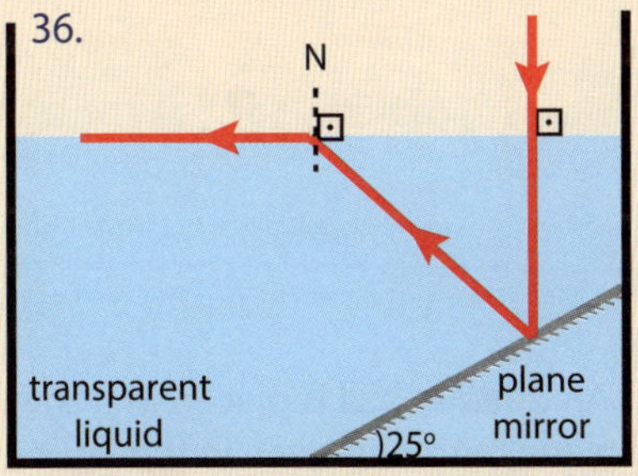

The path followed by the light ray is given in the figure.

Find the refractive index of the liquid.

($\sin 50 = 0.77$, $\sin 90 = 1$, $n_{air} = 1$)

37. Some amount of water is poured in a container. A person looking at normal direction to the container observes as if the half of the container is full.

Find the real height of water in the container.

($n_{glass} = 1.5$, $n_{air} = 1$)

38. The image formed by a lens is 3 times larger, upright and 120 cm away from the lens.

a. What is the type of the lens?

b. Find the focal length of the lens.

39. The focal lengths of a converging and a diverging lens are equal and 20 cm. The separation distance between the lenses is 60 cm. An object is placed 40 cm away from the converging lens.

Find the distance between the object and the image that will be formed by the diverging lens.

35. a. convex; b. 180 cm / 36. 1.3 / 37. 2.5 units /38. a. converging; b. 60 cm /39. 90 cm

Multiple Choice Questions

1.

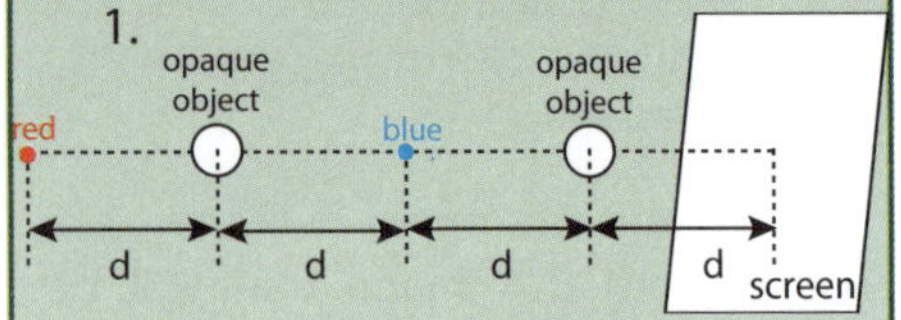

A screen, two identical opaque sources, a red and a blue point light sources are placed as given in the figure.

Which is the order of colors observed from center outwards on the screen?

A. Magenta / Red /Black

B. Magenta / Blue / Black

C. Black / Red / Magenta

D. Grey / Red / Yellow

E. Black / Blue / Magenta

2.

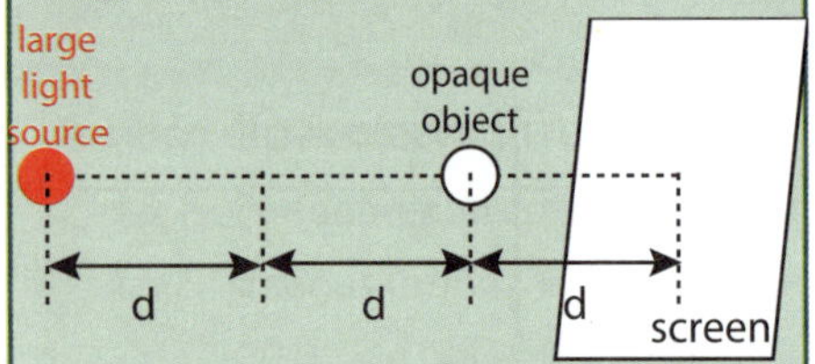

A large light source and an opaque object have the same size. They are placed as placed in the figure.

Find the ratio of area of penumbra to umbra; P/U.

A. 1 B. 2 C. 3 D. 4 E. 5

3.

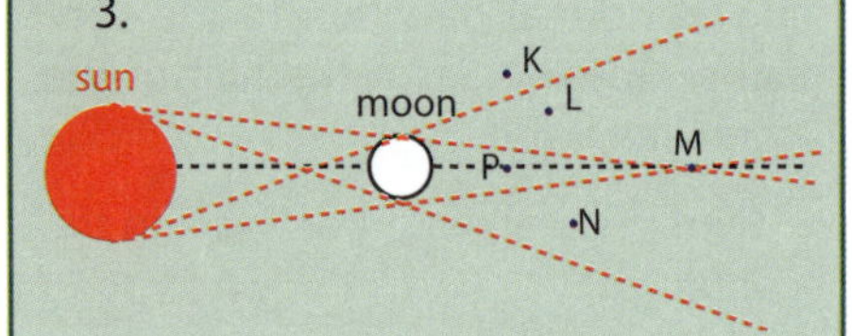

At which point a total eclipse is observed?

A. K B. L C. M D. N E. P

4.

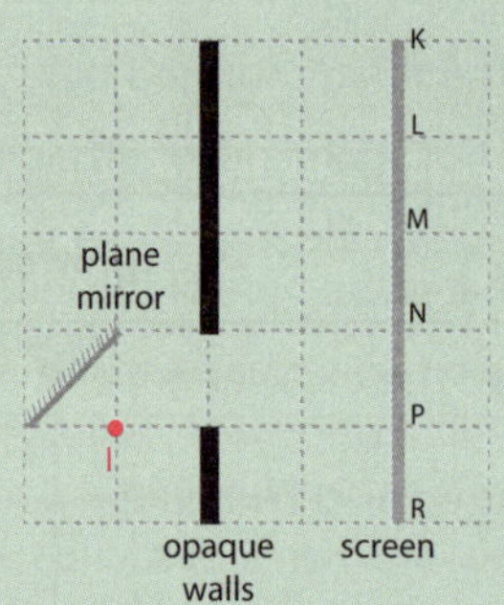

A plane mirror, point light source I are placed behind opaque walls as given in the figure.

Which part(s) of the screen will be illuminated by the light source?

A. KL B. LN C. LP D. LR E. KR

5.

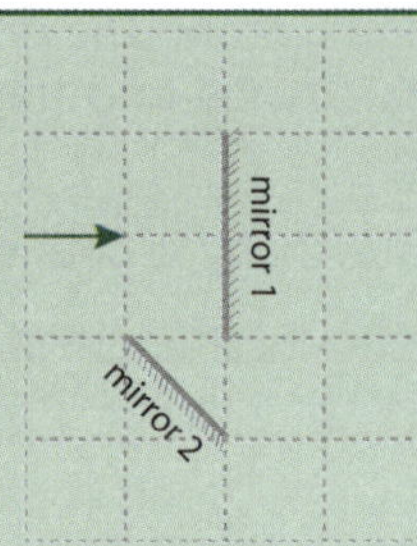

Image of the arrow given in the figure is first formed mirror 1, then by mirror 2.

Which of the followings is the final image of the arrow?

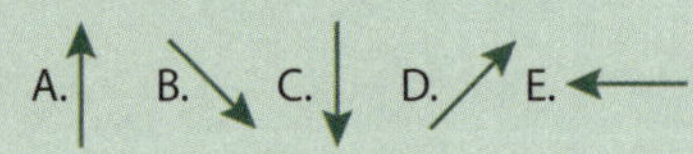

6.

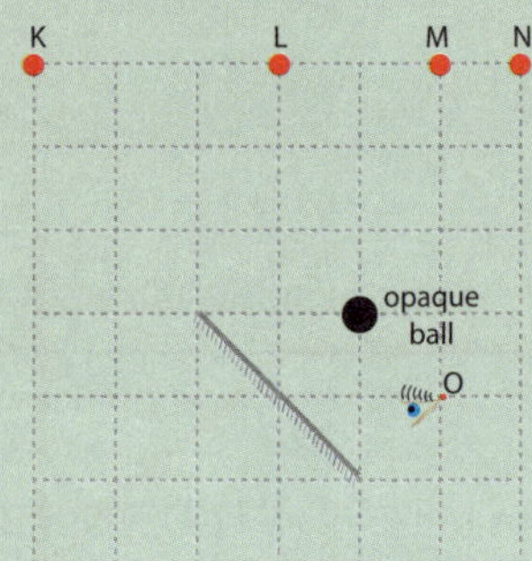

An opaque ball is placed in front of a plane mirror as given in the figure.

Which points can be seen by the observer looking into the mirror?

A. Only M B. L and N

C. M and N D. K and N

E. K, L and M

1.E 2.C 3.E 4.D 5.A 6.B

7.

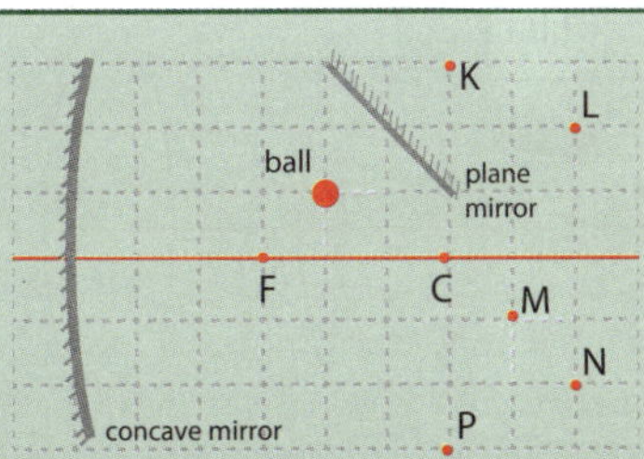

The image of the ball is first formed by the plane mirror, than by the concave mirror.

Which is the final image of the object?

A. K   B. L   C. M   D. N   E. P

8.

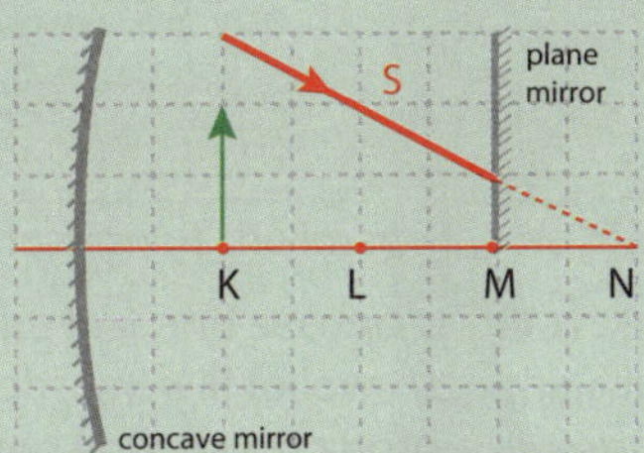

Light ray S is first reflected from the plane mirror, then reflected back on itself from the concave mirror. An image of arrow at point K is formed by concave mirror only.

Which are the properties of the image of the arrow at point K.

A. Real, upright, smaller

B. Virtual, inverted, larger

C. At infinity

D. Real, smaller, between K and L

E. Real, larger, beyond L

9.

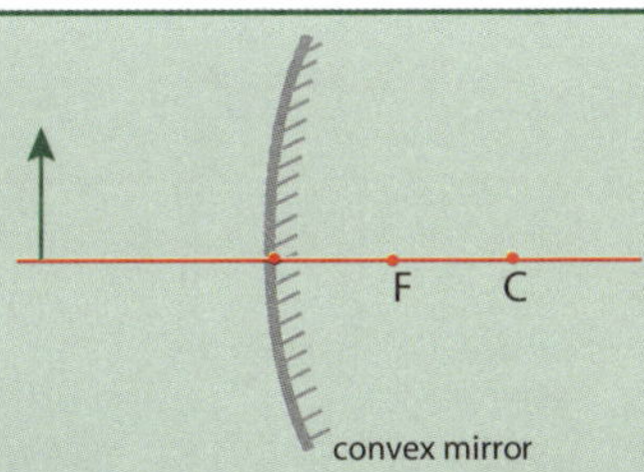

An object is placed in front of a convex mirror.

How will the height of the image , H_i and the distance of the image to the mirror, D_i, change when the object is approached to the mirror?

	H_i	D_i
A.	increases	decreases
B.	stays the same	increases
C.	decreases	increases
D.	decreases	decreases
E.	increases	stays the same

10.

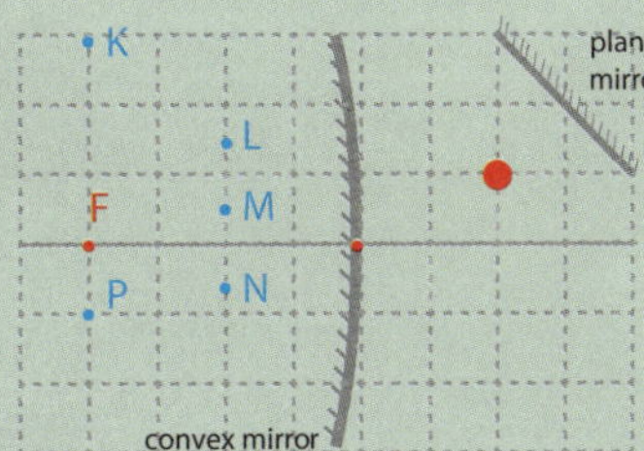

A plane mirror and a convex mirror are placed as shown in the figure. The image of the ball is first formed by the plane mirror, then by the convex mirror.

Which is the final image of the object?

A. K   B. L   C. M   D. N   E. P

11.

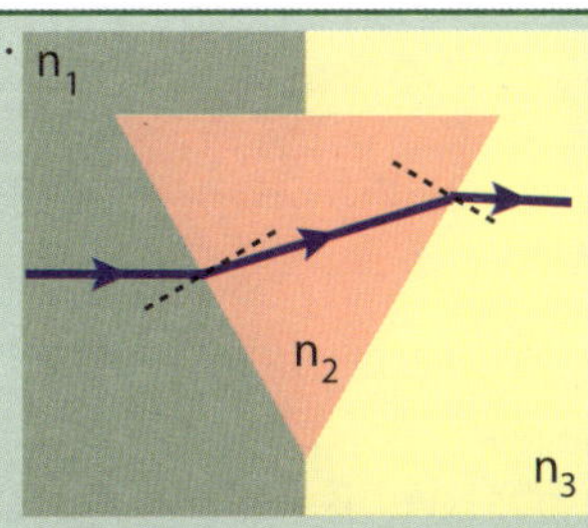

The path followed by a light ray in three different media is given in the figure.

What is the relationship between the refractive indices of the media, n_1, n_2, n_3.

A. $n_1 < n_2 < n_3$ B. $n_2 < n_1 < n_3$

C. $n_3 < n_2 < n_1$ D. $n_1' < n_3 < n_2$

E. $n_2 < n_3 < n_1$

12.

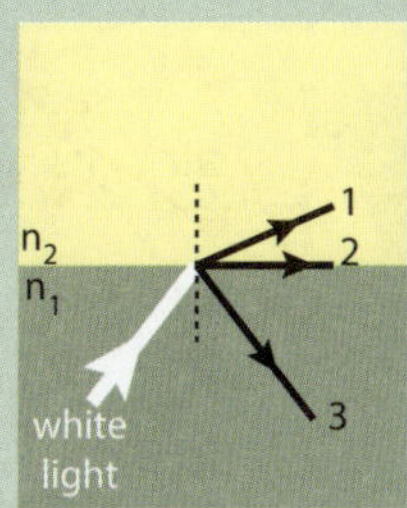

White light incident to the interface of transparent media 1 and 2 is dispersed as it is changing its medium.

Which of the followings can be the colors of the rays 1, 2, 3?

	1	2	3
A.	orange	blue	red
B.	yellow	violet	orange
C.	red	violet	yellow
D.	red	green	blue
E.	blue	green	red

7. E   8. C   9. A   10. B   11. A   12. D

13.

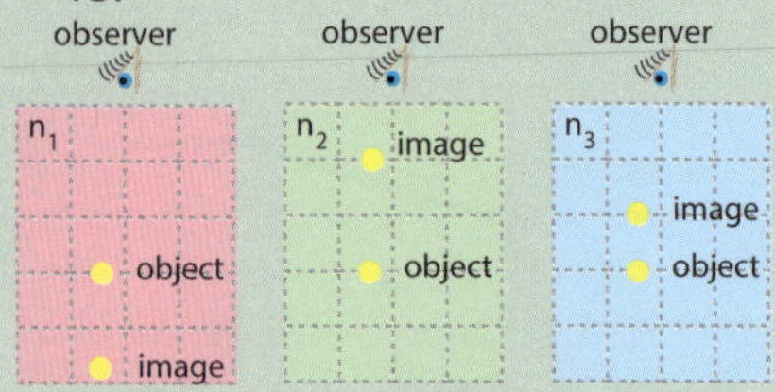

An observer is looking at perpendicular direction to an object which is placed in three different transparent media. The images observed are shown in the figure.

Which is the relationship between refractive indices of the media?

A. $n_1 = n_2 < n_3$ B. $n_2 < n_1 < n_3$

C. $n_3 < n_2 = n_1$ D. $n_1 < n_3 < n_2$

E. $n_2 < n_3 < n_1$

14.

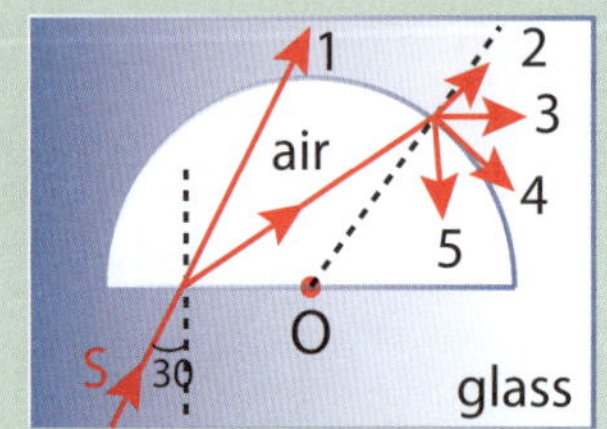

Light ray S propagates from glass to a hemispherical air medium.

Which is the correct path followed by this light ray?

(critical angle for glass-air is 42°)

A. 1 B. 2 C. 3 D. 4 E. 5

15.

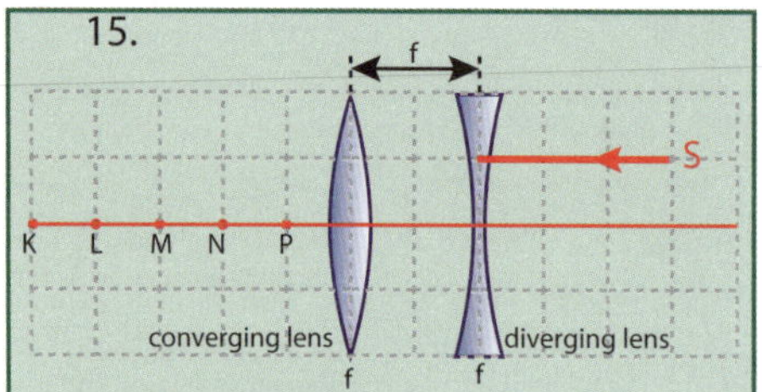

Light ray S is parallel to the principal axes of the diverging and converging lenses.

From which point(s) will it intersect with the principal axis as a result of refractions from the lenses?

A. K

B. L

C. Between M and N

D. Between K and L

E. Between N and P

16.

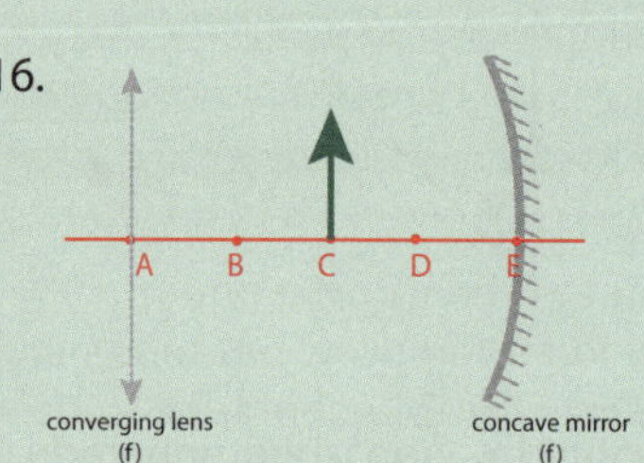

An object is placed between a concave mirror and a converging lens with equal focal lengths. The image of the object is formed first by concave mirror, then by converging lens.

Which is the properties of the final image formed?

(A, optical center of the lens; C, center of the mirror)

A. At infinity

B. Real, upright, same size

C. Virtual, inverted, smaller

D. Real, point

E. Real, upright, larger

17.

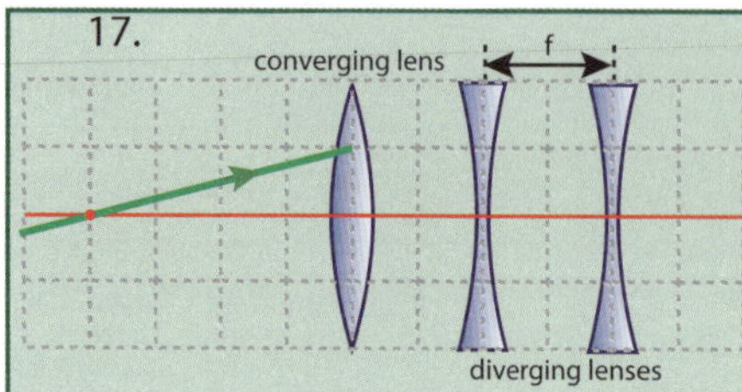

Focal length of the lenses given in the figure are equal to each other and f. A light ray is incident to 2F point of the converging lens.

Which is the path followed by the light ray as it leaves the system?

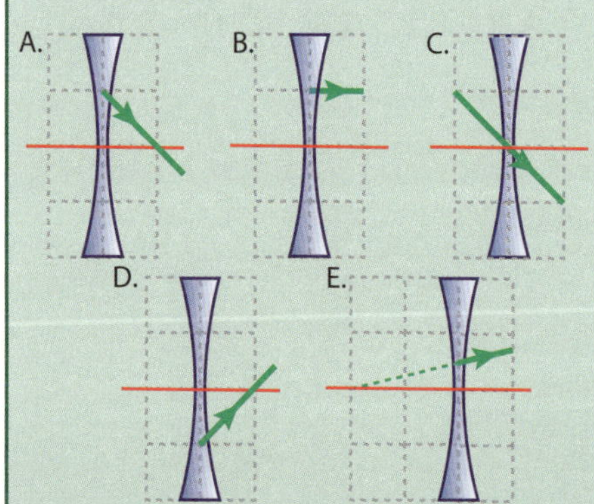

18.

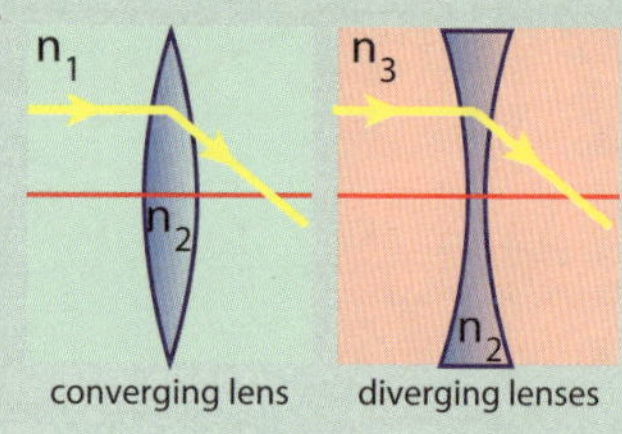

A light ray parallel to the principal axis of the converging and diverging lenses follows the path given in the figure.

A. $n_2 < n_1 < n_3$ B. $n_1 < n_2 < n_3$

C. $n_3 < n_2 < n_1$ D. $n_1 = n_3 < n_2$

E. $n_2 = n_3 < n_1$

13. D 14. B 15. B 16. B 17. E 18. B

19.

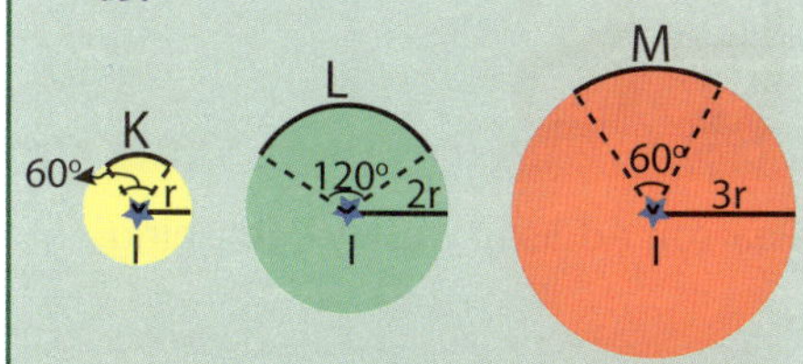

A light source is placed to the center of the spheres.

What is the relationship between the illuminations at the lids K, L and M?

A. $\Phi_K < \Phi_L = \Phi_M$

B. $\Phi_K = \Phi_L < \Phi_M$

C. $\Phi_K > \Phi_M > \Phi_L$

D. $\Phi_K = \Phi_L = \Phi_M$

E. $\Phi_K = \Phi_M < \Phi_M$

20.

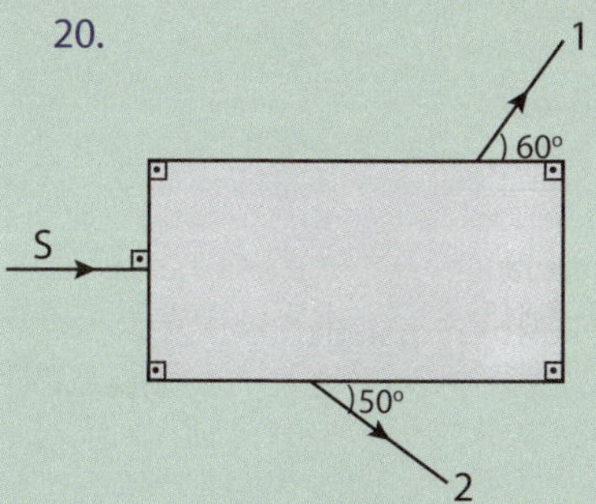

There is a plane mirror in the rectangular box. Light ray S is being reflected as light ray 1. When the mirror is rotated by α, S is being reflected as 2.

Find angle of rotation, α.

A. 125 B. 55 C. 135 D. 60 E. 145

21.

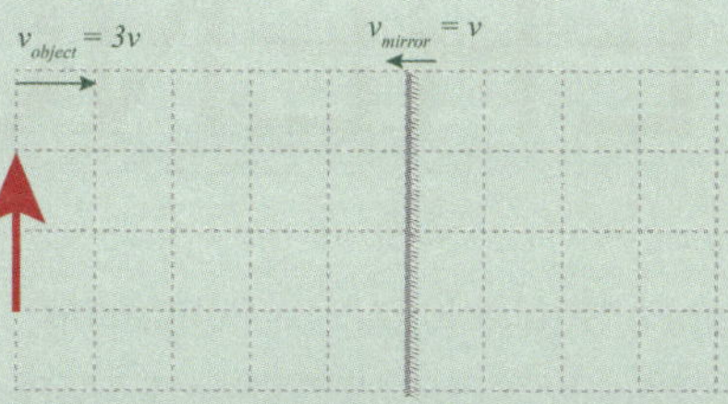

As the object is moving with a velocity of 3v, the plane mirror is moving towards the object with a velocity of -v as given in the figure.

What is the velocity of the image of the object in terms of v?

A. 2v B. 5v C. -2v D. -5v E. -v

22. An object is placed 60 cm away from a convex mirror with a focal length of 40 cm.

Find the position of the image of the object.

A. 6 B. 12 C. 18 D. 24 E. 30

23.

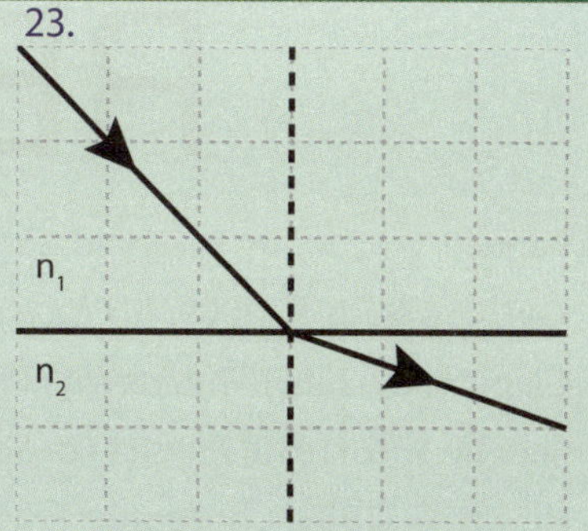

The path followed by a light ray as it is passing from transparent medium 1 to transparent medium 2 is given in the figure.

Find the ratio of the refractive indices; n_1 / n_2.

A. $\frac{1}{2}$ B. $\frac{1}{\sqrt{5}}$ C. $\frac{2}{3}$ D. $\frac{1}{\sqrt{10}}$ E. $\frac{1}{4}$

24. The image formed by a thin lens is 3 times larger, real and inverted. The distance between the image and the object is 160 cm.

Find the focal length of the lens.

A. 30 B. 35 C. 40 D. 45 E. 50

19. E 20. A 21. C 22. D 23. B 24. A

REFERENCES

Randall, D.K., Jones, B. & Field, S., (2007) *College Physics - A Strategic Approach.* USA: Pearson Addison Wesley Publishing

Kerr, G. & Ruth, P, (2009) *IB Physics* (3rd Edition) Australia: IBID Press Publishing

Giancoli, D. (2005) *Physics With Principles and Applications* (6th Edition) USA: Pearson Addison Wesley Publishing

Serway, A. S. (1990) *Physics For Scientist and Engineers* (3rd Edition) USA: Sounders College Publishing

Hewitt, G. (1989) *Conceptual Physics* (6th Edition) USA: Harper and Collins Publishing

Nolan, P. J. (1994) *Fundamentals of College Physics* Volume 1 (2nd Edition) USA: Wm. C. Brown Publishers

http://www.scienceclarified.com/As-Bi/Barometer.html

http://www.chem.science.unideb.hu/Pharm/P03ChangesofState.pdf

http://www.preservearticles.com/201012231569/effect-of-pressure-on-the-melting-point-of-solids.html

http://chemistry.about.com/od/lecturenotesl3/a/Regelation.htm

http://www.iapws.org/faq1/boil.htm

http://prezi.com/ucui2rfnodib/atomic-model-timeline/

http://www.supraconductivite.fr/en/#supra-intro

http://en.wikipedia.org/wiki/Faraday_cage

http://www.nobelprize.org/educational/physics/vacuum/experiment-1.html

http://www-outreach.phy.cam.ac.uk/camphy/electron/electron5_1.htm

http://hyperphysics.phy-astr.gsu.edu/hbase/tables/rstiv.html

E. Generalic, http://glossary.periodni.com/glossary.php?en=Volta,+Alessandro

http://www.rpc.com.au/information/developing-countries/battery-disposal.html

http://www.ehow.com/list_5846105_negative-effects-dry-cell-battery.html

http://standby.lbl.gov/faq.html

http://www.theregister.co.uk/2014/07/08/standby_consumes_more_power_than_canada_iea/

https://www.energystar.gov/index.cfm?c=about.vampires

http://www.buzzle.com/articles/how-an-electric-shock-affects-the-human-body.html

http://www.allaboutcircuits.com/vol_1/chpt_3/8.html

http://www.howstuffworks.com/electromagnet.htm

http://stars.astro.illinois.edu/aurora.html

http://www.phils.com.au/magneticfield.htm

http://electricalnotes.wordpress.com/2012/02/17/effects-of-high-voltage-transmission-lines-on-humans-and-plants/

http://science.howstuffworks.com/transport/engines-equipment/maglev-train.htm

http://www.doodlesandjots.com/2012/02/28/ripple-effect/

http://www.colorado.edu/physics/phys1230/phys1230_fa01/topic14.html

http://www.nuffieldfoundation.org/practical-physics/hand-stroboscope

http://science.howstuffworks.com/radar1.htm

http://www.tutorvista.com/content/science/science-i/sound/echoes.php

http://www.ehow.com/facts_5712580_difference-sound-insulation-sound-absorption.html

http://www.acoustics.salford.ac.uk/feschools/waves/reflect.php

http://hyperphysics.phy-astr.gsu.edu/hbase/sound/reson.html

http://science.howstuffworks.com/resonance-info.htm

http://www.sound-physics.com/Sound/Resonance-NaturalFrequency/

http://www.police.uci.edu/em/EarthquakeSafetyTop10.pdf

http://www.thehartford.com/home-insurance/earthquake-safety

http://environment.nationalgeographic.com/environment/natural-disasters/earthquake-safety-tips/

http://science.howstuffworks.com/environmental/earth/geophysics/question142.htm

http://rmsa.karnatakaeducation.org.in/sites/rmsa.karnatakaeducation.org.in/files/documents/Light/Light_Resource_Material_Subject_Teacher_Forum_September_2011.html#__RefHeading__10893_902763810

http://yollisclassblog.blogspot.com.tr/2010/08/science-light-lesson.html

http://www.passmyexams.co.uk/GCSE/physics/lunar-and-solar-eclipse.html

http://www.space.com/15771-solar-eclipse-safe-observing-tips.html

http://2011sec2lss.wikispaces.com/Applications+of+Total+Internal+Reflection

http://www.thefeaturedcreature.com/2010/09/take-aim-with-archer-fish.html

http://www.bbc.co.uk/nature/20068607

http://www.liveaquaria.com/product/prod_display.cfm?c=954+1600&pcatid=1600

http://science-edu.larc.nasa.gov/EDDOCS/Wavelengths_for_Colors.html#violet

http://scribol.com/environment/10-most-incredible-eyes-in-the-animal-kingdom/9

http://www.nhm.ac.uk/nature-online/earth/rock-minerals/diamonds/diamond-formation/

http://hyperphysics.phy-astr.gsu.edu/hbase/electric/dcex6.html

https://learn.sparkfun.com/tutorials/measuring-internal-resistance-of-batteries/internal-resistance

https://www.wikihow.com/Use-a-Compass

https://www.newscientist.com/article/dn28494-animal-magnetic-sense-comes-from-protein-that-acts-as-a-compass/

http://www.scienceclarified.com/As-Bi/Barometer.html

http://www.chem.science.unideb.hu/Pharm/P03ChangesofState.pdf

http://www.preservearticles.com/201012231569/effect-of-pressure-on-the-melting-point-of-solids.html

http://chemistry.about.com/od/lecturenotesl3/a/Regelation.htm

http://www.iapws.org/faq1/boil.htm

http://hyperphysics.phy-astr.gsu.edu/hbase/Sound/timbre.html

http://www.physicsclassroom.com/class/sound/Lesson-2/Pitch-and-Frequency

http://www.physicsclassroom.com/class/sound/Lesson-2/Intensity-and-the-Decibel-Scale

https://www.conserve-energy-future.com/causes-and-effects-of-noise-pollution.php

https://www.s-cool.co.uk/gcse/physics/uses-of-waves/revise-it/uses-of-sound-waves

https://www.famousscientists.org/abu-nasr-al-farabi/

https://plato.stanford.edu/entries/al-farabi/#MathMusi

https://www.selcuk.edu.tr/farabi/birim/web/sayfa/ayrinti/1373/tr

https://byjus.com/physics/loudness-of-sound/

My Notes